# Contents

Software Development for Engineers

with C, Pascal, C++, Assembly Language,
Visual Basic, HTML, JavaScript and Java

William J. Buchanan, BSc, CEng, PhD

Senior Lecturer
Department of Electrical and Electronic Engineering
Napier University
Edinburgh
UK

Newnes

This book is dedicated to the memory of my Grandmother

Newnes
An imprint of Butterworth-Heinemann
Linacre House, Jordan Hill, Oxford OX2 8DP
225 Wildwood Avenue, Woburn MA 01801-2041
A division of Reed Educational and Professional Publishing Ltd

A member of the Reed Elsevier plc group

OXFORD AUCKLAND BOSTON
JOHANNESBURG MELBOURNE NEW DELHI

First published 1997
Transferred to digital printing 2001

© William Buchanan 1997

British Library Cataloguing in Publication Data
A catalogue record for this book is available from the British Library

Library of Congress Cataloguing in Publication Data
A catalogue record for this book is available from the Library of Congress

ISBN 0 340 70014 9

For information on all Newnes publications
visit our website at www.newnespress.com

FOR EVERY TITLE THAT WE PUBLISH, BUTTERWORTH-HEINEMANN
WILL PAY FOR BTCV TO PLANT AND CARE FOR A TREE.

 **Preface**

Specialization in software development is becoming a thing of the past. Previously many software developers specialized on software languages such as FORTRAN, C and Pascal. This was mainly because these languages allowed access to all the required functionality. In modern times with the move towards graphical user interface programming a developer must choose not only the required software language(s) but also the required set of development tools for a specific purpose. Typical decision might be to:

- Minimize development time;
- Create a usable interface (such as DOS, or Microsoft Windows or X-Windows, and so on);
- Operate within critical timings (such as the use of fast code, or that DOS programs generally operate faster than Microsoft Windows programs, or that compiled programs generally work faster than interpreted programs);
- Integrate with other software or systems (such as the integration with previous written software, different operating systems or with precompiled libraries);
- Maintain the long-term development of the program (typical questions might be: will there be updates to the development tools; will the development company still be around in a few years?, and so on).

Typical modern development languages are C/C++, Visual Basic, Ada (especially in military applications), Java and Delphi. This book introduces C/C++ which can be used in C/C++ and Java development applications. Pascal is useful in developing Delphi and Ada applications. Visual Basic is used to write Microsoft Windows applications, and 80X86 Assembly Language programming is useful in writing extremely fast sections of code and in appreciating the operation of the PC.

The main objective of the text is to provide a single source of reference and learning material for most of the main technical programming languages. It can be used by undergraduates through a course of study from first year to final and from introductory tutorial work to advanced user interfaces and project work. It can also be used by professional developers with a knowledge of one or more of the software development language who wish to learn some, or all, of the others, or how these languages can be used in 'real-life' applications.

The text splits into nine main sections:

Part A: Pascal/C programming – gives an introduction to structured software development using Pascal and C.
Part B: C++ programming – gives an introduction to object-oriented design with C++.
Part C: 80x86 Assembly Language programs – gives an introduction to Assembly Language programming and PC architecture.
Part D: Visual Basic programming – gives an introduction to the development of graphical user interfaces for Microsoft Windows.
Part E: HTML and Java programs – show how to develop WWW-based pages and gives an introduction to Java.
Part F: DOS.
Part G: Windows 3.
Part H: Windows 95.
Part I: UNIX.

The text uses C and Pascal to provide a basic grounding in software development. These are used to show structured software development concepts, such as repetition, decision making and modular development. The more advanced concepts of object-oriented design is introduced with the C++ development. The Visual Basic section contains program examples which can be used to develop graphical user interface programs.

Many software development job advertisements now specify the requirement for a mixture of software languages on possibly several different operating systems. Software development has thus evolved to the point where it is possible to integrate different software tools to produce the required system. The user interface of a program might be developed using a graphical programming language such as Visual Basic and various specialized modules within the program could be developed in C/C++. In summary, in a changing employment market:

*'it is **essential** to become multi-skilled in different*
areas and applications'.

Author email: w.buchanan@napier.ac.uk
WWW page: http://www.eece.napier.ac.uk/~bill_b
Source code: http://www.eece.napier.ac.uk/~bill_b/soft.html

PART A

C/Pascal
C++
Assembly Language
Visual Basic
HTML/Java
DOS
Windows 3.x
Windows 95
UNIX

 **Introduction**

1.1 Introduction

Software development has grown over the years from simple BASIC programs written on small hobby computers to large software systems that control factories. Many applications that at one time used dedicated hardware are now implemented using software and programmable hardware. This shift in emphasis has meant that, as a percentage, an increasing amount of time is spent on software and less on hardware development.

Electrical, electronic and software engineers require a great deal of flexibility in their approach to system development. They must have an understanding of all levels of abstraction of the system, whether it be hardware, software or firmware. The system itself could range from a small 4-bit central heating controller to a large industrial control system. In the development of any system the engineer must understand the system specification from its interface requirements, its timing requirements, its electrical characteristics, and so on.

The software that runs on a system must be flexible in its structure as the developer could require to interrogate memory addresses for their contents or to model a part of the system as an algorithm. For this purpose the programming languages C and Pascal are excellent in that they allow a high-level of abstraction (such as algorithm specification) and allow low-level operations (such as operations on binary digits). They have a wide range of applications, from commerce and business to industry and research, which is a distinct advantage as many software languages have facilities that make them useful only in a particular environment. For example, in the past, business and commercial applications used COBOL extensively, whereas engineering and science used FORTRAN.

1.2 Hardware, software and firmware

A system consists of hardware, software and firmware, all of which interconnect. Hardware is 'the bits that can be touched', that is, the components, the screws and nuts, the case, the electrical wires, and so on. Software is the programs that run on programmable hardware and change their operation depending on the inputs to the system. These inputs could be taken from a keyboard, interface hardware or from an external device. The program itself cannot exist without some form of programmable hardware such as a microprocessor or controller. Firmware is a hardware device that is programmed using software. Typical firmware devices are EEPROMs (Electrically Erasable Read Only Memories), and interface devices that are programmed using registers.

In most applications, dedicated hardware is faster than hardware that is running software, although systems running software programs tend to be easier to modify and require less development time.

1.3 Basic computer architecture

The main elements of a basic computer system are a central processing unit (or microprocessor), memory, and input/output (I/O) interfacing circuitry. These are interconnected by three main buses: the address bus; the control bus; and the data bus, as illustrated in Figure 1.1. External devices such as a keyboard, display, disk drives, and so on, can connect directly onto the data, address and control buses, or connect through I/O interfacing circuitry.

Memory normally consists of RAM (random access memory) and ROM (read only memory). ROM stores permanent binary information, whereas RAM is a non-permanent memory and loses its contents when the power is taken away. RAM memory is used to run application programs and to store information temporarily.

The microprocessor is the main controller of the computer. It fetches binary instructions (known as machine code) from memory, it then decodes these into a series of simple actions and carries out the actions in a sequence of steps. These steps are synchronized by a system clock.

The microprocessor accesses a memory location by putting its address on the address bus. The contents at this address are placed on the data bus and the microprocessor reads the data from the data bus. To store data in memory the microprocessor places the data on the data bus. The address of the location in memory is then put on the address bus and the data is then read from the data bus into the required memory address location.

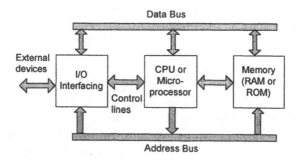

Figure 1.1 Block diagram of a simple computer system

1.4 Compiling, linking and producing an executable program

A microprocessor only understands binary information and operates on a series of binary commands known as machine code. It is extremely difficult to write large programs in machine code, so that high-level languages are used instead. A low-level language is one which is similar to machine code and normally involves the usage of keyword macros to replace machine code instructions. A high-level language has a syntax that is almost like written English and thus makes a program easy to read and to modify. In most programs the actual operation of the hardware is invisible to the programmer.

A compiler changes the high-level language into machine code. High-level languages include C, BASIC, COBOL, FORTRAN and Pascal; an example of a low-level language is 80386 Assembly Language.

Figure 1.2 shows the sequence of events that occur to generate an executable program from a C or Pascal source code file (the filenames used in this example relate to a PC-based system). An editor creates and modifies the source code file; a compiler then converts this source code into a form which the microprocessor can understand, that is, machine code. The file produced by the compiler is named an object code file code (note that Turbo Pascal does not produce an object code file). This file cannot be executed as it does not have all the required information to run the program. The final stage of the process is linking, which involves adding extra machine code into the program so that it can use devices such as a keyboard, a monitor, and so on. A linker links the object code file with other object code files and with libraries to produce an executable program. These libraries contain other object code modules that are compiled source code.

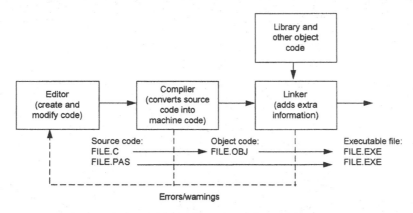

Figure 1.2 Edit, compile and link processes

If compilation or linking steps generate errors or warnings then the source code must be modified to eliminate them and the process of compilation/linking begins again. Warnings in the compile/ link process do not stop the compiler or linker from producing an output, but errors will. All errors in the compilation or linking stage must be eliminated, whereas it is only advisable to eliminate warnings.

1.5 C compilation

Borland C++ Version 3.0 is an integrated development package available for PC-based systems. It contains an editor, compiler, linker and debugger (used to test programs). The editor creates and modifies source code files and is initiated by running BC.EXE. Figure 1.3 shows a main screen with a source code file PROG1_1.C.

Figure 1.4 shows the compile menu options within this package. A source code file is compiled by selecting Compile to OBJ. If there are no errors an object code file is produced (in this case PROG_1.OBJ). This is linked using Link EXE file (producing the file PROG_1.EXE). A compile and link process can also be initiated using the Make EXE file option. Programs are run from the Run menu option.

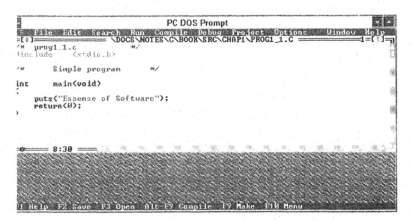

Figure 1.3 Borland C++ Version 3.0 main screen

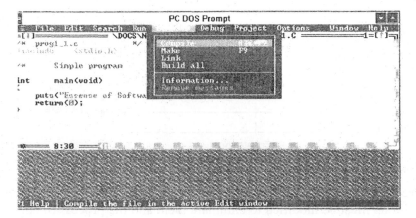

Figure 1.4 Borland C++ Version 3.0 compile menu options

1.6 Pascal compilation

Turbo Pascal Version 5.0 is an integrated development package available for PC-based systems. It contains an editor, compiler, linker and debugger (used to test programs). The editor creates and modifies source code files and is initiated by running TURBO.EXE. Figure 1.5 shows a main screen with a source code file PROG1_1.PAS.

Figure 1.6 shows the compile menu options within this package. A source code file is compiled by selecting Compile. If there are no errors then an executable program is produced. If the destination is given as Memory then it does not save the executable file to the disk but runs it from memory. If the destination is to the Disk then an executable file will be produced (producing the file PROG_1.EXE). The destination can be toggled by pressing the ENTER key while the line cursor is on the Destination option. A program is run from the Run menu option.

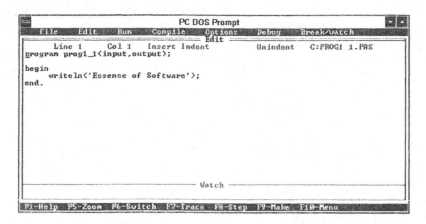

Figure 1.5 Turbo Pascal Version 5.0 main screen

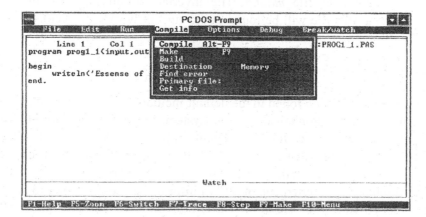

Figure 1.6 Turbo Pascal Version 5.0 compile menu options

1.7 Introduction to C

This section gives a brief introduction to ANSI-C.

1.7.1 Pre-processor

The pre-processor acts on programs before the compiler. It uses commands that have a number-sign symbol ('#') as the first non-blank character on a line. Figure 1.7 shows its main uses, which are: including special files (header files) and defining various macros (or symbolic tokens). The #include directive includes a header file and #define defines macros. By placing these directives near the top of a source code file then all parts of the program have access to the information contained in them.

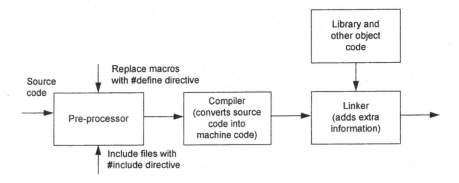

Figure 1.7   Operations on the program to produce an executable file

For example, the pre-processor directive:

```
#include "main.h"
```

includes the header file *main.h*. The inverted commas inform the pre-processor that this file will be found in the current working directory, while the directive

```
#include <stdio.h>
```

includes the file *stdio.h* found in the default include directory. This directory is normally set-up automatically by the system. For example, Turbo C Version 2.0 stores its header files, by default, in the directory \TC\INCLUDE and Borland C uses \BORLANDC\INCLUDE. Typically, header files on a Unix system are stored in the /usr/include directory.

To summarize, inverted commas (" ") inform the pre-processor to search for the specified header file in the current directory (or the directory specified in the pathname). The chevron characters (<>) inform the pre-processor to search in the default include directory. It is not advisable to include any other file apart from header files. These have a '.h' file extension (although this is not obligatory). Standard header files are used in conjunction with functions contained in libraries. They do not contain program code, but have information relating to functions. A given set of functions, such as maths or I/O, has a header file associated with it. Table 1.1 gives typical header files and their functionality.

A macro replaces every occurrence of a certain token with another specified token. The following examples show substitutions using the #define directive.

```
#define    PI                3.14
#define    BEGIN             {
#define    END               }
#define    _sqr(X)           ((X) * (X))
#define    SPEED_OF_LIGHT    3e8
```

Typically, as a matter of programming style, the definitions of constants, such as π, are given in uppercase characters.

Table 1.1 Typical header files

Header file	Comment
ctype.h	character classification and conversion
math.h	maths functions
stddef.h	defines several common data types and macros
stdio.h	Input/Output (I/O) routines, such as input from keyboard, output to display and file handling (*stdio* is a contraction of **standard input/output**)
stdlib.h	miscellaneous routines
string.h	string manipulation functions
time.h	time functions

1.7.2 Structure

Normally programs are split into a number of sub-tasks named functions. These are clearly distinctive pieces of code that perform particular operations. The main function (main()) is the basic routine for controlling the flow of the program and calls other sub-functions.

C Program 1.1 is a simple program which uses the puts() function to display the text "Essence of Software". The puts() function is a standard function used to output text to the display; the header file associated with it is *stdio.h*. This header file is included using the #include directive.

The statement terminator (;) is used to end a line of code (or statement) and braces ({}) show the beginning ({) and end (}) of a block of code. Comments are inserted in the program between a start comment identifier (/*) and an end identifier (*/).

⬡ C
Program 1.1

```
/* Simple program */
#include <stdio.h>

int    main(void)
{
    puts("Essence of Software");
    return(0);
}
```

All C programs have a main() function which defines the entry point into the program and, by means of calling functions, controls general program flow. It can be located anywhere in the source code program, but is normally placed near the top of the file it is located in (making it easier to find). The int keyword preceding main() defines that the program returns a value to the operating system (or calling program). In this case, the return value is 0 (return(0)). Normally, a non-zero return value is used when the program has exited due to an error; the actual value of this gives an indication of why the program has exited. The void within the parenthesis of main() defines that there is no communication between the program and the operating system (that is, no values are passed into the program). Figure 1.8 shows the basic structure of a C program.

```
/* This is a comment */   comment

int main(void)
{
int    var1, var2;    variable
float var3, var4;     declaration

    statement;
    statement;
    return(0);        main
}                     function
```

Figure 1.8   C program structure

1.7.3 Data types

Variables within a program can be stored as either numbers or characters. For example, the resistance of a copper wire would be stored as a number (a real value) and the name of a component (such as, "R1") would be stored as characters. Table 1.2 gives the four basic data types which define the format of variables.

There are three basic extensions for the four types; these are:

```
short      long      unsigned
```

Table 1.2 Basic data types

Type	Usage
char	single character 'a', '1', and so on
int	signed integer
float	single-precision floating point
double	double-precision floating point

An integer is any value without a decimal point; its range depends on the number of bytes used to store it. A floating point value is any number and can include a decimal point; this value is always in a signed format. Again, the range depends on the number of bytes used.

Integers normally take up 2 or 4 bytes in memory, depending on the compiler implementation. This gives ranges of −32 768 to 32 767 (a 2-byte int) and −2 147 483 648 to 2 147 483 647 (a 4-byte int), respectively.

1.7.4 Declaration of variables

A program uses variables to store data. Before the program can use a variable, its name and its data type must first be declared. A comma groups variables of the same data type. For example, if a program requires integer variables num_steps and bit_mask, floating point variables resistor1 and resistor2, and two character variables char1 and char2, then the following declarations can be made:

```
int        num_steps,bit_mask;
float      resistor1,resistor2;
char       char1,char2;
```

C Program 1.2 is a simple program that determines the equivalent resistance of two resistors of 1000 Ω and 500 Ω connected in parallel. It contains three floating point declarations for the variables resistor1, resistor2 and equ_resistance.

Program 1.2
```
/* Program to determine the parallel equivalent        */
/* resistance of two resistors of 1000 and 500 Ohms    */

#include <stdio.h>

int        main(void)
{
float resistor1, resistor2,equ_resistance;

    resistor1=1000.0;
    resistor2=500.0;
    equ_resistance=1.0/(1.0/resistor1+1.0/resistor2);
    printf("Equivalent resistance is %f\n",equ_resistance);
    return(0);
}
```

It is also possible to assign an initial value to a variable at the point in the program at which it is declared; this is known as variable initialization. C Program 1.3 gives an example of this with the declared variables `resistor1` and `resistor2` initialized to `1000.0` and `500.0`, respectively.

⬡ c
Program 1.3

```
/* Program to determine the parallel equivalent        */
/* resistance of two resistors of 1000 and 500 ohms    */

#include <stdio.h>

int      main(void)
{
float resistor1=1000.0, resistor2=500.0,equ_resistance;

    equ_resistance=1.0/(1.0/resistor1+1.0/resistor2);
    printf("Equivalent resistance is %f \n",equ_resistance);

    return(0);
}
```

1.7.5 Keywords

ANSI-C has very few reserved keywords (only 32); these cannot be used as program identifiers. C is case-sensitive and thus they must be used in lowercase. From these simple building blocks large programs can be built. The following gives a list of the keywords.

auto	do	for	return	switch
break	double	goto	short	typedef
case	else	if	signed	union
char	enum	int	sizeof	unsigned
const	extern	long	static	void
continue	float	register	struct	volatile
default				while

Functions are sections of code that perform a specified operation. They receive some input and produce an output in a way dictated by their functionality. These can be standardized functions which are inserted into libraries or are written by the programmer. ANSI-C defines some standard functions which provide basic input/output to/from the keyboard and display, mathematical functions, character handling, and so on. They are grouped together into library files and are not an intrinsic part of the language. These libraries link into a program to produce an executable program.

1.8 Introduction to Pascal

This section gives a brief introduction to Turbo Pascal.

1.8.1 Constant declarations

Pascal uses the `const` keyword to defined constant numeric values. The following examples show constant declarations for π and the speed of light (which is 3×10^8).

```
const      PI=3.14;
           SPEED_OF_LIGHT=3e8;
```

In Pascal the case of the characters is ignored but, as a matter of programming style, the definition of constants, such as π, is given in uppercase characters.

1.8.2 Structure

Normally programs are split into a number of sub-tasks named procedures or functions. These are clearly distinctive pieces of code that perform particular operations. The main program is the basic routine to control the flow of a program and calls other sub-functions.

Pascal Program 1.1 is a simple program which uses the `writeln` procedure to display the text 'Essence of Software'. The `writeln` procedure is a standard procedure which is used to output text to the display.

The statement terminator (`;`) is used to end a line of code (or statement) and the keywords `begin` and `end` define the beginning and end of a block of code. Comments are inserted into the program between a start comment identifier (`(*`) and an end identifier (`*)`).

All Pascal programs have a main program which defines the entry point into the program and, by means of calling functions and procedures, controls general program flow. In most cases it is be located at the end of the source code file.

Program 1.1
```
program prog1_1(input,output);
(* Simple program *)
begin
     writeln('Essence of Software');
end.
```

Figure 1.9 shows the basic structure of a Pascal program. Each program has a program header which is defined with the `program` keyword. After this the program variables are declared. In this case the variables declared are *var1*, *var2* (which are integers) and *var3*, *var4* (which are real values). The main program is defined after the variable declaration and can be identified between the `begin` and `end` keywords. The final `end` keyword has a full-stop after it.

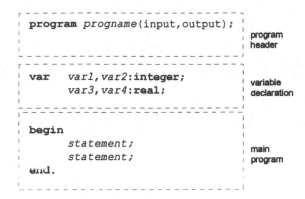

```
program progname(input,output);          program
                                          header

var    var1,var2:integer;                 variable
       var3,var4:real;                     declaration

begin
       statement;                          main
       statement;                          program
end.
```

Figure 1.9 Pascal program structure

1.8.3 Data types

Variables within a program can be stored as either numbers or characters. For example, the resistance of a copper wire would be stored as a number (a real value) and the name of a component (such as, 'R1') would be stored as characters. Table 1.3 gives the four basic data types which define the format of variables.

Table 1.3 Basic Pascal data types

Type	Usage	Range
char	single character 'a', '1', and so on	Character range
integer	signed integer	−32 768 to 32 767
real	single-precision floating point	2.9×19^{-39} to 1.7×10^{38}
boolean	boolean type	true or false

Other data types used in Turbo Pascal include:

`shortint, longint, byte, word` and `double`

An `integer` is any value without a decimal point; its range depends on the number of bytes used to store it. A floating point value is any number and can include a decimal point; this value is always in a signed format. Again, the range depends on the number of bytes used.

The integer type uses 2 bytes in memory. This gives ranges of −32 768 to 32 767 (a 2-byte `int`) and −2 147 483 648 to 2 147 483 647 (a 4-byte `longint`), respectively.

1.8.4 Declaration of variables

A program uses variables to store data. Before the program can use a variable, its name and its data type must first be declared. A comma groups variables of the same data type. For example, if a program requires integer variables num_steps and bit_mask, floating point variables resistor1 and resistor2, and two character variables char1 and char2, then the following declarations can be made:

```
var     num_steps,bit_mask:integer;
var     resistor1,resistor2:float;
var     char1,char2:char;
```

Pascal Program 1.2 is a simple program that determines the equivalent resistance of two resistors of 1000 Ω and 500 Ω connected in parallel. It contains three floating point declarations for the variables resistor1, resistor2 and eq_resistance.

Program 1.2
```
program prog1_2(input,output);
(* Program to determine the parallel equivalent    *)
(* resistance of two resistors of 1000 and 500 Ohms *)

var     resistor1,resistor2,equ_resistance:real;

begin
    resistor1:=1000;
    resistor2:=500;

    equ_resistance:=1/(1/resistor1+1/resistor2);
    writeln('Equivalent resistance is ',equ_resistance);
end.
```

1.8.5 Keywords

Turbo Pascal has 52 reserved keywords; these cannot be used as program identifiers and can be in upper- or lower-case. Large programs can be built from these simple building blocks. The following gives a list of the keywords.

absolute	and	array	begin	case
const	constructor	destructor	div	do
downto	else	end	external	file
for	forward	function	goto	if
implementation	in	inline	interface	interrupt
label	mod	nil	not	object
of	or	packed	procedure	program
record	repeat	set	shl	shr
string	then	to	type	unit
until	uses	var	virtual	while
with	xor			

Functions and procedures are sections of code that perform a specified operation. They receive some input and produce an output in a way dictated by their functionality. These can be standardized functions which are inserted into libraries or are written by the programmer. Turbo Pascal defines some

16 *Introduction*

standard functions which provide basic input/output to/from the keyboard and display, mathematical functions, character handling, and so on. They are grouped together into library files and are not an intrinsic part of the language. These libraries link into a program to produce an executable program.

1.9 Exercises

1.9.1   Determine the errors in the C Programs 1.4 to 1.6 or the Pascal Programs 1.3 to 1.5. Each program has a single error. Enter them into the compiler and after the error has been corrected, run them.

Program 1.4
```
#include    <stdio.h>
/*      Simple program     */

int     main(void)
{
   puts("This is a sample program")
   return(0);
}
```

Program 1.5
```
#include    <stdio.h>
/*      Simple program        */
int     main(void)
{
   puts("This is another sample program");
   return(0);
```

Program 1.6
```
#include    <stdio.h>
/*      Simple program

int     main(void)
{
   puts("And another one");
   return(0);
}
```

Program 1.3
```
program prog1_3(input,output)
(*      Simple program     *)
begin

   writeln('This is sample program');

end.
```

Program 1.4
```
program prog1_4(input,output);
```

```
(*      Simple program     *)
begin

  writeln('This is sample program');
```

Program 1.5
```
program prog1_5(input,output);
(*      Simple program
begin

  writeln('This is sample program');

end.
```

Worksheet 1

W1.1   Which programming language is your choice for software development (tick one):

C   [   ]
Turbo Pascal   [   ]
Mixture of C and Pascal   [   ]

> Reason for choice:

W1.2   If you are running the compiler over the network then what is your login name:

> Login name:

W1.3   Locate and run Pascal or C compiler:

> How is it selected:

W1.4   Enter either C Program 1.1 or Pascal Program 1.1 and save this to a file on floppy disk as `PROG1_1.PAS` (for the Pascal file) or `PROG1_1.C` (for the C file).

> Completed successfully: YES/ NO
> Notes:

W1.5   Compile the program and note any messages that the compiler gives.

```
Program compiled successfully:          YES/ NO
Notes:
```

W1.6   If there are errors in the program then compare the entered file with the program listing and try to identify how they differ. The compiler should identify the location of the error (note look also at the line before). Then recompile.

```
Program compiled successfully:          YES/ NO
Notes:
```

W1.7   After the program has been successfully compiled, run the program and determine its output.

```
Program output:
```

W1.8   Enter either C Program 1.2 or Pascal Program 1.2 and save file on floppy disk as PROG1_2.PAS (for the Pascal file) or PROG1_2.C (for the C file).

```
Completed successfully:                 YES/ NO
Program output:
```

W1.9   Using C Program 1.2 or Pascal Program 1.2 determine the equivalent resistance for two parallel resistors. Use this program and by changing the resistor values complete Table W1.1.

Table W1.1   Equivalent resistance

Resistor1 (Ω)	Resistor2 (Ω)	Equivalent resistance (Ω)
1000	1000	
25	100	
1e6 (1MΩ)	1e6	
150	50	

Input/Output

2.1 Introduction

Every program has some form of output and normally an input. Figure 2.1 shows some examples of input and output devices. The input could be from a keyboard, a file, input/output ports, a mouse, and so on. Output can be sent to devices such as displays, printers, hard-disks, and so on. Typically, engineers also communicate with devices such as ADC/DACs, LEDs, interface adapters, IC programmers, and so on.

The default input device is normal from a keyboard and the default output from a display. Most programs prompt the user to enter data from the keyboard. This data is then processed and the results displayed to the screen. The user can then enter new data and so the cycle continues.

Most operating systems also allows a redirection of the input or output. For example, a text file can act as an input to a program and the printer as the output.

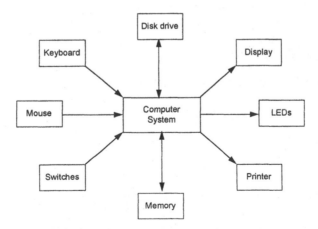

Figure 2.1   Input/output devices

2.2 Pascal input/output

Compared with C, Pascal has a very limited set of input/output statements (for this reason this chapter contains many more pages on the C input/output statements). Pascal uses the keyboard as the standard input and the display as the standard output. The two statements which are used to control this input and output are `read` and `write`.

2.2.1 writeln

The statements used to output data from a program to the screen are `writeln` and `write`. The `write` statement does not move the cursor to a newline once the data has been printed, whereas the `writeln` will. The standard format is:

`writeln`(*'text',argn1, arg2 ... argn*);	outputs a text string defined by *'text'* and values given by arguments *arg1..argn*. The output is appended with a new-line.
`write`(*'text',argn1, arg2 ... argn*);	outputs a text string defined by *'text'* and values given by arguments *arg1..argn*. The output is not appended with a new-line.

A text string is enclosed within quotes (' ') and can be printed at any place in the `write` statement. Values will be printed in a format defined by their type. For example an integer will be displayed without a decimal point, a very large or small real value will be displayed in exponent form. The actual format of the value to be printed can be modified using the colon modifier, the standard format is:

`value:n`	prints `value` with n spaces used to print the variable
`value:n:m`	prints `value` with n spaces and m places after the decimal point

Note that all printed values are right justified.

2.2.2 readln

The statements used to input data into a program from the keyboard are `read` and `readln`. The `read` statement does not move the cursor to a new-line once the data has been entered, whereas the `readln` will. The standard format is:

`readln` (*arg1,arg2...argn*)	reads values from the keyboard and loads them into the arguments *arg1*, *arg2*, and so on.
`read` (*arg1,arg2...argn*)	reads values from the keyboard and loads them into the arguments *arg1*, *arg2*, and so on.
ch= `readkey()`	reads a single character from the keyboard into *ch*

Pascal Program 2.1 shows a simple example of a program which uses input/output statements

⬡ Pascal
Program 2.1

```pascal
program prog2_1(input,output);
var      voltage,current,resistance:real;
begin
     writeln('Enter voltage and current');
     readln(voltage,current);
     resistance:=voltage/current;
     writeln('Resistance is ' ,resistance:8:3, ' Ohms');
end.
```

2.3 C input/output

The standard input/output (I/O) functions in C are not intrinsic (built-in) to the language, but are stored in libraries that are linked into the program. The #include pre-processor directive includes the header files associated with them. Input/Output functions use *stdio.h*. In order to allow all parts of the source code access to the functions defined in the header file the pre-processor directive is located near the top of the file in which it is used. The compiler will then initiate extra error checking whenever any of the standard I/O functions are used. C Program 2.1 shows a program which includes the file *stdio.h*.

⬡ C
Program 2.1

```c
#include        <stdio.h>

int    main(void)
{
   printf("Enter a value of resistance");
   return(0);
}
```

2.3.1 C standard output (`printf()`, `puts()` and `putchar()`)

There are three basic output functions in C, these are:

printf(*"format"*, *arg1*, *arg2* ... *argn*) outputs a formatted text string to the output in a form defined by *"format"* using the arguments *arg1..argn*

puts(*"string"*) outputs a text string to the standard output and appends it with a new line

putchar(*ch*) outputs a single character (*ch*) to the standard output

The printf() function sends a formatted string to the standard output (the display). This string can display formatted variables and special control characters, such as new lines ('\n'), backspaces ('\b') and tabspaces ('\t'); these are listed in Table 2.1.

The puts() function writes a string of text to the standard output and no formatted variables can be used. At the end of the text, a new line is automatically appended.

The parameters passed into printf() are known as arguments; these are separated commas. C Program 2.1 contains a printf() statement with only one argument, that is, a text string. This string is referred to as the message string and is always the first argument of printf(). It can contain special control characters and/or parameter conversion control characters.

Conversion control characters describe the format of how the message string uses the other arguments. If printf() contains more than one argument then the format of the output is defined using a percent (%) character followed by a format description character. A signed integer uses the %d conversion control characters, an unsigned integer %u. A floating point value uses the %f conversion control characters, while scientific notation uses %e. Table 2.2 lists the main conversion control characters.

Table 2.1 Special control (or escape sequence) characters

Characters	Function	Characters	Function
\"	Double quotes (")	\b	Backspace (move back one space)
\'	Single quote (')	\f	Form-feed
\\	Backslash (\ )	\n	New line (line-feed)
\nnn	ASCII character in octal code, such as \041 gives '!'	\r	Carriage return
\0xnn	ASCII character in hexadecimal code, such as \0x41 gives an 'A'	\t	Horizontal tab spacing
\a	Audible bell		

Table 2.2 Conversion control characters

Operator	Format	Operator	Format
%c	single character	%s	string of characters
%d	signed decimal integer	%o	unsigned octal integer
%e	scientific floating point	%%	prints % character
%f	floating point	%x	unsigned hexadecimal integer
%u	unsigned decimal integer	%g	either floating point or scientific notation

Figure 2.2 shows an example of the printf() statement with four arguments. The first argument is the message string followed by the parameters to be printed in the message string. In this case the parameters are val1, val2 and ch; val1 is formatted in the message string as a floating point (%f), val2 as an integer (%d) and ch as a character (%c). Finally, a new line character ('\n') is used to force a new line on the output.

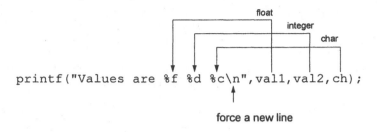

Figure 2.2 An example `printf()` statement

A numerical value is output to a given specification using a precision specifier. This specifies the number of characters used to display the value and the number of places after the decimal point. The general format of a floating point value is:

%m.nX

where *m* is the width of the value (the number of digits including the decimal point), *n* is the number of digits following the decimal point, and *X* is the format type (f for float). The general format of a string or integer is:

%mX

where *X* is the format type (c for character, s for string or d for integer) and *m* is the width of the output. Table 2.3 gives a few examples.

Table 2.3 Example of conversion control modifiers

Format	Function
%.3f	format floating point value with 3 decimal places and a default width
%8.3f	format floating point with 8 reserved spaces and 3 places after the decimal point such as 32.453
%10d	format integer for 10 reserved spaces such as 23
%3o	format octal integer number for 3 hexadecimal characters
%10.6e	format exponent format with 6 decimal places

2.3.2 C standard input (`scanf()`, `gets()` and `getchar()`)

The keyboard is normally the standard input to a program. As with the output functions, the input functions are not part of the standard language and are contained in a standard C library. Definitions (or prototypes) of these functions are found in the header file *stdio.h*. By including this header file a degree of error checking is initiated at compilation. The compiler checks, among other things, the data types of the parameters passed into the functions. It is thus less likely that there will be any run-time errors.

There are three main input functions, these are:

scanf (*"format",&arg1,&arg2..&argn*)	reads formatted values from the keyboard in a format defined by *format* and loads them into the arguments *arg1*, *arg2*, etc.
gets (*string*)	reads a string of text from the keyboard into *string* (up to a new line)
ch= getchar ()	reads a single character from the keyboard into *ch*

If a numeric or a character variable is used with the scanf () function an ampersand (&) precedes each parameter in the argument list (there are exceptions and these will be discussed in Chapters 6 and 7). This prefix causes the memory address of the variable to be used as a parameter and not the value. This allows scanf () to change the value of the variable (this will also be explained in more detail in Chapter 6). For now, it should be assumed that an ampersand precedes all simple numerical and character data types when using scanf (). The general format of the scanf () function is scanf (format, &arg1, &arg2...).

The first argument *format* is a string that defines the format of all entered values. For example, "%f %d" specifies that *arg1* is entered as a float and *arg2* as an integer. This string should only contain the conversion control characters such as %d, %f, %c, %s, etc., separated by spaces. Figure 2.3 shows an example of the scanf () function reading a float, an integer and a character into the variables val1, val2 and ch.

The gets (str) function reads a number of characters into a variable (in this case str); these characters are read until the ENTER key is pressed. The getchar () function reads a single character from the input. This character is returned via the function header and not through the argument list.

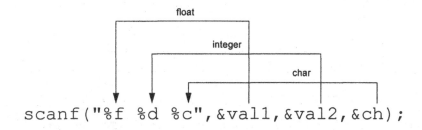

Figure 2.3 An example of the scanf () statement

2.4 Examples

This section contains some practical electronics-related examples of C and Pascal programs. Each of the test runs relate to a run of the C program (although most of the Pascal test runs are almost identical).

2.4.1 Fahrenheit to centigrade conversion

Temperature is typically measured in either centigrade or Fahrenheit. The conversion from Fahrenheit to centigrade is:

$$C = \frac{5}{9}.(F - 32) \quad °C$$

Program 2.2 converts from an entered value of Fahrenheit (faren) into centigrade (cent). and Test run 2.1 shows a sample run for an entered value of 80°F. The resultant value is displayed with 8 places reserved for the answer and 2 decimal places. This is specified in C as %8.2f and in Pascal with :8:2 after the variable.

Program 2.2

```c
#include <stdio.h>

int    main(void)
{
float faren,cent;

    puts("Program to convert Fahrenheit to centigrade");
    printf("Enter a temperature (in Fahrenheit)");
    scanf("%f",&faren);

    cent=5.0/9.0*(faren-32);

    printf("Temperature is %8.2f deg C\n",cent);

    return(0);
}
```

Program 2.2

```pascal
program prog2_2(input,output);

var     faren,cent:real;

begin
    writeln('Program to convert Fahrenheit to centigrade');
    writeln('Enter a temperature (in Fahrenheit)');
    readln(faren);

    cent:=5/9*(faren-32);

    writeln('Temperature is ',cent:8:2,' deg C');

end.
```

Test run 2.1

```
Program to convert Fahrenheit to centigrade
Enter a temperature (in Fahrenheit)  80
Temperature is    26.67 deg C
```

2.4.2 Gradient of a straight line

The equation of a straight line is:

$$y = mx + c$$

where m is the gradient of the line and c is the point at which the line cuts the y-axis. If two points on the line are known, (x_1,y_1) and (x_2,y_2) then m can be calculated by:

$$m = \frac{y_2 - y_1}{x_2 - x_1}$$

and the c value can be calculated from:

$$c = y - mx$$
$$= y_1 - mx_1$$

Program 2.3 determines the gradient of a straight line for entered value of x_1,y_1 and x_2,y_2 (Note that the solution of the value for c will be left as an exercise). Test run 2.2 is a sample test run.

Program 2.3

```c
#include <stdio.h>

int    main(void)
{

float x1,x2,y1,y2,m;

   puts("Program to determine the gradient");
   puts("of a straight line");

   printf("Enter x1, y1 >> ");
   scanf("%f %f", &x1,&y1);

   printf("Enter x2, y2 >> ");
   scanf("%f %f",&x2,&y2);

   m=(y2-y1)/(x2-x1);

   printf("Gradient is %8.2f",m);

   return(0);
}
```

Program 2.3

```pascal
program  prog2_3(input,output);

var      x1,x2,y1,y2,m:real;
```

```
begin
   writeln('Program to determine the gradient');
   writeln('of a straight line');

   write('Enter x1, y1 >> ');
   readln(x1,y1);
   write('Enter x2, y2 >> ');
   readln(x2,y2);

   m:=(y2-y1)/(x2-x1);

   writeln('Gradient is ',m:8:2);
end.
```

Test run 2.2
```
Program to determine the gradient
of a straight line
Enter x1, y1 >> 3 4
Enter x2, y2 >> 5 6
Gradient is     1.00
```

2.4.3 Force of attraction

The gravitational force between two objects of mass m_1 and m_2 of a distance d apart is given by:

$$F = \frac{G.m_1.m_2}{d^2} \quad N$$

where G is a gravitation constant and is equal to 6.67×10^{-11} $m^3.kg^{-1}sec^{-2}$. Program 2.4 determines the gravitation force and Test run 2.3 shows a test run for the gravitation force between an apple and the earth. The parameters used are:

$$m_{earth} = 6 \times 10^{24} \text{ kg}$$
$$m_{apple} = 0.1 \text{ kg}$$
$$r_{earth} = 6370000 \text{ m}$$

The resultant gravitation force is 0.99 N, which is similar to the calculation using:

$$F = ma$$
$$= 0.1 \times 9.81 \quad N$$
$$= 9.81 \text{ N}$$

The gravitation force constant (G) has been defined, in C, with the #define statement and, in Pascal, with a const.

Program 2.4

```c
#define  G  6.67e-11

#include <stdio.h>

int   main(void)
{
float    force,m1,m2,distance;

   puts("Program to force between two objects");

   printf("Enter mass of first object (kg)");
   scanf("%f",&m1);

   printf("Enter mass of second object (kg)");
   scanf("%f",&m2);

   printf("Enter distance between objects (m)");
   scanf("%f",&distance);

   force=G*m1*m2/(distance*distance);

   printf("Force is %8.2f N\n",force);
   return(0);
}
```

Program 2.4

```pascal
program prog2_4(input,output);
const    G=6.67e-11;

var      force,m1,m2,distance:real;

begin
   writeln('Program to force between two objects');

   writeln('Enter mass of first object (kg)');
   readln(m1);

   writeln('Enter mass of second object (kg)');
   readln(m2);

   writeln('Enter distance between objects (m)');
   readln(distance);

   force:=G*m1*m2/(distance*distance);

   writeln('Force is ',force:8:2,' N');
end.
```

Test run 2.3
```
Program to force between two objects
Enter mass of first object (kg)  6e24
Enter mass of second object (kg) 0.1
Enter distance between objects (m) 6.67e-11
Force is     0.99 N
```

2.4.4 Capacitive reactance

The reactance of a capacitor depends upon the applied frequency. At low frequencies the reactance is extremely high and at high frequencies it is low.

The reactance (X_C) of a capacitor, of capacitance C (Farads), at an applied frequency f (Hertz) is be given by:

$$X_C = \frac{1}{2\pi f C} \quad \Omega$$

Figure 2.4 shows a schematic of this arrangement. There is one output variable (X_C), two input variables (f and C) and a single constant (π). Program 2.5 shows a sample program and test run 2.4 is a sample run. In C a constant is declared with the #define pre-processor option and Pascal uses the const keyword.

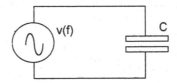

Figure 2.4   Capacitor connected to sinusoidal voltage source

Program 2.5
```
/*     Program to calculate capacitive reactance     */
#include    <stdio.h>
#define  PI 3.14159

int      main(void)
{
float      freq,cap,X_c;

    puts("Enter frequency and capacitance");
    scanf("%f %f",&freq,&cap);

    X_c=1.0/(2.0*PI*freq*cap);

    printf("Capacitive Reactance is %6.3f ohms\n",  X_c);
    return(0);
}
```

Program 2.5
```
program  prog2_2(input,output);

const    PI=3.14157;

var      freq,cap,X_c:real;

begin
     writeln('Enter frequency and capacitance');
     readln(freq,cap);

     X_c:=1/(2*PI*freq*cap);

     writeln('Capacitive reactance is ',X_c:8:3,' ohms');
end.
```

⌨ **Test run 2.4**
```
Enter frequency and capacitance
10e3 1e-6
Capacitive Reactance is 15.916 ohms
```

2.4.5 Impedance of an RL series circuit

The magnitude of the impedance of an RL series circuit (modulus $|Z|$) is given by the equation:

$$|Z| = \sqrt{R^2 + X_L{}^2} \quad \Omega$$

and the angle of the impedance (argument $\langle Z \rangle$) is given by:

$$\langle Z \rangle = \tan^{-1} \frac{X_L}{R}$$

Figure 2.5 shows a schematic of an RL series circuit. Program 2.3 determines the magnitude and the angle of the impedance using entered values of resistance (R), inductance (L) and frequency (freq). In C the inverse tangent ($\tan^{-1}$) function is defined (or prototyped) in *math.h* and is named atan(), whereas, Pascal uses the arctan() function. Both these functions return the inverse tangent in radians. The program converts the returned value into degrees by scaling it by $\pi/180$.

The program uses the square root function; in C this function is (sqrt()) which is prototyped in the *math.h* header file. This inclusion helps the compiler check the format of the values sent to the function as it checks the general syntax of the function call. It also informs the compiler that the value returned is a floating point (this will be discussed in greater detail in a later chapter). In Pascal it is also named sqrt().

Test run 2.5 is a sample output using entered values $R = 100\,\Omega$, $L = 100\,\text{mH}$ and *frequency* $= 1\,\text{kHz}$. The impedance has a magnitude of $118.10\,\Omega$ and an angle of $32.14°$.

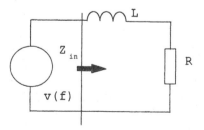

Figure 2.5 RL series circuit

⬡ C

Program 2.6

```
/*      Program to determine the impedance of a         */
/*      series RL circuit                                */

#include      <stdio.h>
#include      <math.h> /*required for sqrt() and atan() */

#define  PI 3.14159

int       main(void)
{
float     R,L,freq,Xl,Zin_mag,Zin_angle;

    printf("Enter R, L and frequency >>");
    scanf("%f %f %f",&R,&L,&freq);

    Xl= 2 * PI * freq * L;
    Zin_mag = sqrt(R*R+Xl*Xl);
    Zin_angle= atan(Xl/R)*180.0/PI;
        /* atan is arc tan and returns radians    */
        /* 180/PI converts to degrees             */

    printf("Zin mag %6.2f ohm, angle %6.2f degrees\n",Zin_mag,Zin_angle);
    return(0);
}
```

⬡ Pascal

Program 2.6

```
program  prog2_3(input,output);

const     PI=3.14159;

var       R,L,freq,Xl,Zin_mag,Zin_angle:real;

begin
    write('Enter R, L and frequency >>');
    readln(R,L,freq);
    Xl:= 2 * PI * freq * L;
    Zin_mag := sqrt(R*R+Xl*Xl);
    Zin_angle:= arctan(Xl/R)*180.0/PI;
        { arctan is arc tan and returns radians   }
        { 180/PI converts to degrees              }
    writeln('Zin mag ', Zin_mag:6:2, ' ohm, angle ',Zin_angle:6:2,' deg');
end.
```

🖥 **Test run 2.5**

```
Enter R, L and frequency >> 100 10e-3 1000
Zin mag 118.10 ohm, angle  32.14 degrees
```

2.4.6 Resistors in parallel

Program 2.7 determines the equivalent resistance of three resistors connected in parallel. Figure 2.6 gives a schematic diagram of this set-up. The resistors connected are R_1, R_2 and R_3 and the equivalent input resistance is R_{equ}. Test run 2.6 shows a run with values of 250, 500 and 1000 Ω.

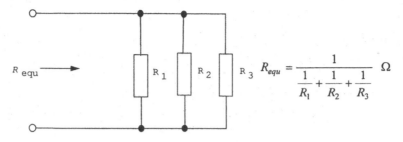

Figure 2.6 Three resistors connected in parallel

Program 2.7

```c
/*      Program to determine equivalent resistance   */
#include <stdio.h>

int       main(void)
{
float     R1,R2,R3,R_equ;

   puts("Program to determine equivalent resistance");
   puts("of three resistors connected in parallel");
   puts("Enter three values of resistance >>");
   scanf("%f %f %f",&R1,&R2,&R3);

   R_equ=1.0/(1.0/R1+1/R2+1/R3);

   printf("R1=%8.3f, R2=%8.3f and R3=%8.3f ohms\n",   R1,R2,R3);
   printf("Equivalent resistance is %8.3f ohms\n",R_equ);
   return(0);
}
```

Program 2.7

```pascal
program   prog2_4(input,output);

var       R1,R2,R3,R_equ:real;
begin
   writeln('Program to determine equivalent resistance');
   write('of three resistors connected in parallel');
   write('Enter three values of resistance >>');
   readln(R1,R2,R3);

   R_equ:=1.0/(1.0/R1+1/R2+1/R3);
   writeln('R1=',R1,'R2=',R2,' and R3=',R3, ' ohms');
   writeln('Equivalent resistance is ',R_equ,' ohms');
end.
```

🖳 **Test run 2.6**
```
Program to determine equivalent resistance
of three resistors connected in parallel
Enter three values of resistance >>
1000 500 250
R1=1000.000, R2= 500.000, R3= 250.000 ohms
Equivalent resistance is 142.857 ohms
```

2.4.7 Bit operations

Program 2.8 relates to digital electronics and illustrates the power of C and

Pascal when dealing with low-level bit operators. The C program uses &, | , ^ and ~ bitwise operators to create AND, OR, EX-OR, NAND and NOR Boolean functions. Pascal uses the operators: and, or, xor and not. The NAND and NOR functions are generated by inverting the AND and OR operations.

The C program allows the entry of hexadecimal values using the %x format descriptor in scanf() statement. Test run 2.7 shows a run with test values.

Program 2.8

```
/*     Program to bitwise AND, OR, NAND, NOR and EX-OR    */
/*     two hexadecimal values                             */

#include <stdio.h>

int    main(void)
{
int    value1, value2;

       /* & - bitwise AND operator    */
       /* | - bitwise OR  operator    */
       /* ^ - bitwise EX-OR operator  */
       /* ~ - bitwise NOT operator    */

   printf("Enter two hex values >>> ");
   scanf("%x %x",&value1,&value2);

   printf("Values ANDed is %x\n",value1 & value2);
   printf("Values ORed is %x\n",value1 | value2);
   printf("Values Ex-ORed is %x\n",value1 ^ value2);
   printf("Values NANDed is %x\n",~(value1 & value2));
   printf("Values NORed is %x\n",   ~(value1 | value2));
   return(0);
}
```

Test run 2.7

```
Enter two hex values >>> E215 C431
Values ANDed is c011
Values ORed is e635
Values Ex-ORed is 2624
Values NANDed is 3fee
Values NORed is 19ca
```

The bit patterns used in the test run are 1110 0010 0001 0101 (E215h) and 1100 0100 0011 0001 (C431h). To verify the program the hexadecimal equivalents of these values are operated on by the Boolean operators and the results checked against the test run results.
The AND operation gives the following:

HEX	BINARY
E215	1110 0010 0001 0101
C431	1100 0100 0011 0001
C011	1100 0000 0001 0001

The OR operation gives the following:

HEX	BINARY
E215	1110 0010 0001 0101
C431	1100 0100 0011 0001
E635	1110 0110 0011 0101

The EX-OR function gives the following:

HEX	BINARY
E215	1110 0010 0001 0101
C431	1100 0100 0011 0001
2624	0010 0110 0010 0100

The inverse of AND (NAND) will be 0011 1111 1110 1110 (3FEEh); the inverse of the OR (NOR) is 0001 1001 1100 1010 (19CAh). These results are identical to these in test run 2.16. Thus the test has been successful.

The Pascal Program 2.8 allows the input of the values as an integer. Test run 2.8 shows a sample test run.

Program 2.8
```
program prog2_8(input,output);

var    value1, value2:integer;

begin
   writeln('Enter integer values >>> ');
   readln(value1,value2);
   writeln('Values ANDed is ',value1 and value2);
   writeln('Values ORed is ',value1 or value2);
   writeln('Values Ex-ORed is ',value1 xor value2);
   writeln('Values NANDed is ',not(value1 and value2));
   writeln('Values NORed is ',   not(value1 or value2));
end.
```

Test run 2.8
```
Enter integer values >>>
25 11
Values ANDed is 9
Values ORed is 27
Values Ex-ORed is 18
Values NANDed is -10
Values NORed is -28
```

Worksheet 2

W2.1 Enter two programs from the chapter and verify that their output conforms with the sample test runs.

Completed successfully:	YES/NO

W2.2 Modify program 2.3 so that it also calculates the value of c. Use this program to complete Table W2.1.

Table W2.1 Straight lines calculations

x_1	y_1	x_2	y_2	m	c
3	3	6	5		
7	1	−1	4		
1000	500	10	40		
−100	3	−5	−9		
5	−10	−10	10		

W2.3 Write a program which calculates the magnitude of a complex number of x+jy (or in another form x+iy) and complete Table W2.2 (note that the first row has been completed). The magnitude is given by:

$$Mag = \sqrt{x^2 + y^2}$$

Table W2.2 Magnitude

x	y	Mag
3	4	5
50	70	
−9	−9	
100	100	
0.1	0.5	
30	−10	

W2.4 Write a program which calculates the angle of a complex number of x+jy (or in another form x+iy) and complete Table W2.3. The angle is given by:

$$Angle = \tan^{-1}\left(\frac{y}{x}\right) \text{ radians}$$

Table W2.3 Angle

x	y	Angle (radians)
3	4	0.9273
–9	–9	
100	100	
0.1	0.5	
30	–10	

Worksheet 3

W3.1 Modify the program written in W2.4 so that it converts the angle to degree and complete Table W3.1. An angle converted from radians to degree using:

$$Angle \ (degrees) = \frac{180}{\pi} . Angle(radians)$$

Table W3.1 Angle

x	y	Angle (degrees)
3	4	53.13
–9	–9	
100	100	
30	–10	

W3.2 Write a program which determines the equivalent resistance of three parallel resistors. Use this program to complete Table W3.2.

Table W3.2 Equivalent parallel resistance

R1 (Ω)	R2 (Ω)	R3 (Ω)	R_equ (Ω)
1000	1000	1000	
200	100	50	
1.2K	1K	800	
1M	0.5M	250K	

W3.3 Modify the program in Question W3.2 so that the user enters the applied voltage to the parallel resistors and the program determines the current in each of the resistor and the input current. Use this program to complete Table W3.3. Note that the current in each of the resistors is simply the applied voltage divided by each of the resistors. A sample test is given in Test run 2.9.

💻 **Test run 2.9**

```
Enter resistor values R1, R2 and R3>> 1000 500 250
Enter applied voltage >> 10
Equivalent resistance is  142.86 ohms
I1=0.010 A, I2=0.020 A, I3=0.040 A, I=0.070 A
```

Table W3.3 Current flow

V_{in}	$R1\,(\Omega)$	$R2\,(\Omega)$	$R3\,(\Omega)$	$R\_equ\,(\Omega)$	$I\,(A)$	$I_{R1}\,(A)$	$I_{R2}\,(A)$	$I_{R3}\,(A)$
10	1000	1000	1000					
1	200	100	50					
5	1.2K	1K	800					
100	1M	0.5M	250k					

W3.4 Modify the program in Question W3.3 so that is displays the current in milliAmps (mA). Note to convert to mA then multiply the value by 1000. A sample test is given in Test run 2.10.

💻 **Test run 2.10**

```
Enter resistor values R1, R2 and R3>> 1000 500 250
Enter applied voltage >> 10
Equivalent resistance is  142.86 ohms
I1=10 mA, I2=20 mA, I3=40 mA, I=70 mA
```

---

3 Selection Statements

3.1 `if...else` statements

A decision is made with the `if` statement. It logically determines whether a conditional expression is TRUE or FALSE. For a TRUE, the program executes one block of code; a FALSE causes the execution of another (if any). The keyword `else` identifies the FALSE block. In C, braces (`{ }`) are used to define the start and end of the block. In Pascal, the `begin` and `end` keywords are used.

Relationship operators, include:

- Greater than (`>`).
- Less than (`<`).
- Greater than or equal to (`>=`).
- Less than or equal to (`<=`).
- Equal to (in C, it is `==` and, in Pascal, it is `=`).
- Not equal to (in C, it is `!=` and, in Pascal it is `<>`).

These operations yield a TRUE or FALSE from their operation. Logical statements (`&&`, `||`, `!`) can then group these together to give the required functionality. These are:

- AND (in C, it is `&&`, and, in Pascal it is `and`);
- OR (in C, it is `||`, and, in Pascal it is `or`);
- NOT (in C, it is `!`, and, in Pascal it is `not`).

If the operation is not a relationship, such as bitwise or an arithmetic operation, then any non-zero value is TRUE and a zero is FALSE. The following is an example syntax of the `if` statement. If the statement block has only one statement then, in C, the braces (`{ }`) can be excluded (in Pascal the `begin` and `end` can be excluded).

Pascal Syntax
```
if (expression)
begin
    statement block
end;
```

C Syntax
```
if (expression)
{
    statement block
}
```

The following is an example format with an `else` extension.

Pascal Syntax
```
if (expression)
begin
    statement block1
end
else
begin
    statement block2
end;
```

C Syntax
```
if (expression)
{
    statement block1
}
else
{
    statement block2
}
```

It is possible to nest `if..else` statements to give a required functionality. In the next example, *statement block1* is executed if `expression1` is TRUE. If it is FALSE then the program checks the next expression. If this is TRUE the program executes *statement block2*, else it checks the next expression, and so on. If all expressions are FALSE then the program executes the final `else` statement block, in this case, *statement block 4*:

Pascal Syntax
```
if (expression1) then
begin
    statement block1
end
else if (expression2) then
begin
    statement block2
end
else if (expression3) then
begin
    statement block3
end
else
begin
    statement block4
end;
```

C Syntax
```
if (expression1)
{
    statement block1
}
else if (expression2)
{
    statement block2
}
else if (expression3)
{
    statement block3
}
else
{
    statement block4
}
```

Figure 3.1 shows a diagrammatic represention of this example statement.

3.1.1 Examples

This section contains some C and Pascal example programs.

Quadratic equations
Some electrical examples require the solution of a quadratic equation. The standard form is:

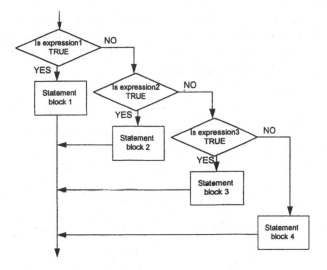

Figure 3.1 Structure of the compound `if` statement

$$ax^2 + bx + c = 0$$

The solution of x in this equation is given by:

$$x_{1,2} = \frac{-b \pm \sqrt{b^2 - 4ac}}{2a}$$

This can yield three possible types of results:

1. if b^2=4ac, there will be a single real root (x=-b/2a)
2. else, if b^2>4ac, there will be two real roots:

$$x_1 = \frac{-b + \sqrt{b^2 - 4ac}}{2a}, \qquad x_2 = \frac{-b - \sqrt{b^2 - 4ac}}{2a}$$

3. else, the roots will be complex:

$$x_1 = \frac{-b}{2a} + j\frac{\sqrt{4ac - b^2}}{2a}, \qquad x_2 = \frac{-b}{2a} - j\frac{\sqrt{4ac - b^2}}{2a}$$

Program 3.1 determines the roots of a quadratic equation. In this program the `if..else` statement is used to determine if the roots are real, complex or singular. The value passed to the square-root function (`sqrt()`) should be tested to determine if it is negative. If it is, it may cause the program to terminate as the square root of a negative number cannot be calculated (it is numerically invalid). The program may also terminate if a is zero as this causes a divide by zero error (the trap for this error is left as a tutorial question). Note that, in C, the sqrt() function is prototyped in the *math.h*, thus the *math.h* header is included in the program.

⬡ C

Program 3.1

```c
/*      Program to determine roots of a quadratic equation*/
#include    <stdio.h>
#include    <math.h>

int     main(void)
{
float   a,b,c,real1,real2,imag;

    puts("Program to determine roots of a quadratic equation");
    printf("Enter a,b and c >>>");

    scanf("%f %f %f",&a,&b,&c);

    printf("Equation is %.2fx*x + %.2fx + %.2f\n",a,b,c);
    if ((b*b)==(4*a*c))
    {     /*      singular root            */
       real1=-b/(2*a);
       printf("Root is %.2f\n",real1);

    } else if ((b*b)>(4*a*c))
    {     /*      real roots              */
       real1=(-b+sqrt( (b*b)-4*a*c )) /(2*a);
       real2=(-b-sqrt( (b*b)-4*a*c )) /(2*a);
       printf("Roots are %.2f, %.2f\n",real1,real2);
    } else
    {     /*      complex roots           */
       real1=-b/(2*a);
       imag=sqrt(4*a*c-b*b)/(2*a);
       printf("Roots are %.2f +/- j%.2f\n",real1,imag);
    }
    return(0);
}
```

⬡ Pascal

Program 3.1

```pascal
program if1(input,output);
(* Program to determine roots of a quadratic equation *)

var a,b,c,real1,real2,imag:real;

begin
    writeln('Program to determine roots of a quadratic equation');
    write('Enter a, b and c >');
    readln(a,b,c);

    writeln('Equation is ',a:6:2,'x*x+',b:6:2,'x+',c:6:2);

    if ((b*b)=(4*a*c)) then
    begin
       real1:=-b/(2*a);
       writeln('Root is ',real1:6:2);
    end
    else if ((b*b)>=(4*a*c)) then
    begin
       real1:=(-b+sqrt( (b*b)-4*a*c))/(2*a);
       real2:=(-b-sqrt( (b*b)-4*a*c))/(2*a);
```

```
        writeln('Roots are ',real1:6:2,real2:6:2);
   end
   else
   begin
      real1:=-b/(2*a);
      imag:=sqrt(4*a*c-b*b)/(2*a);
      writeln('Roots are ',real1:6:2,'+/-j',imag:6:2);
   end;
end.
```

Three test runs 3.1, 3.2 and 3.3 test each of the three types of roots that occur. In Test run 3.1 the roots of the equation are real. In Test run 3.2 the roots are complex, i.e. in the form x+jy. In Test run 3.3 the result is a singular root.

Test run 3.1
```
Program to determine roots of a quadratic equation
Enter a,b and c >>> 1 1 -2
Equation is 1.00x*x + 1.00x + -2.00
Roots are 1.00, -2.00
```

Test run 3.2
```
Program to determine roots of a quadratic equation
Enter a,b and c >>> 2 2 4
Equation is 2.00x*x + 2.00x + 4.00
Roots are -0.50 +/- j1.32
```

Test run 3.3
```
Program to determine roots of a quadratic equation
Enter a,b and c >>> 1 2 1
Equation is 1.00x*x + 2.00x + 1.00
Root is -1
```

Electromagnetic (EM) waves

Program 3.2 uses the if statement to determine the classification of an EM wave given its wavelength. Figure 3.2 illustrates the EM spectrum spanning different wavelengths. The classification of the wave is determined either by the frequency or the wavelength (normally radio and microwaves are defined by their frequency, whereas other types by their wavelength). For example, an EM wave with a wavelength of 10 m is classified as a radio wave, a wavelength of 500 nm as visible light and a wavelength of 50 cm is in the microwave region. Test run 3.4 shows a sample run.

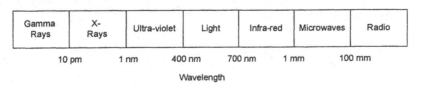

Figure 3.2 EM spectrum

⬡ C

Program 3.2

```c
/*      Program to determine type of EM wave for a given wavelength */
#include <stdio.h>
int       main(void)
{
float     lambda;

   printf("Enter wavelength>>>");
   scanf("%f",&lambda);

   printf("Electromagnetic wave is ");
   if       (lambda<1e 11)     puts("Gamma Ray !!!");
   else if  (lambda<1e-9)      puts("X-ray");
   else if  (lambda<400e-9)    puts("Ultaviolet");
   else if  (lambda<700e-9)    puts("LIGHT");
   else if  (lambda<1e-3)      puts("Infrared");
   else if  (lambda<1e-1)      puts("Microwave");
   else                        puts("Radio wave");

   return(0);
}
```

⬡ Pascal

Program 3.2

```pascal
program if2(input,output);
(*      Program to determine type of EM wave for a given wavelength *)
var     lambda:real;

begin
   write('Enter wavelength >>>');
   readln(lambda);

   write('Electomagnetic wave is ');

   if (lambda<1e-11) then writeln('Gamma Rays !!!')
   else if (lambda<1e-9) then writeln('X-rays')
   else if (lambda<400e-9) then writeln('Ultra-violet')
   else if (lambda<700e-9) then writeln('LIGHT')
   else if (lambda<1e-3) then writeln('Infrared')
   else if (lambda<1e-1) then writeln('Microwaves')
   else writeln('Radio waves');
end.
```

🖳 **Test run 3.4**

```
Enter wavelength>>> 1e-10
Electromagnetic wave is X-ray
```

EM waves can also be specified by their frequency. Program 3.3 allows the user to enter the frequency of the wave, and the program then determines the wavelength using the formula:

$$\lambda = \frac{c}{f}$$

where c is the speed of light and f the frequency of the wave.

Program 3.3

```c
/*    Program to determine type of wave for an entered frequency */
#include    <stdio.h>
#define  SPEED_OF_LIGHT 3e8

int       main(void)
{
float lambda,freq;

   printf("Enter frequency>>>");
   scanf("%f",&freq);

   lambda=SPEED_OF_LIGHT/freq;

   printf("Wavelength is %.2e m. EM wave is ", lambda);
   if        (lambda<1e-11)      puts("Gamma Rays !!!");
   else if   (lambda<1e-9)       puts("X-rays");
   else if   (lambda<400e-9)     puts("Ultaviolet");
   else if   (lambda<700e-9)     puts("LIGHT");
   else if   (lambda<1e-3)       puts("Infrared");
   else if   (lambda<0.3e-1)     puts("Microwave");
   else                          puts("Radio waves");

   return(0);
}
```

Program 3.3

```pascal
program if3(input,output);
(*    Program to determine type of wave for an entered frequency *)
const SPEED_OF_LIGHT=3e8;

var lambda,freq:real;

begin

   write('Enter frequency >>>');
   readln(freq);

   lambda:=SPEED_OF_LIGHT/freq;

   write('Wavelength is ',lambda:6:2,' EM wave is ');

   if (lambda<1e-11) then writeln('Gamma Rays !!!')
   else if (lambda<1e-9) then writeln('X-rays')
   else if (lambda<400e-9) then writeln('Ultra-violet')
   else if (lambda<700e-9) then writeln('LIGHT')
   else if (lambda<1e-3) then writeln('Infrared')
   else if (lambda<1e-1) then writeln('Microwave')
   else writeln('Radio waves');
end.
```

Test run 3.5 shows a sample run.

Test run 3.5

```
Enter frequency>>> 10e9
Wavelength is 3.0e-02 m. EM wave is Microwave
```

Series/ parallel resistances

Program 3.4 determines the equivalent resistance of two resistors connected either in series or parallel. The C program uses scanf() to get the two resistance values and getchar() to select the circuit configuration. Problems can occur when using getchar() after scanf() due to new-line characters being stored in the keyboard buffer. The statement fflush(stdin) has been inserted into the program in order to clear the buffer before getchar() is called (stdin represents the standard input device, that is, the keyboard). The C program also uses tolower() to convert the entered circuit selection character to lower-case (this is protoyped in the *ctype.h*). In Turbo Pascal there is no lowercase conversion so the uppercase equivalent is used (upcase).

⬡ C
Program 3.4

```
/*    Program to determine the equivalent resistance of two*/
/*    resistors connected either in series or parallel     */

#include <stdio.h>
#include <ctype.h>   /* required for tolower() function    */

int      main(void)
{
float    R1,R2,R_equ;
char     ch;

   printf("Enter two resistance values >>");

   scanf("%f %f",&R1,&R2);

   fflush(stdin); /* flush keyboard buffer                */
   printf("Do you require (s)eries or (p)arallel >>");
   ch=getchar();

   if (tolower(ch)=='s')   /* convert character to lowercase     */
   {
      R_equ=R1+R2;
      printf("Equivalent series resistance is %8.2f ohms",R_equ);
   }
   else if (tolower(ch)=='p')
   {
      R_equ=(R1*R2)/(R1+R2);
      printf("Equivalent parallel resistance is %8.2f ohms",R_equ);
   }
   else puts("Invalid entry");

   return(0);
}
```

⬡ Pascal
Program 3.4

```
program if4(input,output);

(* Program to determine the equivalent resistance of two   *)
(* resistors either connected in series of parallel        *)

var     R1,R2,R_equ:real;
        ch:char;
begin
   writeln('Enter two resistor values >>');
   readln(R1,R2);
```

```
    writeln('Do you require (s)eries or (p)arallel >>');
    readln(ch);

    if (upcase(ch)='S') then (* convert character to uppercase *)
    begin
        R_equ:=R1+R2;
        writeln('Equivalent series resistance is ',R_equ,' Ohms');
    end
    else if (upcase(ch)='P') then
    begin
        R_equ:=(R1*R2)/(R1+R2);
        writeln('Equivalent parallel resistance is ',R_equ,' Ohms');
    end
    else writeln('Invalid entry');
end.
```

3.2 `switch` **statement**

The `case` statement (and `switch` in C) is used when there are multiple decision to be made. It is normally used to replace the `if` statement when there are many routes of execution the program execution can take. The syntax of `case`/`switch` is as follows.

Pascal Syntax
```
case (constant) of
const1: statement(s);
const2: statement(s);
   :        :
end;
```

C Syntax
```
switch (expression)
{
    case const1: statement(s) : break;
    case const2: statement(s) ; break;
      :              :
    default:        statement(s) ; break;
}
```

In Pascal, the `case` statement simply selects which one of the constants (`const1`, `const2`, and so on) matches the `constant` value. In C, the `switch` statement checks the `expression` against each of the constants in sequence (the constant must be an integer or character data type). When a match is found the statement(s) associated with the constant is(are) executed. The execution carries on to all other statements until a `break` is encountered or to the end of `switch`, whichever is sooner. If the `break` is omitted, the execution continues until the end of `switch`.

If none of the constants matches the `switch` expression a set of statements associated with the default condition (`default:`) is executed.

3.2.1 Examples

Resistor colour code
Resistors are normally identified by means of a colour code system, as outlined in Table 3.1. Program 3.5 uses a `case`/`switch` statement to determine the colour of a resistor band for an entered value.

Table 3.1 Resistor colour coding system

Digit	Colour	Multiplier	Digit	Colour	Multiplier
	SILVER	0.01	4	YELLOW	10 K
	GOLD	0.1	5	GREEN	100 K
0	BLACK	1	6	BLUE	1 M
1	BROWN	10	7	VIOLET	10 M
2	RED	100	8	GREY	
3	ORANGE	1 K	9	WHITE	

⬡ C

Program 3.5

```c
/*      Program to determine colour code for a single    */
/*      resistor band digit                              */
#include <stdio.h>

int   main(void)
{
int colour;

    printf("Enter value of colour band(0-9)>>");
    scanf("%d",&colour);

    printf("Resistor colour band is ");

    switch (colour)
    {
        case 0: printf("BLACK");    break;
        case 1: printf("BROWN");    break;
        case 2: printf("RED");      break;
        case 3: printf("ORANGE");   break;
        case 4: printf("YELLOW");   break;
        case 5: printf("GREEN");    break;
        case 6: printf("BLUE");     break;
        case 7: printf("VIOLET");   break;
        case 8: printf("GREY");     break;
        case 9: printf("WHITE");    break;
    }
    return(0);
}
```

⬡ Pascal

Program 3.5

```pascal
program case1(input,output);
(*      Program to determine colour code for a single    *)
(*      resistor band digit                              *)

var colour:integer;

begin
    write('Enter value >>');
    readln(colour);

    write('Resistor colour band is ');

    case (colour) of
    0: write('BLACK');
    1: write('BROWN');
    2: write('RED');
    3: write('ORANGE');
    4: write('YELLOW');
    5: write('GREEN');
```

```
    6: write('BLUE');
    7: write('VIOLET');
    8: write('GREY');
    9: write('WHITE');
    end;

end.
```

Test run 3.6 shows a sample run.

```
Enter value of colour band(0-9)>> 3
Resistor colour band is ORANGE
```

Program 3.6 uses #define directives to define each of the resistor colour bands. There may be a clash with these defines if other header files contain these definitions. If this occurs change the defines to RES_BLACK, RES_BROWN, etc.

A default: has been added to catch any invalid input (such as less than 0 or greater than 9).

Program 3.6
```c
/* Program to determine colour code for resistor band digit   */
#include <stdio.h>
#define  BLACK     0
#define  BROWN     1
#define  RED       2
#define  ORANGE    3
#define  YELLOW    4
#define  GREEN     5
#define  BLUE      6
#define  VIOLET    7
#define  GREY      8
#define  WHITE     9

int   main(void)
{
int   colour;

    printf("Enter value of colour band(0-9)>>");
    scanf("%d",&colour);
    printf("Resistor colour band is ");

    switch (colour)
    {
        case BLACK: printf("BLACK");        break;
        case BROWN: printf("BROWN");        break;
        case RED:   printf("RED");          break;
        case ORANGE:printf("ORANGE");       break;
        case YELLOW:printf("YELLOW");       break;
        case GREEN: printf("GREEN");        break;
        case BLUE:  printf("BLUE");         break;
        case VIOLET:printf("VIOLET");       break;
        case GREY:  printf("GREY");         break;
        case WHITE: printf("WHITE");        break;
        default:    printf("NO COLOUR");    break;
    }
    return(0);
}
```

Program 3.6

```pascal
program case2(input,output);
(* Program to determine colour code for resistor band digit   *)

const BLACK=0;     BROWN=1;      RED=2;      ORANGE=3;
      YELLOW=4;    GREEN=5;      BLUE=6;     VIOLET=7;
      GREY=8;      WHITE=9;
var colour:integer;

begin
     write('Enter value >>');
     readln(colour);

     write('Resistor colour band is ');

     case (colour) of
      BLACK:   write('BLACK');
      BROWN:   write('BROWN');
      RED:     write('RED');
      ORANGE:  write('ORANGE');
      YELLOW:  write('YELLOW');
      GREEN:   write('GREEN');
      BLUE:    write('BLUE');
      VIOLET:  write('VIOLET');
      GREY:    write('GREY');
      WHITE:   write('WHITE');
      else write('INVALID');
      end;
end.
```

Resistance of a conductor

The resistance of a cylindrical conductor is a function of its resistivity, cross-sectional area and length. These parameters are illustrated in Figure 3.3. The resistance is given by:

$$R = \frac{\rho \cdot l}{A} \quad \Omega$$

where

ρ = resistivity of the conductor (Ω.m);
l = length of the conductor (m);
A = cross-sectional area of the conductor (m^2).

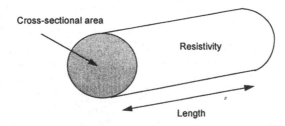

Figure 3.3 Cylindrical conductor

Program 3.7 determines the resistance of a cylindrical conductor made from either silver, manganese, aluminium or copper. The resistivities of these materials have been defined using #define macros (in C) or const (in Pascal).

The user enters the conductor type as a character ('c', 'a', 's' or 'm') which can either be in upper or lowercase format as the tolower() function converts the entered character to lowercase (this is protyped in *ctype.h*). When an invalid character is entered the default condition of the case statement is executed, and the text Invalid option is displayed. The program then calls the exit() function; the argument passed to this function is the termination status. A value of 0 describes a normal termination; any other value signals an abnormal program termination.

In C, the printf() statement displays the resistance in scientific format (%e) as values are typically much less than 1 Ω (such as mΩ or μΩ).

Program 3.7

```
/*      Program to determine the resistance         */
/*      of a cylindrical conductor                  */
#include    <stdio.h>
#include    <math.h>
#include    <ctype.h>
#include    <stdlib.h>
/* Define resistivities  */
#define   RHO_COPPER      17e-9
#define   RHO_AL          25.4e-9
#define   RHO_SILVER      16e-9
#define   RHO_MANGANESE   1400e-9
#define   PI              3.14

int       main(void)
{
float     radius,length,area,rho,resistance;
char  ch;

    puts("Type of conductor >>");
    puts(" (c)opper");
    puts(" (a)luminium");
    puts(" (s)ilver");
    puts(" (m)anganese");
    /* get conductor type                           */
    ch=getchar();

    printf("Enter radius and length of conductor >>");
    scanf("%f %f",&radius,&length);

    /* area of conductor                            */
    area=PI*(radius*radius);

    /* convert to lowercase and determine resistivity */
    switch (tolower(ch))
    {
        case 'c': rho=RHO_COPPER;         break;
        case 'a': rho=RHO_AL;             break;
        case 's': rho=RHO_SILVER;         break;
        case 'm': rho=RHO_MANGANESE;      break;
        default: puts("Invalid option");  exit(0); break;
    }
    resistance= rho*length/area;
    printf("Resistance of conductor is %.3e ohm",  resistance);
    return(0);
}
```

Pascal
Program 3.7

```
program case4(input,output);
(*    Program to determine the resistance       *)
(*    of a cylindrical conductor                *)
const RHO_COPPER=17e-9;
      RHO_AL=25.4e-9;
      RHO_SILVER=16e-9;
      RHO_MANGANESE=1400e-9;
var radius,length, area, rho, resistance:real;
    ch:char;

begin
    writeln('Type of conductor >>');
    writeln(' (c)opper');
    writeln(' (a)luminum');
    writeln(' (s)ilver');
    writeln(' (m)anganese');
    readln(ch);

    writeln('Enter radius and length of conductor >>');
    readln(radius,length);

    area:=Pi*(radius*radius);

    case (ch) of
    'c','C': rho:=RHO_COPPER;
    'a','A': rho:=RHO_AL;
    's','S': rho:=RHO_SILVER;
    'm','M': rho:=RHO_MANGANESE;
    else
        begin
            writeln('Invalid option');
            exit;
        end;
    end; { case statement }

    resistance:=rho*length/area;

    writeln('Resistance of conductor is ',resistance:6:2, ' ohms');
end.
```

Test run 3.7 uses an aluminium conductor with a radius of 1 mm and length 1000 m. The resistance is found to be 8.08 Ω.

Test run 3.7

```
Type of conductor >>
(c)opper
(a)luminium
(s)ilver
(m)anganese
a
Enter radius and length of conductor >> 1e-3 1000
Resistance of conductor is 8.09e+00 ohm
```

The Pascal Program 3.7 shows that several constants for a single case option can be used (for example, 'a', 'A'). In C, it is possible to have several `case` options in the `switch` statement. For example, if the `tolower()` function is not used in program 3.7 then the `case` option can be modified so that it includes the upper and lowercase options, as shown in the following code:

```
switch (ch)
{
    case 'C':case 'c': rho=RHO_COPPER;       break;
    case 'A':case 'a': rho=RHO_AL;           break;
    case 'S':case 's': rho=RHO_SILVER;       break;
    case 'M':case 'm': rho=RHO_MANGANESE;    break;
    default: puts("Invalid option");         exit(0);
}
```

3.3 Exercises

3.3.1   Modify Program 3.1 so that it cannot generate a divide by zero error, that is, when a is 0 (zero). Note that if a is 0 then the root will be $-c/b$.

3.3.2   Modify Program 3.2 so that the user can enter the EM wave as a frequency or a wavelength. A sample run is shown in test run 3.8.

> 🖥 **Test run 3.8**
> ```
> Do you wish to enter
> (f)requency or
> (w)avelength  >>> f
> Enter frequency >>> 10e9
> Wavelength is 3.0e-02 Electromagnetic wave is Microwave
> ```

3.3.3   Modify Program 3.2 so it uses the #define statement to define limits for the wavelength, for example:

```
#define GAMMA_RAY_LIMIT 1e-11
```

3.3.4   Capacitance is normally defined as a value and a specified unit, such as pF, nF, μF, mF or F. Write a program in which a capacitance value and the unit are entered and the program displays the actual numerical value in Farads. The capacitance unit should be entered as a character. A sample test run is given in Test run 3.9. Note that pF is 10^{-12} F, pF is 10^{-9} F, μF is 10^{-6} F and mF is 10^{-6} F.

> 🖥 **Test run 3.9**
> ```
> Enter value: 1
> Enter unit p,n,u, or m : u
> Capacitance value is 0.000001 Farads
> ```

3.3.5   Repeat Q3.10 for the value of the resistance. The units entered are either mΩ ('m'), Ω ('1'), kΩ ('k') or MΩ ('M'). Test run 3.11 shows a sample run. Note that mΩ is 10^{-3} Ω, kΩ is 10^{3} Ω and MΩ is 10^{6} Ω.

> 🖥 **Test run 3.10**
> ```
> Enter value: 3.21
> Enter unit m,1,k, or M : k
> Resistance value is 3210  ohms
> ```

Worksheet 4

W4.1   Enter Program 3.1 and use it to complete Table W4.1

Table W4.1   Roots of a quadratic equation

Equation	Root(s)
$x^2 + 21x - 72 = 0$	
$5x^2 + 2x + 1 = 0$	
$25x^2 - 30x + 9 = 0$	
$6x^2 + 9x - 20 = 0$	

W4.2   Modify Program 3.1 so that it cannot generate a divide by zero error, that is, when a is 0 (zero). Note that if a is 0 then the root will be $-c/b$.

Table W4.2   Root of a quadratic equation

Equation	Root
$0x^2 + 4x - 2 = 0$	
$0x^2 + 6x + 6 = 0$	

W4.3   Write a program in which the user enters a value of resistance and the program displays the resistance value in the best possible units. A possible implementation could be:

If the resistance is less than $1\,000\,\Omega$ (1e3) then it is printed as the value in ohms;
else, if it is between $1\,000$ (1e3) and $1\,000\,000\,\Omega$ (1e6) then the value is printed as $k\Omega$;
else, if it is greater than $1\,000\,000$ (1e6) then it is printed in $M\Omega$.

Test run W4.1 shows some sample runs.

🖥️ **Test run W4.1**

```
Enter a value of resistance >> 500
Resistor value is 500 ohms
Enter a value of resistance >> 1200
Resistance value is 1.200 kohms
Enter a value of resistance >> 1.2e6
Resistance value is 1.2 Mohms
```

Worksheet 5

W5.1   **OPTIONAL.** Enter Program 3.5 and test the results.

W5.2   Write a program using the `case` (or `switch`) statement that allows the user to select from a menu of options. These options allow the user to select either the calculation of the equivalent resistance of two series or two parallel resistors. The user should enter a 1 if the series equivalent is required or a 2 if parallel required. Sample run W5.1 shows a sample test run.

🖳 **Test run W5.1**

```
Enter R1 >> 100
Enter R2 >> 100
Select an option
(1) Series resistance
(2) Parallel resistance
Option >> 2
Equivalent resistance is 50.00 ohms
```

W5.3   The `textbackground` function allows the colour of the background to be changed to BLACK, BROWN, RED, ... and `clrscr` clears the screen. Sample program 3.8 shows an example program of their use. Write a program which allows the user to enter a colour and the then program changes the background colour. Table W5.1 shows the colour definitions.

Table W5.1

Colour	Value	Colour	Value
BLACK	0	CYAN	3
BLUE	1	RED	4
GREEN	2	MAGENTA	5

⬡ c

Program 3.8

```c
#include <conio.h>

int   main(void)
{
   textbackground(RED);
   clrscr();
   return(0);
}
```

Program 3.8

```
program temp;

uses crt;

begin
     textbackground(RED);
     clrscr;
end.
```

Test run W5.2

```
Select a background colour >>
(0)BLACK
(1)BLUE
(2)GREEN
(3)CYAN
(4)RED
(5)MAGNETA

Option >> 2
```

OPTIONAL EXERCISE

W5.4 Modify the program written in W5.2 so that the user enters an 's' for series resistance and a 'p' for parallel. Sample run W5.2 shows a sample test run. Note that the program should accept the input characters in uppercase or lowercase.

Test run W5.2

```
Enter R1 >> 100
Enter R2 >> 100

Select an option
(S)eries resistance
(P)arallel resistance
Option >> p

Equivalent resistance is 50.00 ohms
```

4 Repetitive Statements

4.1 Introduction

Iterative, or repetition, allows the looping of a set of statements. There are three forms of iteration:

	Pascal Syntax	**C Syntax**

for loop	`for val:=start to end do`
	`begin`
	`    statement block;`
	`end;`
		`for (start_cond;loop_cond;loop_oper)`
		`{`
		`    statement block;`
		`}`

repeat..until	`repeat`
	
	`    statement block;`
	
	`until (condition);`
		`do`
		`{`
		`    statement block;`
		`} while (condition);`

while	`while (condition)`
	`begin`
	
	`    statement block;`
	
	`end;`
		`while (condition)`
		`{`
		
		`    statement block;`
		
		`}`

4.2 for

Many tasks within a program are repetitive, such as prompting for data, counting values, and so on. The `for` loop allows the execution of a block of code for a given control function or a given number of times. In Pascal the format is:

Pascal Syntax

```
for value:=startval to endval do
begin
      statement block
end;
```

In this case, `value` starts at `startval` and ends at `endval`. Each time round the loop, value will be incremented by 1. If there is only one statement in the block then the begin and end reserved words can be omitted. In C the format is:

Syntax

```
for (starting condition;test condition;operation)
{
    statement block
}
```

where:

starting condition	–	the starting value for the loop;
test condition	–	if test condition is TRUE the loop will continue execution;
operation	–	the operation conducted at the end of the loop.

4.3 Examples

4.3.1 ASCII characters

Program 4.1 displays ASCII characters for entered start and end decimal values. Test run 4.1 displays the ASCII characters from decimal 40 (' (') to 50 ('2').

Program 4.1

```
program for1(input,output);
{ Program to display ASCII characters }
var i,startloop,endloop:integer;

begin
    write('Enter start and end for ASCII char.>> ');
    readln(startloop,endloop);

    writeln('INTEGER  CHARACTER');
    { The function chr() is used to determine the character }
    { associate with the ASCII integer value                }

    for i:=startloop to endloop do
        writeln(i:5,chr(i):10);
end.
```

Program 4.1

```
/*    Program to print ASCII characters       */
#include <stdio.h>
int   main(void)
{
int   i,start,end;

    printf("Enter start and end for ASCII characters >>");
    scanf("%d %d",&start,&end);
```

```
    puts("INTEGER   HEX     ASCII");

    for (i=start;i<=end;i++)
        printf("%5d  %5x  %5c\n",i,i,i);
    return(0);
}
```

💻 **Test run 4.1**

```
Enter start and end for ASCII characters >> 40 50
INTEGER   HEX     ASCII
   40      28       (
   41      29       )
   42      2a       *
   43      2b       +
   44      2c       ,
   45      2d       -
   46      2e       .
   47      2f       /
   48      30       0
   49      31       1
   50      32       2
```

4.3.2 Transient response of an RC circuit

Figure 4.1 illustrates an RC circuit with a voltage step applied at $t=0$. When a voltage step, amplitude E volts, is applied to this circuit it produces an exponential current.

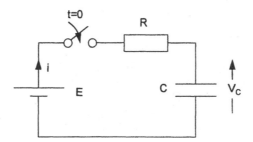

Figure 4.1   RC circuit with step input applied at $t=0$

The following defines the transient current in the circuit.

$$i = \frac{E}{R} e^{-\frac{t}{RC}}$$

and the voltage across the resistor will be:

$$V_R = E e^{-\frac{t}{RC}}$$

Program 4.3 determines the voltage across the resistor at given time intervals. The user enters the end time and the number of time steps required; the program determines the voltage at each of the time steps.

Program 4.2

```pascal
program for3(input,output);
{ Program to determine transient response of an RC circuit }

var R,C,tend,t,Vin,v:real;
    i,tsteps:integer;

begin
    writeln('Program to determine voltage across');
    writeln('resistor is an RC circuit');

    write('Enter R,C >>');
    readln(R,C);

    write('Enter number of time steps and end time');
    readln(tsteps,tend);

    write('Enter voltage step applied>>');
    readln(Vin);

    writeln('   TIME      VOLTAGE');
    t:=0;  { start at time equal to zero }

    for i:=1 to tsteps do
    begin
        v:=Vin*exp(-t/(R*C));
        writeln(t:8:4,v:8:2);
        t:=t+tend/tsteps;
    end;
end.
```

Program 4.2

```c
/*     Program to determine transient response of an RC circuit     */
#include <math.h> /* required for exp() */
#include <stdio.h>

int       main(void)
{
float     R,C,tend,t,E,Vr;
int       tsteps;

    puts("Program to determine voltage across");
    puts("Resistor in an RC circuit");

    printf("Enter R,C >> ");
    scanf("%f %f",&R,&C);

    printf("Enter number of time steps and end time>>");
    scanf("%d %f",&tsteps,&tend); /* enter integer and float*/

    printf("Enter voltage step applied>>");
    scanf("%f",&E);

    puts("   TIME      VOLTAGE");
    for (t=0;t<tend;t+=tend/tsteps)
    {
        Vr=E*exp(-t/(R*C));
        printf("%8.4f %8.2f\n",t,Vr);
    }
    return(0);
}
```

Test run 4.2 shows that the voltage across the resistor starts at a maximum at $t=0$. This is because the voltage across the capacitor is initially zero. As the capacitor charges, the voltage across it will increase until it almost equals the applied voltage. The current in the circuit will also be at a maximum when the step is applied. It will then decay to almost zero at a rate determined by the time constant (which is a product of R and C).

Test run 4.2

```
Program to determine voltage across
Resistor in an RC circuit
Enter R,C >> 1e3 1e-6
Enter number of time steps and end time>> 20 10e-3
Enter voltage step applied>> 10
    TIME          VOLTAGE
    0.0000        10.00
    0.0005        6.07
    0.0010        3.68
    0.0015        2.23
    0.0020        1.35
    0.0025        0.82
    0.0030        0.50
    0.0035        0.30
    0.0040        0.18
    0.0045        0.11
    0.0050        0.07
    0.0055        0.04
    0.0060        0.02
    0.0065        0.02
    0.0070        0.01
    0.0075        0.01
    0.0080        0.00
    0.0085        0.00
    0.0090        0.00
    0.0095        0.00
```

4.3.3 Boolean logic

Program 4.3 is an example of how a Boolean logic function can be analyzed and a truth table generated. The `for` loop generates all the required binary permutations for a truth table. The Boolean function used is:

$$Z = \overline{(A.B) + C}$$

A schematic of this equation is given in Figure 4.2. Test run 4.3 shows a sample run.

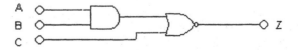

Figure 4.2 Digital circuit

⬡ C

Program 4.3

```c
/*    Program to generate truth table for Boolean function      */

#include <stdio.h>

int   main(void)
{
int   A,B,C,Z;

   puts("Boolean function NOR (AND(A,B),C)");
   puts("    A    B    C      Z");

   for (A=0;A<=1;A++)
      for (B=0;B<=1;B++)
         for (C=0;C<=1;C++)
         {
            Z=!( (A && B) || C);
            printf("%4d %4d %4d %4d\n",A,B,C,Z);
         }

   return(0);
}
```

⬡ Pascal

Program 4.3

```pascal
program for3(input,output);

{ Program to generate truth table for Boolean function      }

var A,B,C,Z:boolean;

begin
     writeln('Boolean function NOR(AND(A,B),C)');
     writeln('    A        B        C        Z');

     for A:=FALSE to TRUE do
         for B:=FALSE to TRUE do
             for C:=FALSE to TRUE do
             begin
                Z:=not( (A and B) or C);
                writeln(A:8,B:8,C:8,Z:8);
             end;
end.
```

💻 **Test run 4.3**

```
Boolean function NOR (AND(A,B),C)
   A    B    C    Z
   0    0    0    1
   0    0    1    0
   0    1    0    1
   0    1    1    0
   1    0    0    1
   1    0    1    0
   1    1    0    0
   1    1    1    0
```

4.4 Exercises

4.4.1   Write a program which prints all the characters from '0' (zero) to 'z' in sequence using a for loop.

4.4.2 Write a program which displays the squares, cubes and fourth powers of the first 15 integers. A sample output is given next.

```
Number   Square  Cube   Fourth
*****************************
1         1       1       1
2         4       8      16
3         9      27      81
       etc
```

4.4.3 Write a program which displays the y values in the formulas given below and with the given x steps.

Equation	Range of x
(i) $y = 4x + 1$	0 to 50 in steps of 5
(ii) $y = \sqrt{x} - 1$	1 to 10, steps of 0.5
(iii) $y = 5x^2 + 3x - 2$	−5 to 5 in steps 0.5

A sample run of the first equation is given next.

```
EQUATION y=4x+1, x goes from 0 to 50 in steps of 5
*********************************************************
 x         y
*********
  0        1
  5       21
 10       41
 15       61
etc
```

4.4.4 Write a program which displays the sine of a number from 0° degrees to 90° in steps of 10°.

4.4.5 Modify Program 4.3 so that it determines the truth table for the following Boolean equation:

$$Z = \overline{(A + B).C}$$

Table 4.1 Truth table

A	B	C	Z
0	0	0	
0	0	1	
0	1	0	
0	1	1	
1	0	0	
1	0	1	
1	1	0	
1	1	1	

4.4.6 Write a program which will determine the impedance of an RL series circuit. The program must calculate the magnitude of the impedance.

Worksheet 6

W6.1   Enter Program 4.1 and use it to complete Table W6.1.

Table W6.1   ASCII characters

Value	Character
34	
35	
36	
37	
38	
64	
65	
66	
67	
68	
69	
70	

W6.2   Write a program which lists the square of the values from 1 to 10.
A sample run in shown in Test run W6.1.

```
      Test run W6.1
Value Square
1       1
2       4
3       9
4       16
5       25
6       36
7       49
8       64
9       81
10      100
```

W6.3   Complete either Exercise 4.4.3(i), Exercise 4.4.4 or Exercise 4.4.5.

4.5 while()

The `while` statement allows a block of code to be executed while a specified condition is TRUE. It checks the condition at the start of the block; if this is TRUE the block is executed, else it will exit the loop. The syntax is

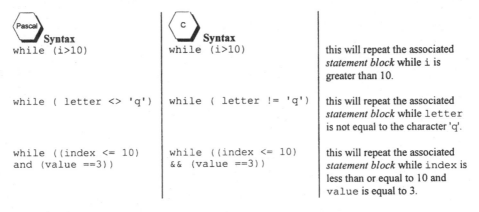

while

Pascal Syntax
```
while (condition)
begin

    statement block;

end;
```

C Syntax
```
while (condition)
{

    statement block;

}
```

If the statement block contains a single statement then the braces may be omitted (although it does no harm to keep them). A few examples are:

Pascal Syntax	**C Syntax**	
`while (i>10)`	`while (i>10)`	this will repeat the associated *statement block* while i is greater than 10.
`while ( letter <> 'q')`	`while ( letter != 'q')`	this will repeat the associated *statement block* while `letter` is not equal to the character 'q'.
`while ((index <= 10) and (value ==3))`	`while ((index <= 10) && (value ==3))`	this will repeat the associated *statement block* while `index` is less than or equal to 10 and `value` is equal to 3.

4.6 repeat..until() and do..while()

In C the `do..while()` statement is similar in its operation to `while()` except that it tests the condition at the bottom of the loop. The repeat..until() in Pascal is also similar to the while() statement but unlike while() the loop quits when the associated condition is true. These loops thus allow *statement block* to be executed at least once. The syntax is:

repeat..until
do..while

Pascal Syntax
```
repeat

    statement block;

until (condition);
```

C Syntax
```
do
{
    statement block;
} while (condition);
```

As with `for` and `while` loops the braces are optional. The `do..while()` and `repeat..until()` loops require a semicolon at the end of the loop, whereas the `while()` does not.

4.7 Examples

4.7.1 Repeating program

Often a user is asked to repeat the program once it has finished calculating values. This is sometimes done by asking the user if they want to repeat (or continue) the program. If the user enters a 'y' character then the program is repeated, else a 'n' character will exit the program. Program 4.4 implements this with a `repeat..until()/do..while()` loop.

⬡ Pascal
Program 4.4

```pascal
program repeat1(input,output);

var   R1,R2,R_equ:real;
   ch:char;

begin
    repeat
          writeln('Enter R1 and R2 >>');
          readln(R1,R2);
          R_equ:=(R1*R2)/(R1+R2);
          writeln('Parallel resistance is ',R_equ:8:2,' ohms');

          writeln('Do you wish to continue (y/n)');
          readln(ch);
      until (ch='n');
end.
```

⬡ C
Program 4.4

```c
#include <stdio.h>

int   main(void)
{
float R1,R2,R_equ;
char  ch;

    do
    {
        printf("Enter R1 and R2 >>");
        scanf("%f %f",&R1,&R2);

        R_equ=(R1*R2)/(R1+R2);
        printf("Parallel resistance is %8.2f ohms\n",R_equ);

        printf("Do you wish to continue (y/n)");
        fflush(stdin); /* flush keyboard buffer */
        ch=getchar();
    } while (ch=='y');

    return(0);
}
```

4.7.2 Limiting ranges of inputs

Most of the values that are entered into a program have a certain range. For example if a user is asked enter their age then the value will always be:

- An integer.
- A positive value.
- Less than 130.

If a user enters an invalid value then the program could either:

- Crash, which typically happens when a program tries to divide by zero or determines the square root of a negative number.
- Gives invalid results.

Thus it is important that the user is stopped from entering values which are invalid. Program 4.5 allows the user to determine the equivalent parallel resistance for two resistors in parallel. The range of value values of entered resistance is between 0Ω and 1MΩ. The repeat .. until() / do .. while() loop is placed around the user entry of each of the values. These loops continue until a user enters a valid value. Test run 4.4 shows a sample run.

Program 4.5

```
program repeat1(input,output);

var R1,R2,R_equ:real;

begin
    repeat
          writeln('Enter R1 >>');
          readln(R1);
          if ( (R1<0) or (R1>1e6)) then
             writeln('Invalid value: re-enter');
    until ( (R1>0) and (R1<1e6) );

    repeat
          writeln('Enter R2 >>');
          readln(R2);
          if ( (R2<0) or (R2>1e6)) then
             writeln('Invalid value: re-enter');
    until ( (R2>0) and (R1<1e6) );

    R_equ:=(R1*R2)/(R1+R2);
    writeln('Parallel resistance is ',R_equ:8:2,' ohms');

end.
```

Program 4.5

```
#include <stdio.h>

int    main(void)
{
float R1,R2,R_equ;

    do
    {
        printf("Enter R1 >>");
```

```
        scanf("%f",&R1);
        if ( (R1<0) || (R1>1e6)) printf("Invalid value: re-enter\n");
    } while ( (R1<0) || (R1>1e6) );

    do
    {
        printf("Enter R2 >>");
        scanf("%f",&R2);
        if ( (R2<0) || (R2>1e6)) printf("Invalid value: re-enter\n");
    } while ( (R2<0) || (R2>1e6) );

    R_equ=(R1*R2)/(R1+R2);
    printf("Parallel resistance is %8.2f ohms\n",R_equ);
    return(0);
}
```

Test run 4.4
```
Enter R1 >> 1e7
Invalid value: re-enter
Enter R1 >> -100
Invalid value: re-enter
Enter R1 >> 100
Enter R2 >> 100
Parallel resistance is      50.00 ohms
```

Program 4.6 gives an example of the determination of acceleration giving the initial and end velocity, and the time difference. The range of velocity values is between 0 and 1000 m/s, and the range of time difference values is between 0 and 60 s. Test run 4.5 shows a sample run.

Program 4.6
```pascal
program repeat3(input,output);

var v1,v2,t,accel:real;

begin
    repeat
        writeln('Enter initial velocity (m/s)>>');
        readln(v1);
        if ( (v1<0) or (v1>1e3)) then
            writeln('Invalid value: re-enter');
    until ( (v1>0) and (v1<1e3) );

    repeat
        writeln('Enter final velocity (m/s)>>');
        readln(v2);
        if ( (v2<0) or (v2>1e3)) then
            writeln('Invalid value: re-enter');
    until ( (v2>0) and (v2<1e3) );

    repeat
        writeln('Enter time (secs)>>');
        readln(t);
        if ( (t<0) or (t>60)) then
            writeln('Invalid value: re-enter');
    until ( (t>0) and (t<60) );

    accel:=(v2-v1)/t;

    writeln('Acceleration is ',accel:8:2,' m/s2');
end.
```

⬡ c

Program 4.6

```c
#include <stdio.h>

int    main(void)
{
float v1,v2,t,accel;

    do
    {
        printf("Enter initial velocity (m/s)>>");
        scanf("%f",&v1);
        if ( (v1<0) || (v1>1e3)) puts("Invalid value: re-enter");
    } while ((v1<0) || (v1>1e3));

    do
    {
        printf("Enter final velocity (m/s)>>");
        scanf("%f",&v2);
        if ( (v2<0) || (v2>1e3)) puts("Invalid value: re-enter");
    } while ((v2<0) || (v2>1e3));

    do
    {
        printf("Enter time (secs)>>");
        scanf("%f",&t);
        if ( (t<0) || (t>60)) puts("Invalid value: re-enter");
    } while ((t<0) || (t>60) );

    accel=(v2-v1)/t;

    printf("Acceleration is %8.2f m/s2\n",accel);
    return(0);
}
```

🖥 **Test run 4.5**

```
Enter initial velocity (m/s)>> -100
Invalid value: re-enter
Enter initial velocity (m/s)>>  1e7
Invalid value: re-enter
Enter initial velocity (m/s)>> 20
Enter final velocity (m/s)>>  10
Enter time (secs)>> 2
Acceleration is     -5.00 m/s2
```

4.8 Exercises

4.8.1   Correct the errors in the following programs:

(a)
```
Program tut1a(input,output)

{ This Program should calculate the current  }
{ flowing in a resistor but has two syntax    }
{ errors and one functional error             }

var    resistance,voltage,ch  :  real;

begin

    writeln('Program to determine current flowing in');
    writeln('a resistor');
```

```
    repeat

        writeln('Enter voltage and resistance');
        readln(Voltage,Resistance);

        { always catch divide by zero errors ! }

        if (resistance=0) then  Writeln('INFINITE CURRENT !!!')
        else writeln('Current is ',voltage/resistance:8:3,' Amps');

        write('Do you wish to continue (y/n)>>');

        Readln(ch);

    until (ch='y');
END.
```

(b)
```
program tut1b(input,output);

{ This Program should calculate the impedance}
{ of a capacitor but has three errors         }

begin

var   frequency,capacitance,Xc  :  real;
      ch :  char;

    writeln('Program to determine impedance');
    writeln('a a capacitor');

    repeat

        writeln('Enter frequency and capacitance');
        readln(Frequency);

        if ( (frequency=0) or (capacitance=0)) then
            Writeln('INFINITE IMPEDANCE !!!')
        else
        BEGIN
            Xc=1/(2*PI*Frequency*Capacitance);
            writeln('Impedance is ',Xc:8:3,' Ohms');
        end;

        write('Do you wish to continue (y/n)>>');
        readln(ch);
    until (ch='n');
end.
```

4.8.2   What will the following sections of code output to the screen.
(i)
```
i:=1;
while (i<10) do
begin
   i:=i+2;
   writeln('i = ',i);
end;
```

(ii)
```
i:=8;
while (i<10) do
begin
   i:=i+2;
```

```
        writeln('i = ',i);
    end;
```

(iii)
```
    i:=1;
    repeat
        i:=i+2;
        writeln('i = ',i);
    until (i<10);
```

(iv)
```
    i:=10;
    repeat
        i:=i-2;
        Writeln('i = ',i);
    until (i<10);
```

(v)
```
    i:=1;
    repeat
        i:=2*i;
        writeln('i = ',i);
    until (i<10);
```

(vi)
```
    i:=1;
    repeat
        i:=2*i;
        writeln('i = ',i);
    until not(i=10);
```

(vii)
```
    i:=1;
    repeat
        i:=2*i;
        writeln('i = ',i);
    until (i>10);
```

4.8.3   Determine the errors in the following programs
(i)

📄 **Program 4.1**
```
/*     Prints the square of the numbers 1 to 10    */
/*     ie 1,4,9..100                               */
#include <stdio.h>

int main(void)
{
int i;
    for (i=1,i<10,i++)
        printf("The square of i is %d,i*i);
    return(0);
}
```

(ii)

📄 **Program 4.2**
```
/*     Prints values from 1 to 100 in power of 3    */
/*     The step used is 0.3                          */
```

```
#include <stdio.h>
int    main(void)
{
int    i;

   while (i != 100)
   {
      printf("%d to the power of three is %d /n", i*i*i);
      i += 0.3;
   return(0);
}
```

(iii)

▤ **Program 4.3**
```
/*   Prints the square of the numbers 1 to 10      */
/*   ie 1,4,9..100                                 */
#include <stdio.h>

int    main(void)
{
int    i;
   for (i=1;i<=10;i++);
      printf("The square of i is %d,i*i);
   return(0);
}
```

(iv)

▤ **Program 4.4**
```
/*   Program to determine input resistance given the   */
/*   input voltage and current                         */
#include <stdio.h>
int        main(void)
{
char   input;

   puts("Program to determine the resistance given");
   puts("input voltage and current");

   while (input == 'y')
   {
      printf("Enter voltage and current >> ");
      scanf("%f %f",voltage,current);
      printf("The resistance is %f",voltage/current);
      puts("Do you wish to continue (y/n)")
      input = getchar();
   }
   return(0);
}
```

(v)

▤ **Program 4.5**
```
#include <stdio.h>

int    main(void)
{
   puts("Program to determine the resistance given");
   puts("input voltage and current");

   while (1)
   {
      printf("Enter voltage and current >> ");
      scanf("%f %f",voltage,current);
```

```
            printf("The resistance is %f",voltage/current);

            puts("Do you wish to continue(y/n)");
            ch=getchar();
            if ((ch=='n') && (ch=='N')) break;
      }
      puts("Program exited");
      return(0);
}
```

4.8.4   Determine the output from the following sections of code (beware of infinite loops).

(i)
```
for (i=0;i<10;i+=2)
    printf("%d ",i);
```

(ii)
```
for (i=1;i<120;i*=3)
    printf("%d ",i);
```

(iii)
```
for (i=19;i<12;i--)
    printf("%d ",i);
```

(iv)
```
i=6;
do
{
    i++;
    printf("%d ",i);
} while (i<10);
```

(v)
```
i=1;
while (1)
{
    i++;
    printf("%d ",i);
    if (i==4) break;
}
```

(vi)
```
i=2
for (;;)
{
    i++;
    printf("%d ",i);
    i*=2;
    if (i==256) break;
}
```

(vii)
```
i=10;
while (i<10)
{
    i--;
    printf("%d ",i);
}
```

(viii)
```
i=10;
do
{
    i--;
    printf("%d ",i);
} while (i<10);
```

(ix)
```
i=1;
while (i>0)
{
    i++;
    printf("%d ",i);
}
```

4.8.5 Write a Program to convert a decimal number to octal (base 8). A sample run is given in Test run 4.6.

```
⌨  Test run 4.6
Enter a decimal number -> 18
The octal equivalent (backwards ) is 22
```

4.8.6 Replace the following `for` loop with a `repeat..until` and with a `while..do` in the examples below.

(i)
```
for i:=1 to 20 do
begin
end;
```

(ii)
```
for j:=-10 do 10 do
begin
end;
```

(iii)
```
for k:= 4 to 100 do
begin
    j:= 4*k;
end;
```

4.8.7 Write a program to convert hexadecimal to decimal. A sample run is shown in Test run 4.7.

```
⌨  Test run 4.7
Number of hex characters> 2
Enter 1 > A
Enter 2 > 1
The hex. equivalent is 31.
```

4.8.8 The drain current, in milliamps, of a sample depletion-type FET is given by:

$$I_D = 6.\left[1 + \frac{V_{GS}}{4}\right]^2 \text{ mA}$$

Write a program which determines the drain current (in mA) for gate-source voltages (V_{GS}) from –5 to 5 V in steps of 0.5 V. A sample run is given in Test run 4.8.

Test run 4.8

VGS (V)	IDS (mA)
−5.00	0.38
−4.50	0.09
−4.00	0.00
−3.50	0.09
−3.00	0.38
−2.50	0.84
−2.00	1.50
−1.50	2.34
−1.00	3.38
−0.50	4.59
0.00	6.00
0.50	7.59
1.00	9.38
1.50	11.34
2.00	13.50
2.50	15.84
3.00	18.38
3.50	21.09
4.00	24.00
4.50	27.09
5.00	30.38

Worksheet 7

W7.1 Modify a program from a previous worksheet so that it prompts the user to if they want to repeat the program.

W7.2 Write a program which will calculate the equivalent resistance of three resistors in parallel. If the user enters an invalid value (such as a negative value) the program will prompt the user to enter a valid input. A sample run is shown in Test run W7.1.

💻 **Test run W7.1**
```
Enter R1> -1
INVALID INPUT PLEASE RE-ENTER
Enter R1> 6
Enter R2> 10
Enter R3> 14
Equivalent resistance is 2.96 ohms
```

W7.3 Write a program which will convert decimal value to hexadecimal (base 16). A sample run is shown in Test run W7.2.

💻 **Test run W7.2**
```
Enter decimal number 17
The hexadecimal equivalent is 11.
Do you wish to continue (y/n)   n
```

W7.4 Write a Program in which you enter the number of bits in a binary number, enter these bits and then calculate the decimal equivalent. A sample run is given in test run W7.3. Note that the valid input for the bits should be only a 0 or a 1.

💻 **Test run W7.3**
```
Enter the number of bits ->   6
Enter bit 1  1
Enter bit 2  1
Enter bit 3  0
Enter bit 4  0
Enter bit 5  1
Enter bit 6  0
The decimal equivalent is 19
```

5 Functions

5.1 Introduction

Functions are identifiable pieces of code with a defined interface. They are called from any part of a program and allow large programs to be split into more manageable tasks, each of which can be independently tested. Functions are also useful in building libraries of routines that other programs use. Several standard libraries exist, such as a maths and input/output libraries.

A function can be thought of as a 'black box' with a set of inputs and outputs. It processes the inputs in a way dictated by its function and provides some output. In most cases the actual operation of the 'black box' is invisible to the rest of the program. A modular program consists of a number of 'black box' working independently of all others, of which each uses variables declared within it (local variables) and any parameters sent to it. Figure 5.1 illustrates a function represented by an ideal 'black-box' with inputs and outputs, and Figure 5.2 shows a main function calling several sub-functions (or modules).

5.2 Arguments and parameters

The data types and names of parameters passed into a function are declared in the function header (its interface) and the actual values sent are referred to as arguments. They can be passed either as values (known as 'passing by value') or as pointers (known as 'passing by reference'). Passing by value involves sending a copy of it into the function. It is not possible to change the value of a variable using this method. Variables can only be modified if they are passed by reference (this will be covered in the next chapter). This chapter looks at how parameters pass into a function and how a single value is returned.

An argument and a parameter are defined as follows:

An 'argument' is the actual value passed to a function.
A 'parameter' is the variable defined in the function header.

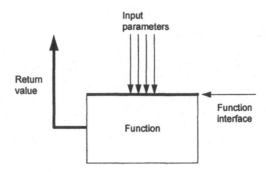

Figure 5.1   An ideal 'black-box' representation of a function

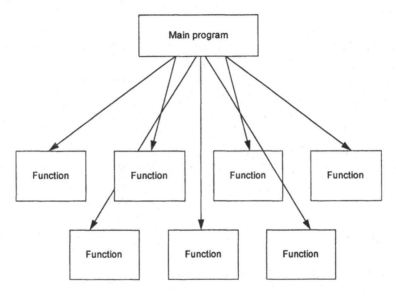

Figure 5.2   Hierarchical decomposition of a program

5.3 C functions

Figure 5.3 shows a program with two functions, `main()` and `function1()`. Function `main()` calls `function1()` and passes three parameters to it; these are passed as values. A copy of the contents of d goes into g, e into h and f into i.

Variables declared within a function are described as local variables. Figure 5.3 shows that d, e and f are local variables within `main()`; g, h, i, j and k are local within `function1()`. These will have no links to variables of the same name declared in other functions. Local variables only exist within a function in which they are declared and do not exist once the program leaves the function. Variables declared at the top of the source file (and not within a function) are defined as global variables. These allow functions, within the source file, to access them. Care must be taken when using global variables for many reasons, one of which is that they tend to lead to programs that are unstructured and difficult to maintain.

In Figure 5.3 the function function1() makes use of the variable a as this is declared as a global variable. This function cannot be modelled as a 'black box' as it can modify a variable which is not passed to it. In a relatively small program this may not create a problem but as the size of the program increases the control of variables can become difficult.

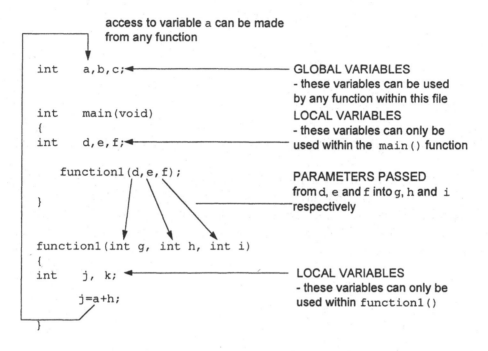

Figure 5.3 Local and global variables

Figure 5.4 shows an example of how things can go wrong with global variables. In this example, a global variable i is used by two functions. Initially, the value of i within the loop in main() will be 0. The function1() function is called within this loop, which uses the variable i within another loop. Each time it is incremented the global variable takes on the incremented value. When the program leaves this function the value of i will have changed (that is, it will be 10). This causes the for() loop in main() to end. If the variable i had been declared locally within both main() and function1() this problem would not have occurred. As a rule, called modules should be self-contained and use only the parameters sent to them.

```
int    i;
int    main(void)
{
int    d,e,f;

   for (i=0;i<5;i++)
   {
       function1(d,e,f);
       ::::::::::::::::::
   }
}
function1(int x, int y, int z)
{
int    j, k;

   for (i=0;i<10;i++)
   {
       ::::::::::::
   }
}
```

Both functions make
use of global variable
i

Figure 5.4   An example of the use of global variables

Program 5.1 contains a function named `print_values()`. This is
called from `main()` and variables a and b are passed into the parameters c
and d, respectively; c and d are local parameters and only exist within
`print_values()`. The values of c and d can be changed with no effect
on the values of a and b.

Program 5.1

```
/*    Simple program that shows parameter passing */
#include <stdio.h>

void    print_values(int a, int b);
                /* ^ function prototype, to be discussed */

int    main(void)
{
int    a=5,b=6;

   print_values(a,b);
   return(0);
}

void    print_values(int c,int d)
{
   printf("The values passed are %d %d \n",c,d);
}
```

5.3.1 Return value

The `return` statement returns a single value from a function to the calling
routine. Programs in previous chapters have used functions that return values. If there are no `return` statements in a function the execution returns
automatically to the calling routine upon execution of the closing brace (i.e.
after the final statement within the function). Program 5.2 contains functions

which will add and multiply two numbers. The function `addition()` uses `return` to send back the addition of the two values to the `main()`.

Program 5.2

```c
#include <stdio.h>
/* Function prototypes, these are discussed in the next section*/
int     addition(int c,int d);
int     multiply(int c,int d);
void    print_values(int c, int d, int sum,int mult);

int   main(void)
{
int    a=5,b=6,summation,multi;

    summation=addition(a,b);
    multi = multiply(a,b);
    print_values(a,b,summation,multi);
    return(0);
}

int     addition(int c,int d)
{
    return(c+d);
}

int     multiply(int c,int d)
{
    return(c*d);
}

void    print_values(int c, int d, int sum,int mult)
{
    printf("%d plus %d is %d \n",c,d,sum);
    printf("%d multiplied %d is %d \n",c,d,mult);
}
```

Figure 5.5 shows a simple structure chart of this program. The function `addition()` is called first; the variables sent are a and b and the return value is put into the variable `summation`. Next, the `multiply()` is called; the variables sent are also a and b and the value returned goes into `multi`. Finally, the function `print_values()` is called; the values sent are a, b, `multi` and `summation`.

A function can have several return points, although it is normally better to have only one return point. This is normally achieved by restructuring the code. An example of a function with two `returns` is shown next. In this example a decision is made as to whether the value passed into the function is positive or negative. If it is greater than or equal to zero it returns the same value, else it returns a negative value (`return(-value)`).

```c
int     magnitude(int value)
{
    if (value >= 0)
        return (value);
    else
        return ( -value);
}
```

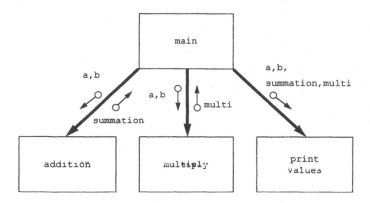

Figure 5.5   Basic structure chart for program 5.2

5.3.2 Function type

Program 5.2 contains functions that return integer data types. It is possible to return any other of C's data types, including `float`, `double` and `char`, by inserting the data type before the function name. If no data type is given then the default return type is `int`. The following gives the general syntax of a function.

```
type_def function_name(parameter list)
{
}
```

C is a flexible language in its structure. It allows the arrangement of functions in any order and even within different files. If the compiler finds a function that has not been defined (or prototyped) then it assumes the return type will be `int`. It also assumes that at the linking stage the linker will be able to find the required function either in the current compiled program, the libraries or other object codes. It is thus important that the function return data type is defined when the compiler is compiling the function; otherwise it will assume that the return type is `int`.

Function declarations (or prototypes) are normally inserted either at the top of each file, locally within a function, or in a header file (the *.h* files). These declarations allow the compiler to determine the return type and the data types of all parameters passed to the function. It thus allows the compiler to test for illegal data types passed to a function in error. For example the following are invalid uses of the function `printf()`, `sqrt()` and `scanf()`. The `printf()` has an incorrect syntax as the first argument should be a format statement (i.e. a string), the `sqrt()` function should be passed a floating point value and the `scanf()` function requires a format string as the first argument.

```
printf(23,"Value is %d\n");
```

```
b=sqrt("1233");
scanf(&val1,&val2);
```

If the *stdio.h* header is not included the compiler does not generate any errors for the incorrect usage of `printf()` and `scanf()`. The same applies to the *math.h* header file and the function `sqrt()`.

When a function prototype is inserted at the top of the file it is a global declaration of that function within the source code file. Otherwise, the declaration can be inserted into the variable declaration lists within a function; this will make the definition local only to the function in which it is defined.

If a function does not return a value then the data type definition for the return value should be `void`. Also, if no parameters are sent to the function then the argument list contains a `void`. The compiler would thus flag an error or warning if any parameters are passed to a function with a `void` argument list or if a returned value is used from a `void` return data type.

Figure 5.6 shows the operation of a function prototype. At the top of the file the prototype declares the parameter types of the arguments passed (in this case, two floats) and the return data type (in this case `float`). The compiler checks all arguments sent to this function to see if they match up with these types. A warning or error is generated if there is a mismatch. The return type is also checked.

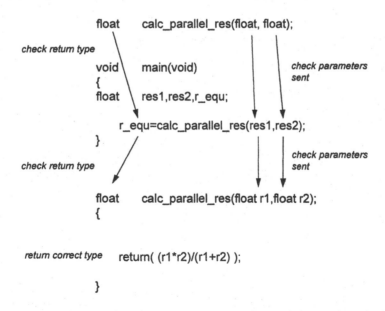

Figure 5.6 Checking conducted by the compiler on function prototypes

Program 5.3 contains a function `power()` which is prototyped at the top of the program. The return value is `double`; the first argument is `double` and the second is `int`. This function uses logarithms to determine the value of x raised to the power of n. The formula used is derived next; the `log()`

function is the natural logarithm and the exp() the exponential function. Both of these require the parameter sent as double and the return type is also double.

$$y = x^n$$

$$\log(y) = \log(x^n)$$

$$\log(y) = n.\log(x)$$

$$y = \exp(n.\log(x))$$

Program 5.3

```
#include <math.h>
#include <stdio.h>

double    power(double x,int n); /* declare function prototype */

int       main(void)
{
double    x,val;
int       n;

  puts("Program to determine the result of a value");
  puts("raised to the power of an integer");

  printf("Enter x to the power of n >>");
  scanf("%f %d",&x,&n);

  val=power(x,n);
  printf("%f to power of %d is %f\n",x,n,val);
  return(0);
}

double    power(double x,int n)
{
  return(exp(n*log(x)) );
}
```

Some compilers give a warning if the return value data type is different from the data type of the variable to which it is assigned. For example, if the data type of the variable x in Program 5.3 is changed to a float then it is good practice to recast the return value to a float, as shown next.

```
int       main(void)
{
float     x,val;
int       n;

  :::::::
  val=(float) power(x,n);
  :::::::
```

5.4 Pascal functions

In Pascal a function returns a single value. It is identified by the function header, which is in the form:

```
FUNCTION  function_name(formal_parameter_list) : result_type;
```

where the parameters are passed through the `formal_parameter_list` and the result type is defined by `result_type`.

Program 5.1 contains two functions named `addition` and `multiply`. In calling the `addition` function and variables a and b are passed into the parameters c and d, respectively; c and d are local parameters and only exist within `addition()`. The values of c and d can be changed with no effect on the values of a and b. This function returns a value back to the main program by setting a value which is the same name as the function.

Program 5.1 also uses a procedure which, in this case, is similar to a function but does not return any values back to the calling program.

Program 5.1

```
program test(input,output);
var    a,b,summation,multi:integer;

function addition(c,d:integer):integer;
begin
   addition:=c+d;
end;

function multiply(c,d:integer):integer;
begin
 multiply:=c*d;
end;

procedure print_values(c,d,sum,mult:integer);
begin
   writeln(c,' plus ',d,' is ', sum);
   writeln(c,' multiplied by ',d, ' is ',mult);
end;

begin
    a:=5;
    b:=6;
    summation:=addition(a,b);
    multi := multiply(a,b);
    print_values(a,b,summation,multi);
end.
```

Figure 5.5 shows a simple structure chart of this program. The function `addition()` is called first; the variables sent are a and b and the return value is put into the variable `summation`. Next, the `multiply()` is called; the variables sent are also a and b and the value returned goes into `multi`. Finally, the function `print_values()` is called; the values sent are a, b, `multi` and `summation`.

Program 5.2 contains functions that return integer data types. It is possible to return any other of C's data types, including `real`, `double` and `character`, by inserting the data type after the function name.

A function can have several return points, although it is normally better to have only one return point. This is normally achieved by restructuring the code. An example of a function with two returns is shown next. In this example a decision is made as to whether the value passed into the function is positive or negative. If it is greater than or equal to zero it returns the same value, else it returns a negative value (`mag=-val`).

```
program test(input,output);

var a:integer;

function mag(val:integer):integer;
begin
    if (val<0) then mag:=-val
    else mag:=val;
end;

begin
    a:=-5;
    writeln('Mag(a) = ',mag(a));
end.
```

Program 5.2 contains a function `power()` which has a return data type of `double`; the first argument is `double` and the second is `integer`. This function uses logarithms to determine the value of x raised to the power of n. The formula used is derived next; the `ln()` function is the natural logarithm and the `exp()` the exponential function.

$$y = x^n$$
$$\ln(y) = \ln(x^n)$$
$$\ln(y) = n.\ln(x)$$
$$y = \exp(n.\ln(x))$$

⬡ Pascal
Program 5.2
```
program pow(input,output);

var    x,val:double;
       n:integer;

function power(x:double;n:integer):double;
begin
    power :=exp(n*ln(x));
end;

begin

    writeln('Program to determine the result of a value');
    writeln('raised to the power of an integer');

    writeln('Enter x to the power of n >>');
    readln(x,n);

    val:=power(x,n);
    writeln(x,' to power of ',n,' is ',val);
end.
```

5.5 Examples

This section contains a few sample C and Pascal programs which use functions.

5.5.1 Tan function

There is no `tan` function in Pascal; thus to overcome this Pascal Program 5.3 contains a tan function and Test run 5.1 shows a sample run.

Program 5.3

```
Program function3(input,output);
(* Program that uses a function to calculate the  *)
(* tan of a number                                *)

var   answer,number: real;

function tan(a:real):real;
begin
   tan:= sin(a)/cos(a);         (* tan is sin over cos *)
end;

begin
   write('Enter number >> ');
   readln(number);

   answer:=tan(number);

   writeln('The tan of ',number:8:3,' is ',answer:8:3);
end.
```

Test run 5.1

```
Enter number >> 1.23
The tan of    1.230 is    2.820
```

C, unlike Pascal, has a tan function, thus to avoid redeclaring it C Program 5.4 has a function called `tangent()`. To improve accuracy the program uses `double` data types. For this reason the `scanf()` and `printf()` functions use the data type `%lf` conversion characters.

Program 5.4

```
/* Program that uses a function to calculate the  */
/* tan of a number                                */

#include <stdio.h>
#include <math.h>

/* Define function prototype */
double   tangent(double a);

int   main(void)
{
double   answer,number;

   printf("Enter number >> ");
```

```
   scanf("%lf",&number);

   answer=tangent(number);

   printf("The tan of %8.3lf is %8.3lf\n",number,answer);
   return(0);
}

double tangent(double a)
{
   return(sin(a)/cos(a));       /* tan is sin over cos */
}
```

5.5.2 Centigrade to Fahrenheit conversion

Pascal Program 5.4 uses two functions to convert from Fahrenheit to centigrade, and vice versa. Test run 5.2 shows a sample test run.

Program 5.4

```
Program function4(input,output);
(* Program that uses a function to convert fahrenheit to centigrade *)

var    fahrenheit,centigrade:real;

function convert_to_centigrade(f:real ):real;
begin
   convert_to_centigrade:=5/9* (f-32);
end;

function convert_to_fahrenheit(c:real )   :real;
begin
   convert_to_fahrenheit:=9/5*c+32;
end;

begin
   write('Enter value in fahrenheit >> ');
   readln(fahrenheit);

   centigrade:=convert_to_centigrade(fahrenheit);

   writeln(fahrenheit:6:3,'deg F is ', centigrade:6:3,' deg C');

   write('Enter value in centigrade');
   readln(centigrade);

   fahrenheit:=convert_to_fahrenhiet(centigrade);

   writeln(centigrade:6:3,' deg C is ',fahrenheit:6:3,' deg F');
end.
```

Test run 5.2

```
Enter value in fahrenheit >> 12
12.000 deg F is -11.111 deg C
Enter value in centigrade >> 60
60.000 deg C is 15.556 deg F
```

Program 5.5

```
/* Program that uses a function to convert fahrenheit to centigrade */
#include <stdio.h>
```

```
/* Function prototypes */
float   convert_to_centigrade(float f);
float convert_to_fahrenheit(float c);

int   main(void)
{

float fahrenheit,centigrade;

   printf("Enter value in fahrenheit >> ");
   scanf("%f",&fahrenheit);

   centigrade=convert_to_centigrade(fahrenheit);

   printf("%8.3f deg F is %8.3f deg C\n",fahrenheit,centigrade);

   printf("Enter value in centigrade");
   scanf("%f",&centigrade);

   fahrenheit=convert_to_fahrenheit(centigrade);

   printf("%8.3f deg C is %8.3f deg F\n",centigrade, fahrenheit);
   return(0);
}

float convert_to_centigrade(float f)
{
   return(5.0/9.0* (f-32));
}

float convert_to_fahrenheit(float c)
{
   return(9.0/5.0*c+32);
}
```

Pascal Program 5.5 and C Program 5.6 uses the two function developed in the previous programs to display a table of values from 0°C to 100°C in steps of 10°C. Test run 5.3 shows a sample run.

(Pascal) **Program 5.5**

```
Program function5(input,output);
{ Program that uses a function to convert centigrade to      *)
{ fahrenheit. Centigrade goes from 0 to 100 in steps of 10.   *)

var   fahrenheit,centigrade:real;
      i: integer;

function convert_to_fahrenheit(c:real ):real;
begin
   convert_to_fahrenheit:=9/5*c+32;
end;

begin
   writeln('Centigrade Farenheit');

   for i:=0 to 10 do
   begin
      centigrade:=10*i;
       fahrenheit:=convert_to_fahrenheit(centigrade);
      writeln(centigrade:8:3,fahrenheit:8:3);
   end;
end.
```

💻 **Test run 5.3**

```
Centigrade Farenheit
    0.000  32.000
   10.000  50.000
   20.000  68.000
   30.000  86.000
   40.000 104.000
   50.000 122.000
   60.000 140.000
   70.000 158.000
   80.000 176.000
   90.000 194.000
  100.000 212.000
```

⬡ **Program 5.6**

```c
/* Program that uses a function to convert   */
/* fahrenheit to centigrade                  */
#include <stdio.h>

/* Function prototypes */
float   convert_to_centigrade(float f);
float convert_to_fahrenheit(float c);

int   main(void)
{
float fahrenheit,centigrade;

   printf("Centigrade Farenheit\n");

   for (centigrade=0;centigrade<=100;centigrade+=10)
   {
      fahrenheit=convert_to_fahrenheit(centigrade);
      printf("%8.3f %8.3f\n",centigrade, fahrenheit);
   }
   return(0);
}

float convert_to_centigrade(float f)
{
   return(5.0/9.0* (f-32));
}

float convert_to_fahrenheit(float c)
{
   return(9.0/5.0*c+32);
}
```

5.5.3 Combinational logic

In this example, the following Boolean equation is processed to determine its truth table.

$$Z = \overline{(\overline{A + B} + (A.C)).C}$$

Figure 5.7 gives a schematic representation of this Boolean function.

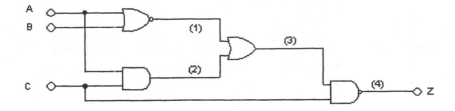

Figure 5.7 Schematic representation of the function $Z = \overline{(\overline{A+B}+(A.C))}.C$

The four nodes numbered on this schematic are:

(1) $\overline{A+B}$
(2) $A.C$
(3) $\overline{A+B}+(A.C)$
(4) $\overline{(\overline{A+B}+(A.C))}.C$

Table 5.1 gives a truth table showing the logical level at each point in the schematic. This table is necessary to check the program results against expected results. Table 5.2 gives the resulting truth table.

Table 5.1 Truth table

B	C	$\overline{A+B}$ (1)	$A.C$ (2)	$\overline{A+B}+(A.C)$ (3)	$(\overline{A+B}+(A.C)).C$	$\overline{(\overline{A+B}+(A.C))}.C$ (4)
0	0	1	0	1	0	1
0	1	1	0	1	1	0
1	0	0	0	0	0	1
1	1	0	0	0	0	1
0	0	0	0	0	0	1
0	1	0	1	1	1	0
1	0	0	0	0	0	1
1	1	0	1	1	1	0

Table 5.2 Truth table

A	B	C	Z
0	0	0	1
0	0	1	0
0	1	0	1
0	1	1	1
1	0	0	1
1	0	1	0
1	1	0	1
1	1	1	0

The permutations of the truth table input variables (i.e. 000, 001, 010, 011, ..., 111) are generated using 3 nested `for` loops. The inner loop toggles C from a 0 to a 1, the next loop toggles B and the outer loop toggles A. The Boolean functions use the logical operators `&&` and `||`. Recall that these operators treat a value of 0 (zero) as FALSE and any other value as TRUE.

Program 5.7

```
#include <stdio.h>
#define      FALSE    0
#define      TRUE     1

/* ANSI C function prototypes */
int      AND(int x,int y);
int      NAND(int x,int y);
int      NOR(int x,int y);
int      OR (int x,int y);
int      NOT(int x);

int      main(void)
{
int      a,b,c,z;

/* Go through all permutations of truth table */
   puts("     A     B     C       Result");
   puts("     **************************");
   for (a=FALSE;a<=TRUE;a++)
      for (b=FALSE;b<=TRUE;b++)
         for (c=FALSE;c<=TRUE;c++)
         {
            z=NAND(OR(NOR(a,b),AND(a,c)),c);
            printf("%6d %6d %6d %6d\n",a,b,c,z);
         }
   return(0);
}

int      AND(int x,int y)
{
   if ( x && y ) return(TRUE); else return(FALSE);
}

int      NAND(int x,int y)
{
   if ( x && y ) return(FALSE);
   else return(TRUE);
}

int      OR(int x,int y)
{
   if ( x || y ) return(TRUE);  else return(FALSE);
}

int      NOR(int x,int y)
{
   if ( x || y ) return(FALSE); else return(TRUE);
}

int      NOT(int x).
{
   if (x) return(FALSE);  else return(TRUE);
}
```

Test run 5.4 shows a sample run of the program. Notice that the results are identical to the truth table generated by analyzing the schematic.

```
      Test run 5.4
     A        B        C       Result
     ****************************
     0        0        0        1
     0        0        1        0
     0        1        0        1
     0        1        1        1
     1        0        0        1
     1        0        1        0
     1        1        0        1
     1        1        1        0
```

Pascal Program 5.6 shows the equivalent Pascal program. As Pascal has reserved keywords for AND, OR and NOT the function names have been changed to reflect the number of inputs they have, such as OR3 for a 3-input OR gate and AND2 for a 2-input AND gate.

Program 5.6

```pascal
program test(input,output);
var a,b,c,z:boolean;

function NOR2(x, y:boolean):boolean;
begin
   if ( x or y ) then nor2:=FALSE
   else nor2:=TRUE;
end;

function OR2(x, y:boolean):boolean;
begin
   if ( x or y) then or2:=TRUE
   else or2:=FALSE;
end;

function AND2(x, y:boolean):boolean;
begin
   if ( x and y) then and2:=TRUE
   else and2:=FALSE;
end;

function NAND2(x, y:boolean):boolean;
begin
   if ( x and y ) then nand2:=FALSE
   else nand2:=TRUE;
end;

begin
   writeln('   A      B      C       Result');
   writeln('   ****************************');

   for a:=FALSE to TRUE do
      for b:=FALSE to TRUE do
         for c:=FALSE to TRUE do
         begin
            z:=NAND2(OR2(NOR2(a,b),AND2(a,c)),c);
            writeln(a:6,b:6,c:6,z:6);
         end;
end.
```

5.5.4 Impedance of a series RL circuit

The magnitude of the impedance of an RL circuit is given by the equation:

$$|Z| = \sqrt{R^2 + X_L^{\,2}} \quad \Omega$$

and the phase angle of this impedance is given by:

$$\langle Z \rangle = \tan^{-1} \frac{X_L}{R}$$

Figure 5.8 gives a schematic of an RL series circuit.

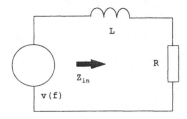

Figure 5.8   RL series circuit

Figure 5.9 gives a structure chart which outlines a basic design for this problem. Inputs are resistance (R), inductance (L) and frequency (f). The program determines the magnitude and phase angle of the impedance. In order to determine these values the reactance of the inductor must be determined using $X_L = 2\pi fL$.

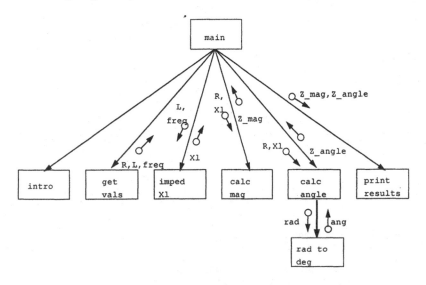

Figure 5.9   Structure chart for a series RL circuit program

⬡ c
Program 5.8
```c
/* Program to determine impedance of an        */
/* RL series circuit                           */

#include <stdio.h>
#include <math.h>

#define PI      3.14159

float calc_mag(float x, float y);
float calc_angle(float x, float y);
float impedance_Xl(float L,float f);
float rad_to_deg(float rad);
void  intro(void);
void  print_results(float mag,float angle);

int    main(void)
{
float     r,l,Xl,Z_mag,Z_angle,freq;

   intro();

   printf("Enter R,L and frequency >>>");

   scanf("%f %f %f",&r,&l,&freq);

   Xl=impedance_Xl(l,freq);

   Z_mag=calc_mag(r,Xl);
   Z_angle=calc_angle(r,Xl);

   print_results(Z_mag,Z_angle);
   return(0);
}

void     intro(void)
{
   puts("Prog to determine impedance of series RL circuit");

}

void     print_results(float mag,float angle)
{
   printf("Impedance is %.2f ohms phase angle %.2f degrees\n",
          mag,angle);
}

float calc_mag(float x, float y)
{
float     z;

   /* determine magnitude for rectangular         */
   /* co-ordinates x+jy                            */
   z=(float) sqrt( x*x + y*y ); /* sqrt() returns a double */

   return(z);
}
float calc_angle(float x, float y)
{
float ang;

   /* determine angle in degrees for rectangular */
   /* co-ordinates x+jy                           */

   ang=(float) atan(y/x); /* atan() returns a double */
   ang=rad_to_deg(ang);
```

```
    return(ang);
}

float impedance_Xl(float L,float f)
{
    /* determine impedance of inductor */
    return(2*PI*f*L);
}

float rad_to_deg(float rad)
{
    /* convert from radians to degrees */
    return(rad*180/PI);
}
```

Test run 5.5 gives a sample run of this program. The input parameters used are resistance 1 kΩ, inductance 1 mH and applied frequency 1 MHz.

Test run 5.5

```
Program to determine impedance of series RL circuit
Enter R,L and frequency >>> 1000 1e-3 1e6
Impedance is 6362.27 ohms phase angle 80.96 degrees
```

Pascal Program 5.7 gives the equivalent Pascal program.

Program 5.7

```
program xlimpedance(input,output);

var r,l,Xl,Z_mag,Z_angle,freq:real;

procedure intro;
begin
   writeln('Prog to determine impedance of series RL circuit');
end;

procedure print_results(mag,angle:real);
begin
   writeln('Impedance is ',mag:8:2, ' ohms phase angle ', angle:8:2,
        ' degrees');
end;
function rad_to_deg(rad:real):real;
begin
 (* convert from radians to degrees *)
   rad_to_deg:=rad*180/PI;
end;


function calc_mag(x, y:real):real;
var   z:real;
begin

   (* determine magnitude for rectangular co-ordinates x+jy *)
   z:=sqrt( x*x + y*y );
   calc_mag:=z;
end;

function calc_angle(x, y:real):real;
var   ang:real;
begin
   (* determine angle in degrees for rectangular *)
   (* co-ordinates x+jy                          *)
```

```
    ang:=arctan(y/x);
    ang:=rad_to_deg(ang);
    calc_angle:=ang;
end;

function impedance_Xl(L, f:real):real;
begin
    (* determine impedance of inductor *)
    impedance_Xl:=2*PI*f*L;
end;


begin
    intro;

    writeln('Enter R,L and frequency >>>');

    readln(r,l,freq);

    Xl:=impedance_Xl(l,freq);
    Z_mag:=calc_mag(r,Xl);
    Z_angle:=calc_angle(r,Xl);

    print_results(Z_mag,Z_angle);
end.
```

5.6 Exercises

5.6.1   Write a program which determines the magnitude of an entered value. The program should use a function to determine this.

5.6.2   Write a program which determines the magnitude and angle of a complex number (in the form $x+iy$, or $x+jy$). The program should use functions to determine each of the values. Complete Table 5.3 using the program (the first row has already been completed).

Table 5.3   Complex number calculation

x	y	Mag.	Angle($^\circ$)
10	10	14.142	45
−10	5		
100	50		
−1	−1		

5.6.3   Write a mathematical function for a factorial calcuation, where:

$$n! = n \times (n-1) \times (n-2) \times \ldots \times 2 \times 1$$

5.6.4   Write mathematical functions for the following

(i) sine_function()
(ii) cosine_function()

The sine and cosine functions should calculated from first principles with:

$$\cos(x) = 1 - \frac{x^2}{2!} + \frac{x^4}{4!} - \frac{x^6}{6!} + \frac{x^8}{8!} - \ldots$$

$$\sin(x) = x - \frac{x^3}{3!} + \frac{x^5}{5!} - \frac{x^7}{7!} + \frac{x^9}{9!} - \ldots$$

The error in the functions should be less than 1×10^{-6}.

5.6.5 Using the functions developed in Exercise 5.6.2 and the standard sine and cosine library functions, write a program which determines the error between the standard library functions and the developed functions. From this, complete Table 5.3.

Table 5.3 Sine and cosine results

Value	Standard cosine function	Developed cosine function	Standard sine function	Developed sine function
2	−0.41614683		0.9092974	
−0.5				
1				
−1				

5.6.6 Write a mathematical function which determines the exponential of a value using the first principles formula:

$$e^x = 1 + \frac{x}{1!} + \frac{x^2}{2!} + \frac{x^3}{3!} + \frac{x^4}{4!} + \ldots$$

Compare the result with the standard exp () library function.

5.6.7 Write Boolean logic functions for the following four digital gates:

```
AND3 (A, B, C)
OR3 (A, B, C)
NAND3 (A, B, C)
NOR3 (A, B, C)
```

Worksheet 8

W8.1   Write a program with separate functions which determine the gradient of a straight line (m) and the point at which a straight line cut the y-axis (c). The entered parameters are two points on the line, that is, (x_1, y_1) and (x_2, y_2). From this program complete Table W8.1 (the first row has already been completed).

Table W8.1   Straight line calculations

x_1	y_1	x_2	y_2	m	c
3	3	4	5	2	−3
−1	5	0	−1		
100	50	−10	−10		
−1	−1	1	3		

W8.2   Write a program which has a function which will only return a real value when the entered value is within a specified range. Examples of calls to this function (which, in this case, is named `get_real`) are given next.

```
inval:=get_real(0,100);    (* in Pascal *)
inval=get_real(0,100);     /* in C */
```

which will only return a value from the function when the entered value is between 0 and 100. A sample run is given next.

⌨ Test run W8.1
```
Enter a value > -1
INVALID INPUT PLEASE RE-ENTER
Enter a value > 6
Success. Bye.
```

W8.3   Modify the program developed and use it to prompt the use to enter a value of current (between 0A and 10A) and voltage (between 0V and 100V). The program will then determine the resistance by dividing the entered voltage by the entered current. The program should contain two functions: `get_real()` and `calc_current()`.

6 Parameter Passing

6.1 Introduction

Parameter passing involves passing input parameters into a module (a function in C and a function and procedure in Pascal) and receiving output parameters back from the module. For example a quadratic equation module requires three parameters to be passed to it, these would be a, b and c. These are defined as the input parameters. The output parameters would be the two roots of the equation, such as root1 and root2. Another parameter could also be passed back to indicate the type of root (such as singular, real or complex). This indication is normally known as a flag. Figure 6.1 illustrates the passing of parameters into and out of a module.

In Pascal the module would be defined as:

```
procedure calc_quadratic(a,b,c:integer; var root1,root2:real;
                         var type:integer)
```

In C a flag is typically passed back through the function header, thus:

```
int calc_quadratic(int a, int b, int c,
                   float *root1, float *root2)
```

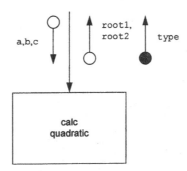

Figure 6.1   Parameter passing

6.2 Pascal parameter passing

Procedures are used in Pascal when parameters need to be passed back to the calling routine. Functions are normally used when there is only one value passed back, whereas a procedure can return any number (or even none). The first part of defining a procedure is to define the input parameters and the output parameters. The input parameters are declared in the parameter list and output parameters are defined with a preceding var keyword. For example, if the input parameters are x and y, and the output parameters are w and z then the following procedure would be defined:

```
procedure  proc1(x,y:real; var w,z:real);
begin

end;
```

In this case only values will be passed into the variables x and y, whereas the values of w and z will be passed back to the calling module. If a procedure requires local variables then these are declared after the procedure header. For example, if the procedure in the last example has two local variables named temp1 and temp2 then it will have the form:

```
procedure  proc1(x,y:real; var w,z:real);
var  temp1,temp2:real;
begin

end;
```

Local variables only exist within the procedure and their contents are lost when the procedure quits.

Program 6.1 shows an example program with a procedure which swaps the values of two variables. In this case a temporary local value (temp) holds one of the values (temp:=x) so that the contents of the other value can be placed in it (x:=y). The temporary value is then put into the other variable (y:=temp). This operation performs a swap.

Program 6.1

```
program proc1(input,output);

var a,b: integer;

procedure swap(var x,y: integer);
var temp: integer;
begin
   temp := x;
   x    := y;
   y    := temp
end;

begin
```

```
      write('Enter two integer values >> ');
      readln(a,b);

      writeln('Values are ',a,' and ',b);
      swap(a,b);
      writeln('Values swapped are ',a,' and ',b);
   end.
```

Program 6.2 calculates the gradient of a straight line given two co-ordinates $(x1,y1)$ and $(x2,y2)$.

⬡ Program 6.2

```
program proc2(input,output);

var x1,y1,x2,y2:real;

procedure get_coord(var x,y:real);
begin
   writeln('Enter x and y coordinate');
   readln(x,y);
end;

function gradient(xstart,ystart,xend,yend:real):real;
begin
   gradient:=(yend-ystart)/(xend-xstart);
end;

begin
   writeln('Start co-ordinate');
   get_coord(x1,y1);
   writeln('End co-ordinate');
   get_coord(x2,y2);
   writeln('The gradient is ',gradient(x1,y1,x2,y2):6:2 );
end.
```

6.3 C parameter passing

One of the most confusion areas of C programming is the usage of pointers. They basically refer to addresses in memory but to the novice they are one of the least liked elements of C programming. In many cases they are extremely useful and allow the programmer to directly access areas of memory.

6.3.1 Introduction

If a company were to send a form to a person and they neglected to inform the person of the correct return address then it would not be possible for the recipient to send back the modified form (unless the person already knows the address). Variables sent to functions operate in a similar manner. If the function does not know where a variable *lives* (its address in memory) then the function cannot change its contents.

A program uses data which is stored by variables. These are assigned to a unique space in memory, the number of bytes they use depends on their data type. For example, a char uses 1 byte, an int will typically take 2 or 4 bytes, and a `float`, typically, 4 or 8 bytes. Each memory location contains one byte and has a unique address associated with it (that is, its binary address). This address is normally specified as a hexadecimal value as this can be easily converted to the actual binary address. The memory map in Figure 6.2 shows how three variables `value1`, `value2` and `ch` could be allocated in memory. This diagram assumes that a `float` uses 4 bytes, and an `int` 2 bytes. The compiler, in this case, has allocated `value1` from addresses `100h` to `103h`, `value2` at `104h` and `105h`, and `ch` is allocated to `106h`. The start of the variable's address in memory can be described as a memory pointer to the variable. A pointer variable is used to store a memory address.

Variables sent to a function can have their contents changed by passing a pointer in the argument list. This method involves sending a memory *address* rather than a copy of the variable's value. A preceding ampersand (`&`) specifies a pointer. This can be thought of as representing *the address of*:

```
&variable_name    {address of variable_name}
```

A pointer to a variable will store the address to the first byte of the area allocated to the variable. An asterisk (*) preceding a pointer is used to access the contents of the location pointed to. The number of bytes accessed will depend on the data type of the pointer. The * operator can be thought of representing *at address*:

```
*ptr   {value stored at address specified by ptr}
```

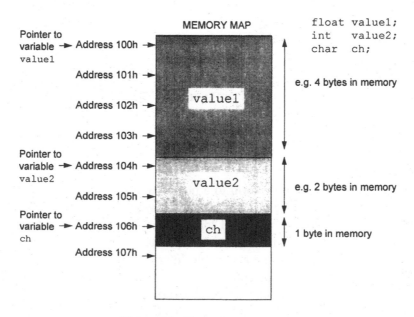

Figure 6.2   Example memory map

Figure 6.3 shows an example memory map. A variable `resistance_1` has the value of `310.0` and is stored at a memory location starting at `107h`. If the data type is a short integer then it will take up 2 bytes in memory (i.e. `107h` and `108h`), if it is a float it may take up 4 bytes in memory (i.e. `107h` to `110h`). The memory map also shows that a pointer `ptr` points to memory location `102h`. The value stored at this location is `15`; `*ptr` accesses its contents. The declaration of the pointer defines the data type of the pointer and thus the number of bytes used to store the value at the address pointed to by the pointer.

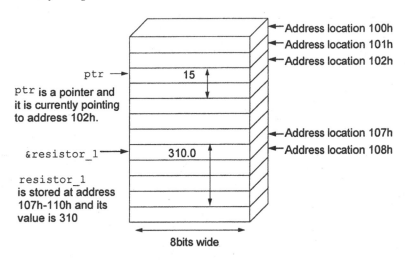

Figure 6.3 Example memory map

6.3.2 Pointers with functions

In the previous chapter it was shown that a single value is passed out of a function through the function header. In order to pass values out through the argument list the address of the variable is passed; that is referred to as *'call by reference'*. To declare a pointer the data type is specified and the pointer name is preceded by an asterisk. The following is the general format:

```
type_def *ptr_name;
```

In this case `ptr_name` is the name of the pointer. The contents of the variable at this address can be accessed using `*ptr_name`. When a function is to modify a variable then a pointer to its address is sent. For example, if the variable to be modified is `value` then the argument passed is `&value`.

Program 6.1 shows an example of a function that swaps the contents of two variables (`a` and `b`). Figure 6.4 shows how the compiler checks the parameters passed to the function and the return type. The function prototype, in this case, specifies that the parameters sent are pointers to integer values and the return type is void. The compiler checks that the parameters sent to the function are integer pointers and that nothing is assigned to the return value from the function.

Program 6.1

```
#include <stdio.h>

void     swap(int *ptr1,int *ptr2);

int      main(void)
{
int      a,b;

    a=5; b=6;
    swap(&a,&b); /* send addresses of a and b */
    printf("a= %d b = %d \n",a,b);
    return(0);
}


void     swap(int *ptr1,int *ptr2)
{
    /* ptr1 and ptr2 are pointers (addresses). */
int      temp;

    temp = *ptr1;
    *ptr1 = *ptr2;
    *ptr2 = temp;

}
```

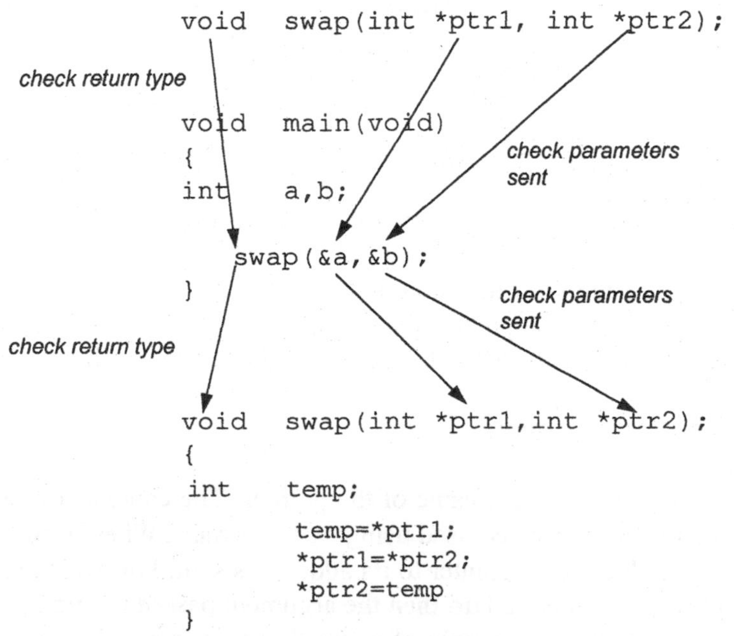

Figure 6.4   Compiler checking for Program 6.1

Program 6.2 calculates the gradient of a straight line given two co-ordinates
($x1$,$y1$) and ($x2$,$y2$). Within the get_coord() function the scanf() do
not require a preceding ampersand; the variable arguments are already
pointers.

Program 6.2

```c
#include <stdio.h>

/* Define function prototypes */
void  get_coord(float *x,float *y);
float gradient(float xstart,float ystart,float xend,float yend);

int    main(void)
{
float x1,y1,x2,y2;

    printf("Start co ordinate >>");
    get_coord(&x1,&y1);
    printf("End co ordinate >> ");
    get_coord(&x2,&y2);
    printf("The gradient is %8.2f\n",gradient(x1,y1,x2,y2));
    return(0);
}

void  get_coord(float *x,float *y)
{
    printf("Enter x and y coordinate");
    scanf("%f %f",x,y); /* x and y are pointers, thus no &'s are required */
}

float gradient(float xstart,float ystart,float xend,float yend)
{
    return((yend-ystart)/(xend-xstart));
}
```

An example of a standard C function that uses call by reference is
scanf() where a pointer is passed for each variable. This allows the function to change its contents.

6.4 Examples

6.4.1 Quadratic equations

C Program 6.2 determines the roots of a quadratic equation. The function
get_values() gets variables a, b and c; these variables are passed as
pointers.

The function to determine the root(s) of a quadratic equation is quad-
ratic_equ(). This returns the root type (such as, singular, real or complex) through the function header and passes the equation root(s) through
the argument list using pointers. The root type returned can be referred to as
a return flag; this flag is set up using an enum declaration. There are three
possible states for this: SINGULAR (a value of 0), REAL_ROOTS (a value
of 1) and COMPLEX_ROOTS (a value of 2). The program then uses the flag

to determine how the root(s) are to be displayed. If the root is singular then
`print_results()` prints a single value of `root1`; else, if the roots are
real, then two values `root1` and `root2` are printed; and if the roots are
complex the function will print the roots in the form `root1 +/-j`
`root2`.

Figure 6.5 gives a basic structure chart of this program. The return flag
from the `quadratic_equ()` function is represented by an arrow with a
circle on the end.

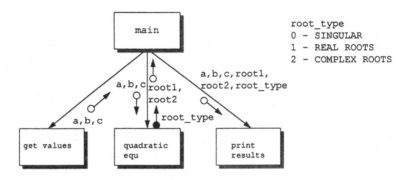

Figure 6.5   Structure chart for Program 6.3

Program 6.3

```c
#include <stdio.h>
#include <math.h>
enum   {SINGULAR, REAL_ROOTS,COMPLEX_ROOTS } quad_roots;

void      get_values(float *ain,float *bin,float *cin);
int       quadratic_equ(float a,float b,float c, float *r1,float *r2);
void      print_results(int r_type,float a,float b,float c, float r1,
             float r2);

int       main(void)
{
float a,b,c,root1,root2;
int       root_type;

   get_values(&a,&b,&c);
   root_type=quadratic_equ(a,b,c,&root1,&root2);
   print_results(root_type,a,b,c,root1,root2);
   return(0);
}

void      get_values(float *ain,float *bin,float *cin)
{
   printf("Enter a, b and c >>");
   scanf("%f %f %f",ain,bin,cin); /*ain,bin and cin are pointers */
}

int       quadratic_equ(float a,float b,float c, float *r1,float *r2)
{
   if (a==0)
   {
     *r1=-c/b;
     return(SINGULAR);
   }
```

```c
    else if ((b*b)>(4*a*c))
    {
        *r1=(-b+sqrt(b*b-4*a*c))/(2*a);
        *r2=(-b-sqrt(b*b-4*a*c))/(2*a);
        return(REAL_ROOTS);
    }
    else if ((b*b)<(4*a*c))
    {
        *r1=-b/(2*a);
        *r2=sqrt(4*a*c-b*b)/(2*a);
        return(COMPLEX_ROOTS);
    }
    else
    {
        *r1=-b/(2*a);
        return(SINGULAR);
    }
}

void    print_results(int r_type,float a,float b,float c,
                float r1,float r2)
{
    printf("Quadratic equation %8.3f x^2 + %8.3f x + %8.3f\n",a,b,c);
    if (r_type==SINGULAR)
        printf("Singular root of %8.3f\n",r1);
    else if (r_type==REAL_ROOTS)
        printf("Real roots of %8.3f and %8.3f\n",r1,r2);
    else
        printf("Complex root of %8.3f +/-j %8.3f\n",r1,r2);
}
```

Pascal Program 6.3 is similar to the C Program 6.3 for the following:

- The flag for the quadratic equation is passed as one of the variables in the parameter of quadratic_equ (rtype).
- The quadratic equation root types are defined as constants using const.

⬡ Pascal

Program 6.3

```pascal
program quad(input,output);

const SINGULAR=0; REAL_ROOTS=1;COMPLEX_ROOTS=2;

var     a,b,c,root1,root2:real;
        root_type:integer;

procedure get_values(var ain,bin,cin:real);
begin
    write('Enter a, b and c >>');
    readln(ain,bin,cin);
end;

procedure quadratic_equ(a,b,c:real; var r1,r2:real; var rtype:integer);
begin
    if (a=0) then
    begin
        r1:=-c/b;
        rtype:=SINGULAR;
    end
    else if ((b*b)>(4*a*c)) then
    begin
        r1:=(-b+sqrt(b*b-4*a*c))/(2*a);
        r2:=(-b-sqrt(b*b-4*a*c))/(2*a);
        rtype:=REAL_ROOTS;
```

```
    end
    else if ((b*b)<(4*a*c)) then
    begin
       r1:=-b/(2*a);
       r2:=sqrt(4*a*c-b*b)/(2*a);
       rtype:=COMPLEX_ROOTS;
    end
    else
    begin
       r1:=-b/(2*a);
       rtype:=SINGULAR;
    end;
end;

procedure print_results(r_type:integer; a,b,c,r1,r2:real);
begin
    writeln('Quadratic equation ',a:8:3, ' x^2 + ', b:8:3, ' x + ',c:8:3);
    if (r_type=SINGULAR) then
       writeln('Singular root of ',r1)
    else if (r_type=REAL_ROOTS)  then
       writeln('Real roots of ', r1:8:3, ' and ',r2:8:3)
    else
       writeln('Complex root of ',r1:8:3,' +/-j ',r2:8:3);
end;

begin
    get_values(a,b,c);
    quadratic_equ(a,b,c,root1,root2,root_type);
    print_results(root_type,a,b,c,root1,root2);
end.
```

Test run 6.1 shows tests for each of the root types.

🖳 Test run 6.1

```
Enter a, b and c >>  2 1 1

Quadratic equation    2.000 x^2 +    1.000 x +    1.000
Complex root of   -0.250 +/-j    0.661

Enter a, b and c >>  1 -2 -3

Quadratic equation    1.000 x^2 +   -2.000 x +   -3.000
Real roots of    3.000 and   -1.000

Enter a, b and c >>  1 2 1

Quadratic equation    1.000 x^2 +    2.000 x +    1.000
Singular root of   -1.000
```

6.4.2 Equivalent parallel resistance

C Program 6.3 uses pointers to determine the equivalent parallel resistance of two resistors. A basic structure chart, given in Figure 6.6, shows that get_values() returns the variables R1 and R2; in order to change their values they are sent as pointers. It also shows that the variables sent to calc_parallel_res() are R1, R2 and R_equ is returned. Variables R1, R2 and R_equ are then passed into print_results().

Pascal Program 6.4 gives the Pascal equivalent of this program.

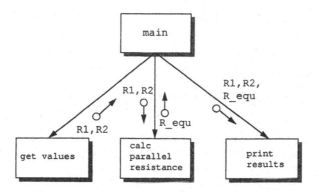

Figure 6.6 Structure chart for program 6.3

Program 6.4

```c
#include <stdio.h>

/* Define function prototypes   */
void    get_values(float *r1,float *r2);
void    get_parallel_res(float r1,float r2,float *r_e);
void    print_results(float r1,float r2,float r_e);

int     main(void)
{
float R1,R2,R_equ;

   get_values(&R1,&R2);
   get_parallel_res(R1,R2,&R_equ);
   print_results(R1,R2,R_equ);
   return(0);
}

void    get_values(float *r1,float *r2)
{
   do
   {
      printf("Enter R1 >>");
      scanf("%f",r1); /*r1 is already a pointer no need for &r1 */
      if (*r1<0) puts("INVALID: re-enter");
   } while (*r1<0);

   do
   {
      printf("Enter R2 >>");
      scanf("%f",r2);
      if (*r2<0) puts("INVALID: re-enter");
   } while (*r2<0);
}
void    get_parallel_res(float r1,float r2,float *r_e)
{
   *r_e=1/(1/r1+1/r2);
}

void    print_results(float r1,float r2,float r_e)
{
   printf("Parallel resistors %8.3f and %8.3f ohm\n",r1,r2);
   printf("Equivalent resistance is %8.3f ohm\n",r_e);
}
```

⬡ *Pascal*

Program 6.4

```pascal
program par_res(input,output);

var     R1,R2,R_equ:real;

procedure get_values(var r1,r2:real);
begin
   repeat
      write('Enter R1 >>');
      readln(r1);
      if (r1<=0) then writeln('INVALID: re-enter');
   until (r1>0);

   repeat
      write('Enter R2 >>');
      readln(r2);
      if (r2<=0) then writeln('INVALID: re-enter');
   until (r2>0);
end;

procedure get_parallel_res(r1,r2:real;var r_e:real);
begin
   r_e:=1/(1/r1+1/r2);
end;

procedure print_results(r1,r2,r_e:real);
begin
   writeln('Parallel resistors ', r1:8:3, ' and ',r2:8:3, 'ohm');
   writeln('Equivalent resistance is ',r_e:8:3, ' ohm');
end;

begin
   get_values(R1,R2);
   get_parallel_res(R1,R2,R_equ);
   print_results(R1,R2,R_equ);
end.
```

Test run 6.2 shows a sample run.

💻 **Test run 6.2**

```
Enter R1 >> 1000
Enter R2 >> 800
Parallel resistors 1000.000 and  800.000 ohm
Equivalent resistance is  444.444 ohm
```

6.5 Exercises

6.5.1   Write a program with a single module that returns both the value of *m* and *c* for a straight line, given passed values of (x_1, y_1) and (x_2, y_2). The equation of the straight line is given by:

$$y = mx + c \qquad \text{where } m = \frac{y_2 - y_1}{x_2 - x_1} \text{ and } c = y_1 - mx_1$$

Example calls for C and Pascal are:

```
calc_line(x1,y1,x2,y2,m,c);      (* in Pascal *)
calc_line(x1,y1,x2,y2,&m,&c);    /* in C      */
```

where x1, y1, x2, y2 are the coordinates, m is the return gradient and c is the returned value for the point at which the line cuts the y-axis.

6.5.2 Write a program which contains a module which is passed two parameters. The module should arrange the values of the parameters so that the first parameter is the largest.

6.5.3 Resistors are normally identified by means of a colour coding system, which is given in Table 6.1. Figure 6.7 shows a 4-band resistor, where the first two bands give a digit, the third a multiplier and the fourth the tolerance. Write a program, with parameter passing, in which the user enters the first three values of the code and the program determines the resistor value. The value range of inputs is between 0 and 9 and A sample run is shown in Test run 6.3.

🖳 **Test run 6.3**
```
Enter band 1 value >> 100
INVALID: re-enter
Enter band 1 value >> 3
Enter band 2 value >> 4
Enter band 3 value >> 2
Resistor value is 3400 ohms
```

Table 6.1   Resistor colour coding

Digit	Colour	Multiplier	No. of zeros
	Silver	0.01	−2
	Gold	0.1	−1
0	Black	1	0
1	Brown	10	1
2	Red	100	2
3	Orange	1k	3
4	Yellow	10k	4
5	Green	100k	5
6	Blue	1M	6
7	Violet	10M	7
8	Grey		
9	White		

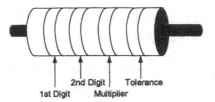

Figure 6.7   4-band resistor colour code

6.5.4   Write a module, with parameter passing, in which a minimum and maximum value are passed to it and the module returns back an entered value which is between the minimum and maximum value.

Example calls for C and Pascal are:

```
get_value(min,max,val);     (* in Pascal  *)
get_value(min,max,&val);    /* in C       */
```

where `min` is the minimum value, `max` is the maximum value and `val` is the return valued.

Use this program with the previous exercise so that the entered values are between 0 and 9.

Worksheet 9

W9.1   Write a program which has a module which converts a complex number in rectangular form into polar form (magnitude and angle). The module should have two values passed to it (x and y) and return two values for the magnitude and angle of the complex number (*mag* and *angle*). These values are determined using:

if $\ z = |Z|\langle\theta\rangle = x + iy$

then

$$|Z| = \sqrt{x^2 + y^2}$$
$$\theta = \tan^{-1}\left(\frac{y}{x}\right) \ \text{radians}$$

Example calls for C and Pascal are:

```
convert_polar(x,y,mag,angle);   (* in Pascal *)
convert_polar(x,y,&mag,&angle);  /* in C */
```

W9.2   Write a program which has a module which converts a complex number in polar form into rectangular form (real and imaginary). The module should have two values passed to it (*mag* and *angle*) and return two values for the real and imaginary parts of the complex number (x and y). These values are determined using:

if $\ z = |Z|\langle\theta\rangle = x + iy$

then

$$x = |Z|\cos\theta$$
$$y = |Z|\sin\theta$$

Example calls for C and Pascal are:

```
convert_rect(mag,angle,x,y);      (* in Pascal *)
convert_rect(mag,angle,&x,&y);   /* in C */
```

7 Arrays

7.1 Introduction

An array stores more than one value, of a common data type, under a collective name. Each value has a unique slot and is referenced using an indexing technique. Figure 7.1 shows a circuit with 5 resistors, which could be declared with a program with 5 simple float (in C) or real (in Pascal) declarations. If these resistor variables were required to be passed into a function all 5 would have to be passed through the parameter list. A neater method uses arrays to store all of the values under a common name (in this case R). Thus a single array variable can then be passed into any function that uses it.

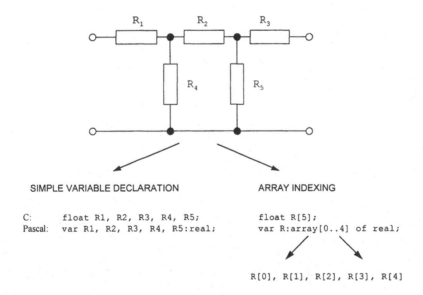

SIMPLE VARIABLE DECLARATION

```
C:      float R1, R2, R3, R4, R5;
Pascal:  var R1, R2, R3, R4, R5:real;
```

ARRAY INDEXING

```
float R[5];
var R:array[0..4] of real;
```

```
R[0], R[1], R[2], R[3], R[4]
```

Figure 7.1 Simple variables against array indexing

7.2 C arrays

The declaration of an array specifies the data type, the array name and the number of elements in the array in brackets ([]). The following gives the standard format for an array declaration.

```
data_type array_name[size];
```

Figure 7.2 shows that the first element of the array is indexed 0 and the last element as `size-1`. The compiler allocates memory for the first element `array_name[0]` to the last array element `array_name[size-1]`. The number of bytes allocated in memory will be the number of elements in the array multiplied by the number of bytes used to store the data type of the array.

The following gives some example array declarations:

```
int circuit[10];   /* allocates space for circuit[0] to circuit[9]   */
float impedance[50]; /*  allocates for impedance[0] to impedance[49] */
```

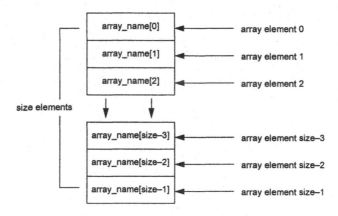

Figure 7.2 Array elements

7.2.1 Pointers and arrays

In C, there is a strong relationship between pointers and arrays. A pointer variable stores a memory address which can be modified, whereas an array name stores a fixed address, set to the first element in the array. The address of the first element of an array named `arrname` is thus `&arrname[0]`. Table 7.1 shows examples of how arrays and pointers use different indexing notations and how it is possible to interchange them.

Figure 7.3 shows two array declarations for `arrname`. Each has five elements; the first is `arrname[0]` and the last `arrname[4]`. The number of bytes allocated to each element depends on the data type declaration. A `char` array uses one byte for each element, whereas an `int` array will typically take 2 or 4 bytes. The array name `arrname` is set to the address of the first element of the array. Each element within the array is referenced with respect to this address.

Table 7.2 gives some examples of array and pointer statements.

Table 7.1 Relationship between arrays and pointers

Using arrays	Using pointers
float arr[10];	float *arr; arr=(float *) malloc(10*sizeof(float));
arr[0]	*(arr)
arr[1]	*(arr+1)
arr[2]	*(arr+2)
arr[9]	*(arr+9)

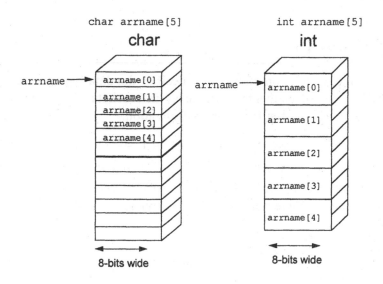

Figure 7.3 Array elements

Table 7.2 Examples of array and pointer statements

Statements	Description
int tmp[100];	declare an array named tmp with 100 elements
tmp[1]=5;	assign 5 to the second element of array tmp
*(tmp+1)=5;	equivalent to previous statement
ptr=&tmp[2];	get address of third element

7.2.2 Passing arrays to C functions

At compilation the compiler reserves enough space for all elements in an array and initializes the array name to the start of it. In order for a function to modify the array the base address is passed through the parameter list. The function itself does not know the maximum number of elements in the array; unless a parameter relating to the maximum number of elements in the array is also passed. It is thus possible to run off the end of an array and access memory not allocated to the array.

The notation used to signify that an array is being passed in a function is square brackets (this signifies that it is a fixed address and not a pointer variable). The following gives an example of array passing.

```
float maximum(int number_of_elements,float arrayname[])
{
float max;
/* number_of_elements maximum number of elements in the array */
/* this function determines maximum value in an array         */
   max=arrayname[0];
   for (i=1;i<number_of_elements;i++)
      if (max<arrayname[i]) max=arrayname[i];
   return(max);
}
```

7.3 Pascal arrays

In Pascal, the declaration of an array specifies the data type, the array name and the number of elements in the array in brackets ([]). The following gives the standard format for an array declaration.

```
var arr_name : array[startval..endval] of datatype;
```

This declares an array of type *datatype* with the first element of *array_name*[*startval*] to the last array element *array_name*[*endval*]. The following gives some example array declarations and assignments.

```
var circuit:array[1..10] of integer;
var impedance:array[1..50] of real;

   circuit[1]:=42;
   impedance[20]:=3.14;
```

7.3.1 Passing arrays to functions

In order to pass an array into a Pascal function or a procedure the array type must first be defined. This is achieved with the type keyword and an array definition is normally done before any of the procedures or functions. For example to define an array type with 100 elements of real values:

```
type arrtype:array[1..100] of real;
```

and an example array is declared with:

```
var  myarr: arrtype;
```

When an array is passed into a function or a procedure, the type definition must be used to define the array type. The following section of code gives an example of array being passing into a function. In this case the array type has been defined as arrtype and the name of the array in the main program is arr1. In the function the array name is arrayin.

```
program maxprog(input,output);

type   arrtype=array[1..100] of real;
var    arr1:arrtype;
       x:real;

function maximum(n:integer; arrayin:arrtype):real;
var max:real;
    i:integer;
begin
(* n is the number of elements in the array             *)
(* This function determines maximum value in an array   *)

   max:=arrayin[1];
   for i:=2 to n do
      if (max<arrayname[i]) then max:=arrayin[i];
   maximum:=max;
end;

begin
   :   :   :
     x:=maximum(10,arr1);
   :   :   :
end.
```

If an array is passed to a procedure, the only way that the procedure can modify the contents of an array so that when its contents are changed when the procedure has completed is to put the var keyword in front of its declaration in the parameter list. The following shows an example of two procedures which can modify the contents of an array. In the first, fill_arr, the function modifies the array passed to it as there is a var in front of it in the parameter list. If there was no var in front of the array name then the contents of the passed array could be modified within this function but when the function was complete then the array which is passed to the fill_arr function (array1) would not have been modified. In the copy_arr procedure, the first array passed (arr1) does not have its contents changed as there is no var keyword in front of it in the parameter list, whereas the second array passed (arr2) has the var keyword in front of it. The section of the code should fill an array with entered values (arr1) and then copy this array into another (arr2).

```
program arrprog(input,output);

type   arrtype=array[1..100] of real;
var    array1, array2:arrtype;

procedure copy_arr(n:integer; arr1:arrtype; var arr2:arrtype);
var        i:integer;
begin
   for i:=1 to n do
      arr2[i]:=arr1[i];
end;

procedure fill_arr(n:integer;var arr:arrtype);
var        i:integer;
begin
   for i:=1 to n do
   begin
```

```
      write('Enter value >>');
      readln(arr[i]);
   end;
end;

begin
   fill_arr(10,array1);
   copy_arr(10,array1,array2);
   :   :   :
end.
```

7.4 Examples

7.4.1 Running average program

Program 7.1 is a 3-point running average program. This type of program has a low-pass filter response and can filter data samples. Figure 7.4 illustrates how the output is a function of the average of three elements in the input array; this is achieved by generating a running average.

The first and last values of the processed array will take on the same values as the input array as there are not three values over which to take an average. Test run 7.1 shows a sample run with 10 entered values.

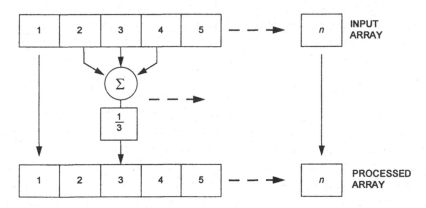

Figure 7.4 Array elements

Program 7.1

```c
#include <stdio.h>

#define     MAX    150    /* maximum values in the array   */
#define     TRUE   1
#define     FALSE  0

void        filter(int n,float array_in[],float array_out[]);
void        get_values(int *n,float array[]);
void        print_values(int n,float array_in[],float array_out[]);

int         main(void)
{
float       input[MAX],output[MAX];/*input values, processed values */
int         nvalues;
```

```
      get_values(&nvalues,input);
      filter(nvalues,input,output);
      print_values(nvalues,input,output);
      return(0);
}

void      filter(int n,float array_in[],float array_out[])
{
int       i;
   array_out[0]=array_in[0];
   array_out[n-1]=array_in[n-1];

   for (i=1;i<n-1;i++)
      array_out[i]=(array_in[i-1]+array_in[i]+array_in[i+1])/3;
}


void      get_values(int *n,float array[])
/* *n is the number of elements in the array */
{
int       i,rtn,okay;

   do
   {
      printf("Enter number of values to be processed >>");
      rtn=scanf("%d",n);
      if ((rtn!=1) || (*n<0) || (*n>MAX))
      {
         printf("Max elements are %d, re-enter\n",MAX);
         okay=FALSE;
      }
      else  okay=TRUE;
   } while (!okay);

   for (i=0;i<*n;i++)
   {
      puts("Enter value >> ");
      scanf("%f",&array[i]);
   }
}

void      print_values(int n,float array_in[],float array_out[])
{
int       i;
   printf("Input     Output\n");
   for (i=0;i<n;i++)
      printf(" %6.3f %6.3f \n",array_in[i],array_out[i]);
}
```

⬡ (Pascal) **Program 7.1**

```
program arr1(input,output);

const ARRAYSIZE=150;

type arrtype = array[1..ARRAYSIZE] of real;

var  input,output:arrtype;
     nvalues:integer;

procedure filter(n:integer;array_in:arrtype;var array_out:arrtype);
var       i:integer;
begin

   array_out[1]:=array_in[1];
   array_out[n]:=array_in[n];
```

```
   for i:=2 to n-1 do
       array_out[i]:=(array_in[i-1]+array_in[i]+array_in[i+1])/3;
end;


procedure get_values(var n:integer;var arr:arrtype);
var    i:integer;
         okay:boolean;
begin
   repeat
       writeln('Enter number of values to be processed >>');
       readln(n);
       if ((n<0) or (n>ARRAYSIZE)) then
       begin
          writeln('Max elements are ',ARRAYSIZE);
          okay:=FALSE;
       end
       else  okay:=TRUE;
   until (okay=TRUE);

   for i:=1 to n do
   begin
       write('Enter value >> ');
       readln(arr[i]);
   end;
end;

procedure print_values(n:integer; array_in,array_out:arrtype);
var       i:integer;
begin

   writeln('Input     Output');
   for i:=1 to n do
       writeln(array_in[i]:6:3,array_out[i]:6:3);
end;

begin
   get_values(nvalues,input);
   filter(nvalues,input,output);
   print_values(nvalues,input,output);
end.
```

💻 **Test run 7.1**

```
Enter number of values to be processed >> 10
Enter value >> 3
Enter value >> -2
Enter value >> 4
Enter value >> 10
Enter value >> 3
Enter value >> 2
Enter value >> 1
Enter value >> 0
Enter value >> 19
Enter value >> 14
Input     Output
  3.000  3.000
 -2.000  1.667
  4.000  4.000
 10.000  5.667
  3.000  5.000
  2.000  2.000
  1.000  1.000
  0.000  6.667
 19.000 11.000
 14.000 14.000
```

7.4.2 Sorting program

Program 7.2 is an example of a sorting program where an array is passed to the `sort` function, which then orders the values from smallest to largest. The algorithm initially checks the first value in an array with all the other values. If the value in the first position is greater than the sampled array value then the two values are swapped.

Figure 7.5 shows an example of how a 6-element array can be sorted to determine the smallest value. In the first iteration the value of 20 is compared with 22. Since 20 is smaller than 22 the values are not swapped. Next, the value of 20 is compared with 12 (the third element), as this is smaller the values are swapped. This now makes 12 the first element. This continues until the last value (15) is tested. At the end of these iterations the smallest value (3) will be the first element in the array. As the first element now contains the smallest value the operation can now continue onto the second element. This is tested against the third, fourth, fifth and sixth elements and so on. The number of iterations required to complete this process will therefore be 15 (5+4+3+2+1).

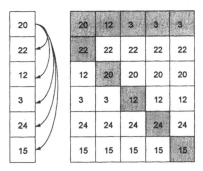

Figure 7.5   Array elements

Program 7.2

```pascal
program  sort_prog(input,output);
const ARRAYSIZE=150;

type arrtype = array[1..ARRAYSIZE] of real;

var   arr:arrtype;
      nvalues:integer;

procedure get_values(var n:integer;var arr:arrtype);
var    i:integer;
       okay:boolean;
begin
   repeat
      writeln('Enter number of values to be processed >>');
      readln(n);
      if ((n<0) or (n>ARRAYSIZE)) then
      begin
         writeln('Max elements are ',ARRAYSIZE);
         okay:=FALSE;
```

```pascal
          end
          else  okay:=TRUE;
    until (okay=TRUE);

    for i:=1 to n do
    begin
       write('Enter value >> ');
       readln(arr[i]);
    end;
end;

procedure print_values(n:integer; array_in:arrtype);
var        i:integer;
begin

    writeln('Ordered values');
    for i:=1 to n do
       writeln(array_in[i]:8:3);
end;

procedure order(var val1,val2:real);
(* val1 is the smallest    *)
var temp:real;
begin
    if (val1 > val2) then
    begin
       temp := val1;
       val1 := val2;
       val2 := temp;
    end;
end;

procedure sort(n:integer; var inarr:arrtype);
var i,j:integer;
begin
    for i:=1 to n-1 do
       for j:=n downto i+1 do
          order(inarr[i],inarr[j]);
end;

begin
    get_values(nvalues,arr);
    sort(nvalues,arr);
    print_values(nvalues,arr);
end.
```

Program 7.2

```c
#include <stdio.h>

#define    MAX   150   /* maximum values in the array   */
#define    TRUE  1
#define    FALSE 0

void    filter(int n,float array_in[],float array_out[]);
void    get_values(int *n,float array[]);
void    print_values(int n,float array_in[],float array_out[]);

int     main(void)
{
float    input[MAX],output[MAX];/*input values, processed values */
int      nvalues;

    get_values(&nvalues,input);
    filter(nvalues,input,output);
    print_values(nvalues,input,output);
```

```
      return(0);
}


void     filter(int n,float array_in[],float array_out[])
{
int      i;
   array_out[0]=array_in[0];
   array_out[n-1]=array_in[n-1];

   for (i=1;i<n-1;i++)
      array_out[i]=(array_in[i-1]+array_in[i]+array_in[i+1])/3;
}


void     get_values(int *n,float array[])
/* *n is the number of elements in the array */
{
int      i,rtn,okay;

   do
   {
      printf("Enter number of values to be processed >>");
      rtn=scanf("%d",n);
      if ((rtn!=1) || (*n<0) || (*n>MAX))
      {
         printf("Max elements are %d, re-enter\n",MAX);
         okay=FALSE;
      }
      else  okay=TRUE;
   } while (!okay);

   for (i=0;i<*n;i++)
   {
      puts("Enter value >> ");
      scanf("%f",&array[i]);
   }
}

void     print_values(int n,float array_in[],float array_out[])
{
int      i;
   printf("Input    Output\n");
   for (i=0;i<n;i++)
      printf(" %6.3f %6.3f \n",array_in[i],array_out[i]);
}
```

Test run 7.2 shows a sample run with 10 entered values.

Test run 7.2
```
Enter number of values be entered >> 10
Enter value >> 3
Enter value >> -2
Enter value >> 4
Enter value >> 10
Enter value >> 3
Enter value >> 2
Enter value >> 1
Enter value >> 0
Enter value >> 19
Enter value >> 14
Ordered values
  -2.000    0.000    1.000    2.000    3.000    3.000    4.000
10.000   14.000   19.000
```

7.4.3 Preferred values

In C, the initialization of an array with values (or characters) is defined between a set of braces ({ }). The following gives the standard format for initializing an array:

```
type arrname[nvalues]={val_0,val_1, .... val_n-1};
```

Program 7.3 determines the nearest preferred resistor value in the range 10 to 100 Ω. An initialized array pref_values[] contains normalized preferred values of 10, 12, 15, 18, 22, 27, 33, 39, 47, 56, 68, 82 and 100 Ω.

In Pascal, it is not possible to initialize an array with values. Thus the set_pref procedure is used to fill up the pref_values array.

The find_nearest_pref function determines the nearest preferred value. Its operation uses the difference between the entered value and an index value in the preferred value array. If the difference is less than the difference between the previous nearest value and the entered value then the current preferred value will take on the current indexed array value. Figure 7.6 shows a basic structure chart for this program (note, the set_pref function has been included for the Pascal equivalent program). Test run 7.3 shows a sample run.

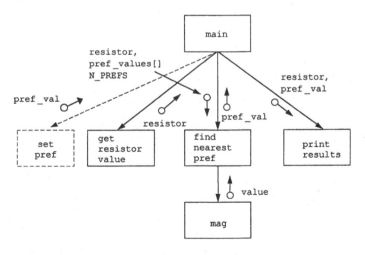

Figure 7.6   Structure chart for Program 7.3

Program 7.3

```
#define   TRUE     1
#define   FALSE    0

#include  <stdio.h>

#define   N_PREF    13

void      get_resistor_value(float *r);
void      find_nearest_pref(float r,float pref_arr[],int n_prf,float *p_val);
```

```
void      print_results(float r,float pref_r);
float     mag(float val);

int       main(void)
{
float     pref_values[N_PREF]={10,12,15,18,22,27,33,39,47,56,68,82,100};
float     resistor,pref_val;

    get_resistor_value(&resistor);
    find_nearest_pref(resistor,pref_values,N_PREF,&pref_val);
    print_results(resistor,pref_val);
    return(0);
}

void      get_resistor_value(float *r)
{
int       rtn,okay;
    /* get a value between 10 and 100 ohms */
    do
    {
        printf("Enter a resistance (10-100 ohm) >> ");
        rtn=scanf("%f",r);
        if ((rtn!=1) || (*r<10) || (*r>100))
        {
            puts("Invalid value, re-enter");
            okay=FALSE;
        }
        else  okay=TRUE;
    } while (!okay);
}

void      find_nearest_pref(float r,float pref_arr[],int n_prf,
                                    float *p_val)
{
int       i;

    *p_val=pref_arr[0];
    for (i=1;i<n_prf;i++)
    {
        if (mag(r-pref_arr[i])<mag(*p_val-pref_arr[i]))
            *p_val=pref_arr[i];
    }
}

void      print_results(float r,float pref_r)
{
    printf("Value entered %8.3f ohm, pref value is %8.3f ohm\n",r,pref_r);
}

float mag(float val)
{  /* Determine the magnitude of val      */
    if (val<0.0) return(-val);
    else return(val);
}
```

🖥 **Test run 7.3**
```
Enter a resistance (10-100 ohm) >> 3
Invalid value, re-enter
Enter a resistance (10-100 ohm) >> 45
Value entered   45.000 ohm, pref value is    47.000 ohm
```

Pascal

Program 7.3

```pascal
const N_PREF=13;

type arrtype = array[1..N_PREF] of integer;

var   pref_values:arrtype;
      resistor,pref_val:integer;


procedure set_pref(var arr:arrtype);
begin
    arr[1]:=10; arr[2]:=12; arr[3]:=15;
    arr[4]:=18; arr[5]:=22; arr[6]:=27;
    arr[7]:=33; arr[8]:=39; arr[9]:=47;
    arr[10]:=56; arr[11]:=68; arr[12]:=82;
    arr[13]:=100;
end;

procedure get_resistor_value(var r:integer);
var   okay:boolean;
begin
   (* get a value between 10 and 100 ohms *)
   repeat
      write('Enter a resistance (10-100 ohm) >> ');
      readln(r);
      if ( (r<10) or (r>100)) then
      begin
         writeln('Invalid value, re-enter');
         okay:=FALSE;
      end
      else  okay:=TRUE;
   until (okay=TRUE);
end;

function mag(val:integer):integer;
begin
   (* Determine the magnitude of val      *)
   if (val<0.0) then mag:=-val
   else mag:=val;
end;

procedure find_nearest_pref(r:integer;pref_arr:arrtype; n_prf:integer; var
p_val:integer);
var   i:integer;
begin
   p_val:=pref_arr[1];

   for i:=2 to n_prf do
   begin
      if (mag(r-pref_arr[i])<mag(p_val-pref_arr[i])) then
         p_val:=pref_arr[i];
   end;
end;

procedure print_results(r,pref_r:real);
begin
   writeln('Value entered ', r:8:3, ', pref value is ', pref_r:8:3,' ohm');
end;


begin
   set_pref(pref_values);
   get_resistor_value(resistor);
   find_nearest_pref(resistor,pref_values,N_PREF,pref_val);
   print_results(resistor,pref_val);
end.
```

7.5 Exercises

7.5.1   Write a function that will arrange an array in descending values. Refer to Program 7.2.

7.5.2   Modify Program 7.3 so that it determines the nearest preferred resistor value between 10 and 100 Ω for the set of preferred values given in Table 7.3.

Table 7.3   Preferred resistor values

10	16	27	43	68
11	18	30	47	76
12	20	33	51	82
13	22	36	56	91
15	24	39	62	100

7.5.3   Write a function which scales an entered real value so that it scales it between 10 and 100 and displays the number of zeros. A sample run is given in Test run 7.4.

🖳   **Test run 7.4**
```
Enter a value >> 32100
Value is 32.1 with 3 zeros
```

Possible algorithm is:

Pascal

```
num_zeros:=0;

while (val<=10)
begin
   val:=val/10;
   num_zeros:=num_zeros+1;
end;
```

C

```
num_zeros=0;

while (val<=10)
{
   val=val/10;
   num_zeros++;
}
```

where, after these codes are complete, the value of val will be between 10 and 100 and the num_zeros will have the number of scaling zeros.

7.5.4   Modify the program in 7.2 so the user can enter any value of resistance and the program will determine the nearest preferred resistor value. Test run 7.5 gives a sample run. Hint: write a function which scales the entered value between 10 and 100 Ω (as written in Exercise 7.5.3) then pass the scaled value to the preferred value's function.

🖳 **Test run 7.5**
```
Enter resistor value >> 42130
Nearest preferred value is 43000 ohms
```

7.5.5   Modify the program in 7.2 so the user can enter any value of resistance and the program will determine the nearest preferred resistor value. Test run 7.5 gives a sample run. Hint: write a function which scales the entered value between 10 and 100 Ω (as written in Exercise 7.5.3) then pass the scaled value to the preferred value's function.

7.5.6   Figure 7.7 shows an alternative representation of the program developed in Section 7.4.1 with the array values represented as time sampled values, where the D represents a single time step delay. In this case the input value `value` is delayed by a single time step (Input[i–1]) and two time steps (Input[i–2]). The representation of the output can be written as:

$$Output[i] = \frac{1}{3}\left(Input[i] + \{Input[i-1] + Input[i-2]\}\right)$$

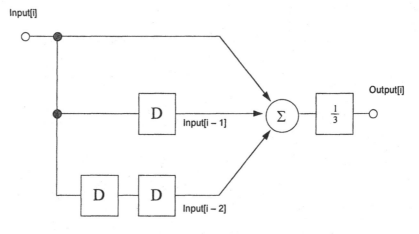

Figure 7.7   Averaging system for time sampling

For the input data of:

Sample	1	2	3	4	5	6	7	8	9	10	11	12
Input	3	4	10	3	−1	−2	−5	−1	4	12	15	20

Determine the output for the following functions:

(i) $Output[i] = \dfrac{1}{2}\left(Input[i] + Input[i-1]\right)$

(ii) $Output[i] = \dfrac{1}{10}\left(2.Input[i] + 6.Input[i-1] + 2.Input[i-2]\right)$

(iii) $Output[i] = \left(Input[i] - Input[i-1]\right)$

Assume that the previous samples to $i=0$ are zero. Test run 7.6 gives a sample run of (i) and Program 7.4 shows a sample program which does not have any functions or procedures.

Test run 7.6

I	Input	Output
0	3.00	3.00
1	4.00	3.50
2	10.00	7.00
3	3.00	6.50
4	-1.00	1.00
5	-2.00	-1.50
6	-5.00	-3.50
7	-1.00	-3.00
8	4.00	1.50
9	12.00	8.00
10	15.00	13.50
11	20.00	17.50

Program 7.4

```c
#include <stdio.h>

int    main(void)
{
float arr1[12]={3,4,10,3,-1,-2,-5,-1,4,12,15,20},arr2[12];
int    i;

    arr2[0]=arr1[0];
    for (i=1;i<12;i++)
    {
        arr2[i]=(arr1[i]+arr1[i-1])/2;
    }

    puts("I  Input Output");

    for (i=0;i<12;i++)
    {
        printf("%2d %8.2f %8.2f\n",i,arr1[i],arr2[i]);
    }
    return(0);
}
```

Program 7.4

```pascal
program arrprog(input,output);

type arrtype=array[1..12] of real;

var arr1,arr2:arrtype;
    i:integer;
```

```
begin
    arr1[1]:=3;  arr1[2]:=4;  arr1[3]:=10;
    arr1[4]:=3;  arr1[5]:=-1;arr1[6]:=-2;
    arr1[7]:=-5; arr1[8]:=-1;arr1[9]:=4;
    arr1[10]:=12; arr1[11]:=15; arr1[12]:=20;

    arr2[1]:=arr1[1];

    for i:=2 to 12 do
    begin
        arr2[i]:=(arr1[i]+arr1[i-1])/2;
    end;

    writeln('I  Input Output');

    for i:=1 to 12 do
    begin
        writeln(i:3,arr1[i]:8:2,arr2[i]:8:2);
    end;
end.
```

Worksheet 10

W10.1   Write a program, using arrays, with a function that will return the largest value entered by the user.

W10.2   Repeat W10.1 with a minimum function.

W10.3   Write a program which will fill an array with values for the function:

$$y = x^2 + 6x - 2$$

for $x = 1$ to 10 (that is, 1, 2, 3, 4, 5, ... 10).

W10.4   Modify the program in W10.3 for a range of -1 to 1, with a step of 0.1 in-between (that is, -1, -0.9, -0.8, ... 0.8, 0.9, 1).

8 Strings

8.1 Introduction

Strings are one-dimensional arrays containing characters. In most cases the number of characters a string has will vary, depending on the input. Thus they must be declared with the maximum number of characters that is likely to occur. In Pascal, strings are relatively easy to use; whereas strings in C are relatively complex, but C has the advantage of having powerful string manipulation routines.

8.2 Pascal strings

Pascal has a special data type reserved for character arrays, which is `string`. The string size can be set as the standard form of an array declaration. For example:

```
var  str1:string[100];   (* string with 100 characters  *)
     str2:string[50];    (* string with 50 characters   *)
```

If the string size parameter is excluded, such as:

```
var  str1:string;          (* string with 255 characters  *)
```

then the string is assumed to have a size of 255 characters.

Table 8.1 lists the routines (functions and procedures) which are used to manipulate strings and Table 8.2 shows the routines which are used to convert from numeric value into strings, and vice-versa.

Program 8.1 shows an example program which uses strings. The assignment operator (:=) is used to assign one string to another (in this case, str1 to str2). The length function is also used to determine the number of characters in the entered string.

A major problem in software development is to guard against incorrect user input. Typically a user may enter a string of characters instead of a numeric value, or a real value instead of an integer. Program 8.2 contains a function (get_int) which overcomes this problem. In this function the user enters an input into a string (inp). This string is then converted into a numeric value using the val function. One of the parameters of the val function is code. If, after it is called, it is a zero then the string has been successfully converted, else either a string or a real value was entered. The user will be told that the input is invalid and will be reprompted for another value (as the `repeat...until` condition is false). The equivalent function to get a real value is:

```
function get_real(msg:string):real;
var      inp:string;
         code:integer, value:real;
begin
    write(msg);
    repeat
        readln(inp);
        val(inp,value,code);
        if (code<>0) then writeln('Invalid input');
    until (code=0);
    get_real:=value;
end;
```

Table 8.1 Pascal array routines

Routine	Description
`function pos(substr:string;` `  str:string)  :byte;`	Searches a string (`str`) for a substring (`substr`) and returns its position.
`procedure Delete(var str : string;` `index : Integer; count : Integer);`	Deletes a substring (`str`) from a string starting at `index` and ending at `index+count`

`function Copy(str : string;`
`index : Integer;`
`count : Integer) : string;`
	Returns a substring of a string (`str`) starting at `index` and ending at `index+count`

`function Concat(s1 [, s2, ..., sn]:` `string): string;`	Concatenates a sequence of strings.

`procedure Insert(source : string;`
`var str : string; index :`
`Integer);`
	Inserts a substring (`str`) into a string (`s`).

Table 8.2 Pascal string conversion routines

Routine	Description

`procedure Val(str : string;`
`  var val;`
`  var code : Integer);`
	Converts a string value (str) to its numeric representation (`val`), as if it were read from a text file with Read.
	`code` is a variable of type Integer. If it is 0 then there is no error else it displays the position of the error.

`procedure Str(x [ : width [ :`
`decimals ]]; var str :`
`string);`
	Converts a numeric value (x) into a string representation (`str`).

Pascal

Program 8.1
```
program str_example(input,output);

var     str1,str2:string;

begin
    writeln('Enter your name >>');
    readln(str1);

    str2:=str1;
    writeln('String 1 is ',str1);
    writeln('String 2 is ',str2);
    writeln('Number of character in name is ',length(str1));
end.
```

Pascal

Program 8.2
```
program str_example(input,output);

var     i:integer;

function get_int(msg:string):integer;
var     inp:string;
        code,value:integer;
begin

    write(msg);
    repeat
        readln(inp);
        val(inp,value,code);
        if (code<>0) then writeln('Invalid input');
    until (code=0);
    get_int:=value;

end;

begin
    i:=get_int('Enter an integer >>');
    writeln('Entered value is ',i);
end.
```

Test run 8.1

```
Enter an integer >> abc
Invalid input
Enter an integer >> 44.4
Invalid input
Enter an integer >> 12
Entered value is 12
```

8.3 C character arrays

In C an array is setup with an array of characters, such as:

```
char str1[100];   /* string with 100 characters  */
char str2[50];    /* string with 50 characters   */
```

The end of a string in C is identified with the ASCII Null or termination character ('\0').   If double quotes are used to define a string, such as "Cap1", then a Null character is automatically appended onto it. If the string is loaded, as an array, with single characters then the Null character must be inserted after the last character. For example, if the string to be loaded is "Cap1" then the array elements would be 'C', 'a', 'p', '1', '\0'. In Figure 8.1 the string "Res-1" is terminated in a Null character (that is, 0000 0000b).

Strings are character arrays with maximum size. The string name is the memory address for the first character in the string. For a declared string of char name[SIZE], the array name name is a fixed address at the start of the string and SIZE the number of characters reserved in memory for the string. Figure 8.1 shows a sample string allocation in memory; the declared string contains a maximum of 11 characters. The maximum number of displayable characters in the stored string will only be 10 as the Null character terminates the string.

As with arrays, the first character of a string declared as char str1[SIZE] is indexed as str1[0] and the last as str1[SIZE-1]. As with arrays, it is possible to overrun the end of a string (especially if the termination character is not present). This can cause data to be read from or written to areas of memory not assigned for this purpose. The dimensioned string should always contain at least the maximum number of entered characters + 1. If a string is read from the keyboard then the maximum number of characters that can be entered is limited by the keyboard buffer. A macro BUFSIZ, defined in *stdio.h*, can be used to determine its size.

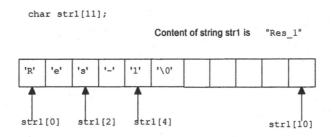

Figure 8.1   Example of string allocation

8.3.1 Standard string functions

There are several string handling functions in the standard library; most are prototyped in *string.h*. Table 8.3 lists these. All the string functions return a value; for example, strlen() returns an integer value relating to the length of a string and the functions strcat(), strupr(), strlwr() and strcpy() return pointers to the resultant string. This pointer can be used, if required, but the resultant string is also passed back as the first argument of these functions. The `strcmp()` function returns a 0 (zero) only if both strings are identical.

C Program 8.1 shows how `sscanf()` scans a string with different data types.

Table 8.3   The main string handling functions

Conversion functions	Header file	Description
`int strcmp(char *str1,char *str2);`	string.h	Function: Compares two strings `str1` and `str2`. Return: A 0 (zero) is returned if the strings are identical, a negative value if `str1` is less than `str2`, or a positive value if `str1` is greater than `str2`.
`int  strlen(char *str);`	string.h	Function: Determines the number of characters in `str`. Return: Number of characters in `str`.
`char  *strcat(char *str1, char *str2);`	string.h	Function: Appends `str2` onto `str1`. The resultant string `str1` will contain `str1` and `str2`. Return: A pointer to the resultant string.
`char  *strlwr(char *str1);`	string.h	Function: Converts uppercase letters in a string to lowercase Return: A pointer to the resultant string
`char  *strupr(char *str1);`	string.h	Function: Converts lowercase letters in a string to uppercase. Return: A pointer to the resultant string.
`char *strcpy(char *str1, char *str2);`	string.h	Function: Copies `str2` into `str1`. Return: A pointer to the resultant string.
`int sprintf(char *str, char *format_str,arg1,....);`	stdio.h	Function: Similar to `printf()` but output goes into string `str`. Return: Number of characters output.
`int sscanf(char *str, char *format_str,arg1,...);`	stdio.h	Function: Similar to `scanf()` but input is from string `str`. Return: Number of fields successfully scanned.

In this case, the user enters a string of text. The first word of the string is read as a string (`res_name`) and the second as a float (`res_values`). The `sscanf()` returns the number of fields successfully scanned; if the number of fields scanned is equal to 2 the variable `okay` is set to `TRUE` and the `do{}while()` loop will thus end. If it is not equal to 2 the `okay` variable is set to `FALSE` and an error message is displayed, the user will then be prompted to re-enter the values.

Program 8.1

```
#define    TRUE       1
#define    FALSE      0
#include   <stdio.h>
#include   <string.h>

int        main(void)
{
char       res_name[BUFSIZ],instr[BUFSIZ],outstr[BUFSIZ];
float      res_value;
int        rtn,okay;

   do
   {
      printf("Enter resistor identifier and the value >>");
      gets(instr);

      rtn=sscanf(instr,"%s %f",res_name,&res_value);
      if (rtn!=2)
      {
         okay=FALSE;
         printf("Invalid input <%s>\n",instr);
      }
      else okay=TRUE;
   } while (!okay);

   printf("Resistor name is %s, value is %8.3f ohm\n",
             res_name,res_value);

   sprintf(outstr,"Resistor name is %s, value is %8.3f ohm\n",
             res_name,res_value);
   puts(outstr);
   return(0);
}
```

Test run 8.2 shows a sample run.

```
Enter resistor identifier and the value >> resistor_1 1430

Resistor name is resistor_1, value is 1430.000 ohm
Resistor name is resistor_1, value is 1430.000 ohm
```

C Program 8.2 uses the string function strcmp(). The do{}while()
loop continues until the user enters the word "exit". Test run 8.3 shows a
sample run.

Program 8.2

```
#include <stdio.h>
#include <string.h>

int        main(void)
{
char       str1[BUFSIZ],str2[BUFSIZ],instr[BUFSIZ];
```

```
    do
    {
        printf("Enter a string >> ");
        gets(str1);

        printf("Enter a string >> ");
        gets(str2);

        if (!strcmp(str1,str2))puts(">>>Strings are identical<<<");
        else puts(">>>Strings differ<<<");

        printf("Do you wish to continue (type 'exit' to quit)>>");
        gets(instr);

    } while (strcmp(instr,"exit"));
    return(0);
}
```

Test run 8.3
```
Enter a string >> this is a string
Enter a string >> this is another string
>>>Strings differ<<<
Do you wish to continue (type 'exit' to quit)>>
Enter a string >> input is zero
Enter a string >> input is zero
>>>Strings are identical<<<

Do you wish to continue (type 'exit' to quit)>> exit
```

8.4 Examples

8.4.1 Counting the number of characters

Program 8.3 contains a function (nochars()) which scans a string and determines the number of occurrences of a given character. It uses gets() to read the string as it accepts spaces between words.

The function nochars() uses pointer arithmetic to read each of the characters in the passed string until a Null character. The getchar() function is used to get the search character. Test run 8.4 shows a sample run.

Program 8.3
```
#include <stdio.h>

/* Find the number of occurrences in a string */

int    nochars(char *str,char ch);

int    main(void)
{
char   str1[BUFSIZ],ch;

    printf("Enter a string >>");
    gets(str1);

    printf("Enter character to find >>");
    ch=getchar();
```

```
    return(0);
}

int   nochars(char str[],char c)
{
int   i=0,no_occ=0,size;
    size=strlen(str);
    for (i=0;i<size;i++)
        if (str[i]==c) no_occ++;

    return(no_occ);
}
```

Test run 8.4
```
Enter a string>> resistor 1 is 100 ohms
Enter a letter to find >> s
Number of occurrences = 4
```

Program 8.3
```
program str_example(input,output);
(* Find the number of occurrences in a string *)

var str1:string;
    ch:char;

function nochars(str:string; c:char):integer;
var no_occ,i:integer;
begin
    no_occ:=0;
    for i:=1 to length(str) do
    begin
        if (c= str[i]) then no_occ:=no_occ+1;
    end;
    nochars:=no_occ; (* no of characters found *)
end;

begin

    writeln('Enter a string >>');
    readln(str1);

    writeln('Enter character to find >>');
    readln(ch);

    writeln('Number of occurrences is ',nochars(str1,ch));
end.
```

8.4.2 Setting up an array of strings

In C, the simplest way of setting up an array of strings is to define a new
string data type. In C program 8.4 a new data type named `string` is defined using the statement `typedef char string[BUFSIZ]`. An array
of strings is then set up using the declaration `string data-
base[MAX_COMP_NAMES]`. Each string in the array can contain a maxi-
mum of `BUFSIZ` characters (note this can be changed to any size if required) and can be accessed using normal array indexing.

Program 8.4

```c
#include <stdio.h>
#include <string.h>

#define     MAX_COMP_NAMES    100

typedef char string[BUFSIZ];

void     get_component_name(int *comp, string data[]);
void     print_component_names(int comp, string data[]);


int      main(void)
{
int      components=0;
string   database[MAX_COMP_NAMES];

   do
   {
      get_component_name(&components,database);

      if (strcmp(database[components-1],"exit")) break;

      print_component_names(components,database);
   } while (components<MAX_COMP_NAMES);
   return(0);
}


void     get_component_name(int *comp, string data[])
{

   printf("Enter component name >>");

   gets(data[*comp]);

   (*comp)++;

}

void     print_component_names(int comp, string data[])
{
int      i;

   puts("Component Names");

   for (i=0;i<comp;i++)
      printf("%d          %s\n",i,data[i]);

}
```

A sample run is given in Test run 8.5. Note that the string "exit" quits the program.

Test run 8.5

```
Enter component name >>resistor 1
0          resistor 1
Enter component name >>resistor 2
Component Names
0          resistor 1
1          resistor 2
Enter component name >>capacitor 1
Component Names
```

```
0          resistor 1
1          resistor 2
2          capacitor 1
Enter component name >>capacitor 4
Component Names
0          resistor 1
1          resistor 2
2          capacitor 1
3          capacitor 4
Enter component name >>inductor 5
Component Names
0          resistor 1
1          resistor 2
2          capacitor 1
3          capacitor 4
4          inductor 5
Enter component name >>exit
```

The equivalent Pascal program is given in Pascal Program 8.4. In this case a
new data type of declared (`str_arr`) which is an array of strings.

(Pascal) **Program 8.4**

```pascal
program str_example(input,output);

const MAX_COMP_NAMES=50;

type  str_arr=array[1..MAX_COMP_NAMES] of string;

var components:integer;
    database:str_arr;

procedure get_component_name(var comp:integer; var data:str_arr);
begin
   write('Enter component name >>');
   comp:=comp+1;
   readln(data[comp]);
end;

procedure print_component_names(comp:integer; data:str_arr);
var        i:integer;

begin
   writeln('Component Names');
   for i:=1  to comp do
      writeln(i,' ',data[i]);
end;

begin
    components:=0;

    repeat
       get_component_name(components,database);
       print_component_names(components,database);
    until ((components=MAX_COMP_NAMES) or (database[components]='exit'));

end.
```

In C, an array of strings can also be initialized using braces. Program 8.5 and
Test run 8.6 show how this is set up.

⬡ C

Program 8.5

```
#include <stdio.h>
#define        MAX_COMP_NAMES     5

typedef  char string[BUFSIZ];

int    main(void)
{
string    database[MAX_COMP_NAMES]={"Res 1","Res 2","Res 3",
                                    "Res 4","Cap 1"};

int       i;

    for (i=0;i<MAX_COMP_NAMES;i++)
       printf("%s\n",database[i]);
    return(0);
}
```

🖥️ **Test run 8.6**

```
Res 1
Res 2
Res 3
Res 4
Cap 1
```

8.4.3 Impedance of a parallel RC circuit

The program in this section determines the impedance of a parallel RC circuit. Figure 8.2 gives a schematic of this circuit. The impedance of this circuit can be found using the product of the impedances over the sum. Thus:

$$Z = \frac{R\dfrac{1}{j\omega C}}{R+\dfrac{1}{j\omega C}} = \frac{R}{j\omega C(R+\dfrac{1}{j\omega C})} = \frac{R}{j\omega CR+1}$$

The magnitude of the impedance is thus:

$$|Z| = \frac{R}{\sqrt{1+(\omega RC)^2}}$$

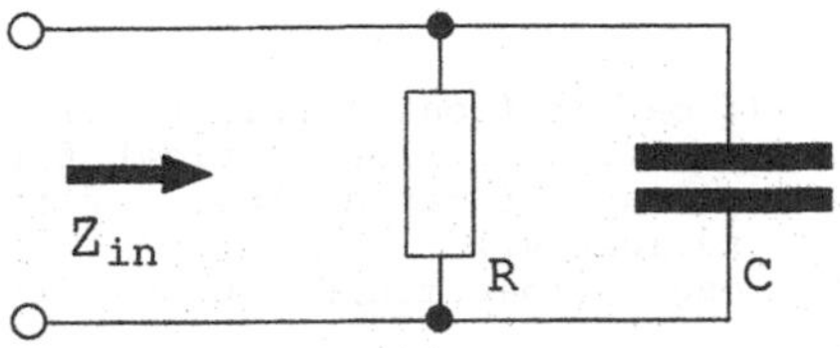

Figure 8.2   Parallel RC circuit

A structure chart for a program which determines this magnitude is given in Figure 8.3 (note that for clarity the parameters passed to the two of the calls to the get_float() function have not been included). The get_parameters() function gets three variables (Res, Cap and freq); parallel_impedance() determines the input impedance and returns it back into the variable Zin. Finally, print_impedance() displays the input parameters and calculated impedance.

The get_float() (or get_real() in Pascal) function gets a value by first putting the entered information into a string and then converting it into a floating point value. If the conversion fails then the user is asked to re-enter the value. It also contains a check for the minimum and maximum value of the entered value.

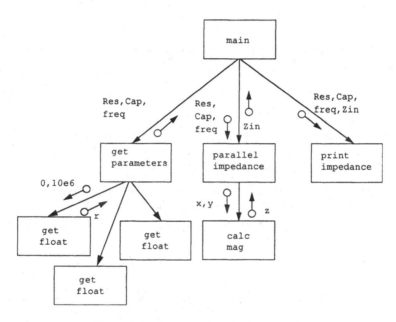

Figure 8.3   Structure chart for Program 8.7

Program 8.6

```
#include <stdio.h>
#define     TRUE     1
#define     FALSE    0
#define     PI       3.14159
#define     MICRO    1e-6

void     get_parameters(float *r,float *c,float *f);
void     parallel_impedance(float r,float c,float f,float *Z);
void     print_impedance(float r,float c,float f,float Z);
float    calc_mag(float x,float y);
void     get_float(char msg[],float min,float max,float *val);

int      main(void)
{
float    Res,Cap,freq,Zin;

    get_parameters(&Res,&Cap,&freq);
    parallel_impedance(Res,Cap,freq,&Zin);
    print_impedance(Res,Cap,freq,Zin);
```

```
      return(0);
}

void     get_parameters(float *r,float *c,float *f)
{
   get_float("Enter resistance  >> ",0,10e6,r);
   get_float("Enter capacitance >> ",0,1,c);
   get_float("Enter frequency   >> ",0,1e7,f);
}

void     get_float(char msg[],float min,float max,float *val)
{
char     instr[BUFSIZ];
int      rtn,okay;
   do
   {
      printf("%s",msg);
      gets(instr);
      rtn=sscanf(instr,"%f",val);
      if ((rtn!=1) || (*val<min) || (*val>max))
      {
         okay=FALSE;
         printf("Invalid input <%s>\n",instr);
      }
      else okay=TRUE;

   } while (!okay);
}

void     print_impedance(float r,float c,float f,float Z)
{
   printf("R=%f ohms C=%f uF f=%f Hz\n",r,c/MICRO,f);
   printf("Zin = %f ohms\n",Z);
}

void     parallel_impedance(float r,float c,float f,float *Z)
{
   *Z=r/(calc_mag(1,2*PI*f*r*c));
}

float calc_mag(float x,float y)
{
   return(x*x+y*y);
}
```

Test run 8.7 shows that the user can enter a value in the incorrect format and the program will re-prompt for another. Notice that the user has entered the strings "none" and "fred"; the program copes with these and re-prompts for an input.

Test run 8.7
```
Enter resistance  >>none
Invalid input <none>
Enter resistance  >>fred
Invalid input <fred>
Enter resistance  >>-100
Invalid input <-100>
Enter resistance >>1000
Enter capacitance   >>1e-6
Enter frequency     >>1e3
R=1000.000 ohms C=    1.00 uF f=1000.000 Hz
Zin = 24.704565 ohms
```

Pascal

Program 8.5

```pascal
program str_example(input,output);
const PI=3.14159; MICRO=1e-6;

var   Res,Cap,freq,Zin:real;

procedure get_real(msg:string; min, max:real; var value:real);
var instr:string;
    code:integer;
    okay:boolean;
begin
   repeat
      write(msg);
      readln(instr);
      val(instr,value,code);
      if (code<>0) then          (* invalid input  *)
      begin
         okay:=FALSE;
         writeln('Invalid input <',instr,'>');
      end
      else okay:=TRUE;

   until (okay=TRUE);
end;

procedure get_parameters(var r,c,f:real);
begin
   get_real('Enter resistance  >> ',0,10e6,r);
   get_real('Enter capacitance >> ',0,1,c);
   get_real('Enter frequency   >> ',0,1e7,f);
end;

procedure print_impedance(r,c,f,Z:real);
begin
   writeln('R=',r,' ohms C=',c/MICRO,' uF f=',f,'Hz');
   writeln('Zin = ',Z,' ohms');
end;

function calc_mag(x,y:real):real;
begin
   calc_mag:=x*x+y*y;
end;

procedure parallel_impedance(r,c,f:real;var Z:real);
begin
   Z:=r/(calc_mag(1,2*PI*f*r*c));
end;

begin
   get_parameters(Res,Cap,freq);
   parallel_impedance(Res,Cap,freq,Zin);
   print_impedance(Res,Cap,freq,Zin);
end.
```

8.5 Exercises

8.5.1   Explain why it is better to input numeric values as a string and then converting it to a numeric value rather than entering it with `scanf()` (in C) or `readln()` (in Pascal).

8.5.2   Write a program that declares the following seven strings.
"RL series", "RC series", "LC series",

"RL parallel", "RC parallel", "LC parallel","EXIT"

Store these strings as a single array of strings named menu by declaring an array of strings. The program should display these strings as menu options using a for loop. Test run 8.8 shows a sample run.

🖥 **Test run 8.8**
```
Menu Options
   RL series
   RC series
   LC series
   RL parallel
   RC parallel
   LC parallel
   EXIT
```

8.5.3   Modify the program in Exercise 8.5.2 so that the user can enter the menu option. The program will display a message on the option selected. Test run 8.9 shows a sample run.

🖥 **Test run 8.9**
```
Menu Options
   RL series
   RC series
   LC series
   RL parallel
   RC parallel
   LC parallel
   QUIT
Enter option >> RL series
   >>> RL series circuit selected
Menu Options
   RL series
   RC series
   LC series
   RL parallel
   RC parallel
   LC parallel
   QUIT
Enter option >> QUIT
```

8.5.4   Modify some programs in previous chapters so that program parameters are entered using the get_float() (or get_real()) function.

8.5.5   Write a function that will capitalize all the characters in a string.

8.5.6   Repeat Exercise 8.5.5 but make the characters lowercase.

8.5.7   Write a function that will determine the number of words in a string.

Worksheet 11

W11.1 Write a program in which the user enters a string of text and then presses the RETURN key. The program will then display the number of characters in the entered text. A sample run is shown in Test run 8.10.

💻 **Test run 8.10**
```
Enter some text >> This is some sample text

This string has 24 characters
```

W11.2 Write a program in which the user enters a string of either "sin", "cos" or "tan" and then a value. The program will then determine the corresponding sine, cosine or tan of the value. A sample run is shown in Test run 8.11.

💻 **Test run 8.11**
```
Enter a math function
    Sin
    Cos
    Tan
    Exit

>> Sin
Enter a value (in degrees) >> 30
Sin of 30 degrees is 0.5
Enter a math function
    Sin
    Cos
    Tan
    Exit

>> Exit
```

W11.3 Modify the program in W11.2 so that the program reads in the entered value as a string and converts it to a real value. The program should re-prompt if the value is invalid.

W11.4 Write a program in which the user enters a number of names, each followed by the RETURN key. The end of the names is signified by entering no characters. The program should store these in an array of strings and then display them to the user. A sample run is shown in Test run 8.12.

🖥 **Test run 8.12**

```
Enter names (or press return for end of list) >>
FRED
BERT
BILL
GARY

The names entered were:
FRED
BERT
BILL
GARY
```

9 File I/O

9.1 Introduction

Information on computers is organized into directories and files. Typically files have a filename followed by a filename extension which identifies the type of file. This file extension is important in some operating systems, such as Windows 95 and Windows NT, as it identifies the application program which is associated with the file. Table 9.1 shows some typical file types.

Table 9.1 Typical file types

File extension	File type	File extension	File type
.c	C program	.jpg	JPEG file (compressed image)
.doc	Word processor	.txt	Text file (ASCII coding)
.bmp	Bitmapped graphics file	.exe	PC executable file
.gif	GIF file (compressed image file)	.obj, .o	Object code
.avi	Motion video	.xls	Microsoft Excell spreadsheet
.ps	Postscript file (printer output)	.hlp	Help files
.au	Compressed sound file	.for	FORTRAN program
.wav	Sound file	.pas	Pascal program
.mpg	Compress motion video	.java	Java program

Files either contain text in the form of ASCII (a text file) or a binary file. A text file uses ASCII characters and a binary file uses the binary digits which the computer uses to store values. It is not normally possible to view a binary file without a special program, but a text file can be viewed with a text editor.

Figure 9.1 shows an example of two files which contain four integer values. The binary file stores integers using two bytes in 2s complement signed notation, whereas the text file uses ASCII characters to represent the values. For example, the value of −1 is represented as 11111111 11111111 in 2s complement. This binary pattern is stored to the binary file. The text file uses ASCII characters to represent −1 (these will be '−'

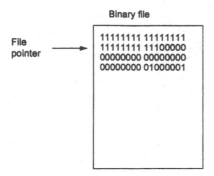

Figure 9.1 File pointer

and '1'), and the bit pattern stored for the text file will thus be 0010 1101 (ASCII '–') and 0011 0001 (ASCII '1'). If a new line is required after each number then a new-line character is inserted after it. Note, there is no new-line character in ASCII and it is typical to represent a new-line with two characters, a carriage return (CR) and a line feed (LF). In C, the new-line character is denoted by '\n'.

The file pointer moves as each element is read/written. Figure 9.2 shows a file pointer pointing to the current position within the file.

The number of bytes used to store each of the elements will depend on the data type of the variable. For example, a long integer will be stored as four bytes, whereas a floating point value can be stored as four bytes (on some systems). The floating point format differs from an integer format; the standard floating point format uses a sign-bit, a significand and an exponent. The end of the file is signified by an EOF character.

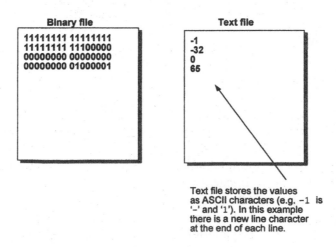

Figure 9.2 Binary and text files

9.2 Pascal file I/O

Pascal has a very basic set of file input/output. The functions used are defined in Table 9.2.

In Pascal, a file pointer is defined either with the text keyword (for a text file) or with:

```
var fptr: file of type;
```

Thus a file pointer for an integer file would use the following:

```
var fptr: file of integer;
```

Pascal Program 9.1 reads in a number of integer values and writes each one in turn to a file ('out.dat'). The program initially assigns the filename to the file pointer (fout) using the assign function. Next the file is created with the rewrite function. Each value that is entered is then written to the file using the write function. The program keeps prompting for values until the user enters a −1 value. When this happens the program will exit the repeat...until loop. The file pointer is then reset to the start of the file with the reset function. Next the values are read back using the read function. Finally, after the −1 value is read-in, the file is closed with the close routine.

Table 9.2 Pascal file I/O functions

Function	Description
assign (*fptr*, *fname*)	Assigns a file point (*fptr*) to a file (*fname*).
rewrite (*fptr*)	Creates and opens a file which has been assigned with *fptr*.
reset (*fptr*)	Opens an existing file which has been assigned with *fptr*.
write (*fptr*, *val*)	Writes a value (*val*) to a file which has been assigned with *fptr*.
read (*fptr*, *val*)	Reads a file which has been assigned with *fptr* and puts the value into *val*.

Program 9.1
```
program example_file(input,output);
var    fout : text;
       val:integer;
```

```
begin
   assign(fout,'out.dat');

   rewrite(fout); (* create new file *)

   repeat
      write('Enter a value');
      readln(val);
      writeln(fout,val);
   until (val=-1);

   reset(fout);   (* set file pointer back to the start of the file *)

   writeln('Values in file are:');

   repeat
      readln(fout, val);
      if (val<>-1) then writeln(val);
   until (val=-1);

   close(fout);
end.
```

9.3 C file I/O

There are 11 main functions used in file input/output (I/O), which are listed below. The `fprint()` and `fscanf()` functions are similar to `printf()` and `scanf()`, but their output goes to a file.

```
file_ptr=fopen(filename,"attributes");
fclose(file_ptr);
fprintf(file_ptr,"format",arg1,arg2,..);
fscanf(file_ptr,"format",arg1,arg2...);
fgets(str,n,file_ptr);
fputs(str,file_ptr);
fputc(ch,file_ptr);
ch=fgetc(file_ptr);
fwrite(ptr,size,n,file_ptr);
fread(ptr,size,n,file_ptr);
feof(file_ptr);
```

A file pointer stores the current position of a read or write within a file. All operations within the file are made with reference to the pointer. The data type of this pointer is defined in *stdio.h* and is named `FILE`.

```
#include <stdio.h>


int      main(void)
{
FILE     *fileptr;

}
```

9.3.1 Opening a file (`fopen()`)

A file pointer is assigned using `fopen()`. The first argument is the file name and the second a string which defines the attributes of the file; these are listed in Table 9.3.

The default mode for opening files is text, but a t attribute can be appended onto the attribute string to specify a text file. For example, the attribute "wt" opens a text file for writing. A binary file is specified by appending a b onto the attribute string. For example, "rb" will open a binary file for reading.

The format of the fopen () function is:

```
file_ptr=fopen("Filename","attrib");
```

If fopen () is completed successfully a file pointer will be returned, and this is initialized to the start of the file. If was not able to open the file then a NULL will be returned. There can be many reasons why a file cannot be opened, such as:

- The file does not exist.
- The file is protected from reading from and/or writing to.
- The file is a directory.

It is important that a program does not read from a file that cannot be opened as it may cause the program to act unpredictably. A test for this condition is given next.

```
int        main(void)
{
FILE      *in;
   if ((in=fopen("in.dat","r"))==NULL)
   {
      printf("File IN.DAT could not be opened");
   }
   ::::::::::::::::::::::::::::::
   return(0);
}
```

Table 9.3 File attributes

Attribute	Function
"r"	open for reading only
"w"	create for writing
"a"	append; open for writing at the end-of-file or create for writing if the file does not exist
"r+"	open an existing file for update (read and write)
"w+"	create a new file for update
"a+"	open for append: open (or create if the file does not exist) for update at the end of the file)

9.3.2 Closing a file (fclose())

Once a file has been used it must be closed before the program is terminated. A file which is not closed properly can cause problems in the file system. The standard format is given next; the return value (rtn) returns a 0 (a zero) on success, otherwise EOF if any errors occur. The macro EOF is defined in *stdio.h*.

```
rtn=fclose(file_ptr)
```

The feof() function detects the end-of-file character. It returns a non-zero value (that is, a TRUE) if the file pointer is at the end of a file, else a 0 is returned. The function shown next uses the feof() function to detect the end of the file and also tests the return from the fscanf() so that an unsuccessful reading from the file is disregarded.

```
int     get_values(int maxval, char fname[],int *n,float arr[])
{
FILE    *in;
int     i=0,rtn;

    if ((in=fopen(fname,"r"))==NULL)
    {
       printf("Cannot open %s\n",fname);
       return(NOFILE);
    }

    while (!(feof(in)))
    {
       rtn=fscanf(in,"%f",&arr[i]);
       if (rtn!=EOF) i++;
       if (i==maxval) break;
    }

    fclose(in);
    *n=i;
    return(!NOFILE);
}
```

9.3.3 Reading from files

Table 9.4 lists the main functions and examples used in reading from files.

Table 9.4   Reading from files

Function	Description
fgetc	The fgetc() function is similar to getchar() and is used to read a single character from a file.
ch=fgetc(file_ptr)	This reads ch from the file specifed by file_ptr. If there is an error in getting the character an EOF will be returned.
fputc	The fputc() function is similar to putchar() and is used to write a single character to a file. The standard format is given next.

`rtn=fputc(ch,file_ptr)`	This writes ch to the file specifed by `file_ptr`. The return (`rtn`) returns the last character written; if there is an error it will return EOF.
fgets	The `fgets()` function is similar to `gets()` and is used to get a string of text from a file up to a new-line character.
`rtn=fgets(str,n,file_ptr)`	It reads from the file specifed by `file_ptr` into `str` with a maximum of n characters. The return (`rtn`) points to a string pointed to by `str` or will return a NULL if the end-of-file (or an error) is encountered.
fscanf	The `fscanf()` function is used with text files and has a similar format to `scanf()` but the input is read from a file.

`rtn=fscanf(file_ptr,`
`"%s %d %d",`
`    str1,&val1,&val2)`
	This function returns the number of fields successfully scanned (`rtn`) or, if there was an attempt to read from the end-of-file, an EOF value is returned.

9.3.4 Writing to files

Table 9.5 lists the main functions used in writing to files.

Table 9.5 Writing to files

Function	Description
fprintf	The `fprintf()` function is used with text files and has a similar format to `printf()` but the output goes to a file.
`rtn=fprintf(file_ptr,` `    "%s %d",str1,value1)`	The return value (`rtn`) returns the number of bytes sent to the file; in the event of an error it returns EOF.
fputc	The `fputc()` function is similar to `putchar()` and is used to write a single character to a file.
`rtn=fputc(ch,file_ptr)`	This writes ch to the file specifed by `file_ptr`. The return (`rtn`) returns the last character written; if there is an error it will return EOF.
fputs	The `fputs()` function is similar to `puts()` and is used to write a string of text to a file. It does not append the output with a new line.
`rtn=fputs(str,file_ptr)`	This outputs `str` to the file specifed by `file_ptr`. The return (`rtn`) returns the last character written; if there is an error it returns EOF.

9.4 Examples

9.4.1 Averages program

C Program 9.1 and Pascal Program 9.2 use text files to determine the average value of a number of floating point values contained in a file. The get_values() function is used to read the values from a file, in this case, *IN.DAT.* This file can be created using a text editor.

Program 9.1

```c
/* Program to determine the average of a file        */
/* containing a number of floating point values      */

#include <stdio.h>

#define     NOVALUES 100 /* max. number of entered values   */
#define     NOFILE   0

int      get_values(int maxvals, char fname[],int *n,float arr[]);
float    calc_average(int nval,float arr[]);
void     display_average(int nval,float arr[],float aver);

int      main(void)
{
float values[NOVALUES],average;
int    nvalues;

   if (get_values(NOVALUES,"IN.DAT",&nvalues,values)==NOFILE)
   return(1);
   average=calc_average(nvalues,values);
   display_average(nvalues,values,average);
   return(0);
}

int   get_values(int maxvals, char fname[],int *n,float arr[])
{
FILE  *in;
int   i=0,rtn;

   if ((in=fopen(fname,"r"))==NULL)
   {
      printf("Cannot open %s\n",fname);
      return(NOFILE);
   }

   while (!feof(in))
   {
      rtn=fscanf(in,"%f",&arr[i]);
      if (rtn!=EOF) i++;
      if (i==maxvals) break;
   }
   fclose(in);

   *n=i;
   return(!NOFILE);
}

float calc_average(int nval,float arr[])
{
int    i;
float running_total=0;
```

```
    for (i=0;i<nval;i++)
        running_total+=arr[i];

    /* note there is no test for a divide by zero */
    return(running_total/nval);
}
void  display_average(int nval,float arr[],float aver)
{
int   i;

    puts("INPUT VALUES ARE:");

    for (i=0;i<nval;i++)
        printf("%8.3f\n",arr[i]);

    printf("Average is %8.3f\n",aver);
}
```

An example of the contents of the *IN.DAT* file are given next.

```
3.240
1.232
6.543
-1.432
```

A sample run using this file is given in Test run 9.1.

🖥 **Test run 9.1**
```
INPUT VALUES ARE:
    3.240
    1.232
    6.543
   -1.432
Average is    2.396
```

⬡(Pascal)
Program 9.2
```
(* Program to determine the average of a file      *)
(* containing a number of floating point values    *)

const NOVALUES=100; (* max. number of entered values *)

type arrtype=array[1..NOVALUES] of real;

var values:arrtype;
    average:real;
    nvalues:integer;

procedure get_values(maxvals:integer; fname:string; var n:integer;
                     var arr:arrtype);
var infile:text;
begin
    n:=0;
    assign(infile,fname);
    reset(infile);
    while ( (not eof(infile)) and (n<maxvals)) do
    begin
        if (not eof(infile)) then n:=n+1;
```

```
            read(infile,arr[n]);
        end;
        close(infile);
end;

function calc_average(nval:integer; arr:arrtype):real;
var i:integer;
    running_total:real;
begin
    running_total:=0;
    for i:=1 to nval do
        running_total:=running_total+arr[i];

    (* note there is no test for a divide by zero *)
    calc_average:=running_total/nval;
end;

procedure display_average(nval:integer; arr:arrtype; aver:real);
var        i:integer;
begin

    writeln('INPUT VALUES ARE:');

    for i:=1 to nval do
        writeln(arr[i]:8:3);

    writeln('Average is ',aver:8:3);
end;

begin
    get_values(NOVALUES,'IN.DAT',nvalues,values);
    average:=calc_average(nvalues,values);
    display_average(nvalues,values,average);
end.
```

9.4.2 Binary read/write

C Program 9.2 and Pascal Program 9.3 is an example of how an array of
floating point values is written to a binary file. In the C program the floating
point values are writen using fwrite() and then read back using
fread(). Note that the NOFILE flags returned from dump_data() and
read_data() are ignored by main(). In the Pascal program the values
are written to the file using write and read using the read routine.

⬡ C
Program 9.2
```c
/* Writes and reads an array of floats      */
/* to and from a binary file                */

#include <stdio.h>
#define    NOFILE      0     /* error flag is file does not exist */
#define    MAXSTRING   100   /* max. number of char's in filename */
#define    MAXVALUES   100   /* max. number of floats in array    */

void    get_filename(char fname[]);
void    get_values(int maxvals, float vals[],int *nov);
int     dump_data(char fname[],float arr[],int nov);
int     read_data(char fname[],float arr[],int *nov);
void    print_values(float arr[],int nov);

int     main(void)
{
```

```
char      fname[MAXSTRING];
float     values[MAXVALUES];
int       no_values;  /* number of values in the array */

   get_filename(fname);
   get_values(MAXVALUES,values,&no_values);
   dump_data(fname,values,no_values);
   read_data(fname,values,&no_values);
   print_values(values,no_values);
   return(0);
}

void      get_filename(char fname[])
{
   printf("Enter file name >>");
   scanf("%s",fname);
}

void      get_values(int maxvals, float vals[],int *nov)
{
int       i;

   do
   {
      printf("Number of values to be entered >>");
      scanf("%d",nov);
      if (*nov>maxvals)
            printf("Too many values: MAX: %d\n",MAXVALUES);
   } while (*nov>MAXVALUES);

   for (i=0;i<*nov;i++)
   {
      printf("Enter value %d >>",i);
      scanf("%f",&vals[i]);
   }
}

int       dump_data(char fname[],float arr[],int nov)
{
FILE      *out;
int       i;

         /* open for binary write */
   if ((out=fopen(fname,"wb"))==NULL)
   {
      printf("Cannot open %s\n",fname);
      return(NOFILE); /* unsuccessful file open */
   }

   for (i=0;i<nov;i++)
      fwrite(&arr[i],sizeof (float),1,out);

   fclose(out);
   return(!NOFILE);
}

int       read_data(char fname[],float arr[],int *nov)
{
FILE      *in;

   *nov=0;     /* number of values in the array */

      /* open for binary read */
   if ((in=fopen(fname,"rb"))==NULL)
   {
      printf("Cannot open %s\n",fname);
      return(NOFILE); /* unsuccessful file open */
   }
```

```
    while (!feof(in))
    {
        if (fread(&arr[*nov],sizeof (float),1,in)==1)
            (*nov)++;

    }

    fclose(in);

    return(!NOFILE);
}

void    print_values(float arr[],int nov)
{
int     i;

    printf("Values are:\n");

    for (i=0;i<nov;i++)
        printf("%d %8.3f\n",i,arr[i]);

}
```

A sample test run is given in Test run 9.2.

🖥 **Test run 9.2**
```
Enter file name >>number.dat
Number of values to be entered >>5
Enter value 0 >>1.435
Enter value 1 >>0.432
Enter value 2 >>-54.32
Enter value 3 >>-1.543
Enter value 4 >>100.01
Values are:
0    1.435
1    0.432
2   -54.320
3   -1.543
4  100.010
```

⬡ (Pascal) **Program 9.3**
```
(*      Writes and reads and array of floats           *)
(*      to and from a binary file                      *)

const MAXVALUES=100; (* max. number of floats in array    *)

type float_file=file of real;
     float_arr=array[1..MAXVALUES] of real;

var  fname:string;
     values:float_arr;
     no_values:integer;

procedure get_filename(var fname:string);
begin
   write('Enter file name >>');
   readln(fname);
end;

procedure get_values(maxvals:integer; var vals:float_arr;var nov:integer);
var i:integer;
begin
```

```
   repeat
      write('Number of values to be entered >>');
      readln(nov);
      if (nov>maxvals) then
                  writeln('Too many values: MAX:',MAXVALUES);
   until ((nov>0) and (nov<maxvals));

   for i:=1 to nov do
   begin
           write('Enter value ',i,' >>');
           readln(vals[i]);
   end;
end;

procedure dump_data(fname:string;arr:float_arr;nov:integer);
var outfile:float_file;
    i:integer;
begin

    assign(outfile,fname); (* assign binary file *)
    rewrite(outfile);

    for i:=1 to nov do
        write(outfile, arr[i]);
    close(outfile);
end;

procedure read_data(fname:string; var arr:float_arr;var nov:integer);
var infile:float_file;
begin
   nov:=0;       (* number of values in the array *)

   assign(infile,fname);
   reset(infile);

   while ( not eof(infile)) do
   begin
      nov:=nov+1;
      read(infile,arr[nov]);
   end;

   close(infile);
end;

procedure print_values(arr:float_arr;nov:integer);
var i:integer;
begin
   writeln('Values are:');

   for i:=1 to nov do
      writeln(i:3,arr[i]:8:3);
end;

begin
   get_filename(fname);
   get_values(MAXVALUES,values,no_values);
   dump_data(fname,values,no_values);
   read_data(fname,values,no_values);
   print_values(values,no_values);
end.
```

9.5 Reading and writing one character at a time

In C a single character is read from a file with `fgetc()` and read back with the `fputc()`. C Program 9.3 shows an example of reading a character, one at a time, from a file and writing it to another file. Pascal Program 9.4 gives the equivalent program. It uses the `read` and `write` routines.

⟨ c ⟩
Program 9.3
```
#include <stdio.h>

int    main(void)
{
FILE   *in,*out;
char   ch;

   in=fopen("in.dat","r");
   out=fopen("out.dat","w");

   while (!feof(in))
   {
      ch=fgetc(in);      /* read character            */
      printf("%c",ch);   /* show character to screen  */
      fputc(ch,out);     /* write character           */
   }
   fclose(in); fclose(out);
   return(0);
}
```

⟨Pascal⟩
Program 9.4
```
program copytext(input,output);
var  inf,outf:text;
     ch:char;
begin
     assign(inf,'in.dat');
     reset(inf);

     assign(outf,'out.dat');
     rewrite(outf);         (* create file               *)

     while (not eof(inf)) do
     begin
          read(inf,ch);     (* read character            *)
          write(ch);        (* show character to screen  *)
          write(outf,ch);   (* write character           *)
     end;
     close(inf); close(outf);
end.
```

9.6 Exercises

9.6.1   Write a program in which the user enters any character and the program will determine the number of occurrences of that character in the specified file. For example:

🖥 **Test run 8.3**
```
Enter filename: fred.dat
Enter character to search for: i
There are 14 occurrences of the character i in the file
fred.dat.
```

9.6.2   Write a program which will determine the number of words in a file. (Hint: count the number of spaces in the file.)

9.6.3   Write a program which will determine the number of lines in a file. A possible method is to count the new-line characters.

9.6.4   Write a program which will get rid of blank lines in an input file and writes the processed file to an output file. Example input and output files are given next.

Input file:

```
Frequency    Impedance (mag)
100             101.1


150             165.1
200             300.5
```

Output file:

```
Frequency    Impedance (mag)
100             101.1
150             165.1
```

Worksheet 12

W12.1   Write a program which will determine the average, the largest and the smallest values of a text file containing floating point values in a text form.

W12.2   Write a program which will count the number of characters in a file. Hint: read the file one character at a time.

W12.3   Write a program which will count the occurrences of the letter 'a' in a file. Hint: read the file one character at a time.

W12.4   Modify Program 9.1 so that the output is written to a file.

10 Structures and Records

10.1 Introduction

A structure (or, in Pascal, a record) is an identifiable object that contains items which define it. These items are linked under a common grouping. For example, an electrical circuit has certain properties that define it. These could be:

- A circuit title.
- Circuit components with identifiable names.
- Circuit components with known values.

For example, a circuit may have a title of "RC Filter Circuit", the circuit components are named "R1", "R2" and "C1" and the values of these are 4320 Ω, 1200 Ω and 1 μF, respectively. The title and the component names are character strings, whereas the component values are floating points. A structure (or record) groups these properties into a single entity. These groupings are referred to as fields and each field is made up of members.

10.2 Records in Pascal

A structure is a type that is a composite of elements that are distinctive and perhaps of different data types. The following is an example of a structure which will store a single electrical component. The record variable declared, in this case, is Component. It has three fields of differing data types; cost (a float), code (an integer) and name (a character string).

```
type
   comp = record
     cost:real;
     code:integer;
     name:string;
   end;

var Component:comp;
```

The dot notation (.) accesses each of the members within the record. For example:

```
Component.cost:=30.1;
Component.code:=32201;
Component.name:='Resistor 1';
```

Pascal Program 10.1 is a simple database program. The database stores a record of a single electrical component which includes its name (Component.name), its cost (Component.cost) and its code number (Component.code). The name is a string, the code a signed integer and the cost a floating point value.

Program 10.1

```
program struct_ex(input,output);
type
   comp = record
      cost:real;
      code:integer;
      name:string;
   end;

var    Component:comp;

begin
    Component.cost:=30.1;
    Component.code:=32201;
    Component.name:='Resistor 1';
    writeln('Name ',Component.name,' Code ',Component.code,
               ' Cost ',Component.cost);
end.
```

Test run 10.1 shows a sample run.

Test run 10.1

```
Name Resistor 1 Code 32201 Cost 30.100000
```

Pascal Program 10.2 contains a procedure to print the record (print_component()). To pass a structure into a function the data type of parameter passed must be defined. For the purpose the type keyword is used to define a new data type, in this case, it is named comp.

Program 10.2

```
program struct_ex(input,output);

type
   comp = record
      cost:real;
      code:integer;
      name:string;
   end;
var     Component:comptype;
```

```
procedure print_component(comp:comptype);
begin
   writeln('Name ',Comp.code, ' Code ',Comp.code,' Cost ',Comp.cost);
end;

begin
    Component.cost:=30.1;
    Component.code:=32201;
    Component.name:='Resistor 1';
    writeln('Name ',Component.name,' Code ',Component.code,
               ' Cost ',Component.cost);
end.
```

Test run 10.2 shows a sample run.

Test run 10.2
```
Name Resistor 1 Code 32201 Cost 30.100000
```

Pascal Program 10.3 uses a function to get data into the record (get_component()). The parameter passed into this function has a var in front of the parameter so that it can be passed back to the calling routine.

Program 10.3
```
program struct_ex(input,output);

type
   comp = record
     cost:real;
     code:integer;
     name:string;
   end;

var    Component:comptype;

procedure get_component(var comp:compType);
begin
   Comp.cost:=30.1;
   Comp.code:=32201;
   Comp.name:='Resistor 1';
end;

procedure print_component(comp:comptype);
begin
   writeln('Name ',Comp.name,' Code ',Comp.code,' Cost ',Comp.cost);
end;

begin
   get_component(Component);
   print_component(Component);
end.
```

Test run 10.3 shows that the results are identical to the previous run.

Test run 10.3
```
Name Resistor 1 Code 32201 Cost 30.100000
```

10.3 Structures in C

A structure is a type that is a composite of elements that are distinctive and perhaps of different data types. The following gives an example of a structure declaration (in this case the strings have been declared with a maximum of 100 characters).

```
struct
{
   char      title[100];
   char      comp1_name[100], comp2_name[100], comp3_name[100];
   float comp1_val, comp2_val, comp3_val;
} circuit;
```

The following is an example of a structure which will store a single electrical component. The structure variable declared, in this case, is Component. It has three fields of differing data types; cost (a float), code (an integer) and name (a character string).

```
struct
{
   float      cost;
   int        code;
   char   name[BUFSIZ];
} Component;
```

The dot notation (.) accesses each of the members within the structure. For example:

```
Component.cost=12.3
Component.code=31004;
strcpy(Component.name,"Resistor 1");
```

C Program 10.1 is a simple database program. The database stores a record of a single electrical component which includes its name (Component.name), its cost (Component.cost) and its code number (Component.code). The name is a string, the code a signed integer and the cost a floating point value.

Program 10.1
```
#include    <stdio.h>
#include <string.h>

int    main(void)
{
struct
{
   float cost;
   int        code;
   char   name[BUFSIZ];
} Component;

   Component.cost=30.1;
```

```
Component.code=32201;
strcpy(Component.name,"Resistor 1");

printf("Name %s Code %d Cost %f\n",
        Component.name,Component.code,Component.cost);
return(0);
}
```

Test run 10.4 shows a sample run.

🖳 Test run 10.4

```
Name Resistor 1 Code 32201 Cost 30.100000
```

Note that the structure could have been set up at initialization using the following:

```
struct
{
    float cost;
    int     code;
    char    name[BUFSIZ];
} Component={30.1, 32201, "Resistor 1"};
```

C Program 10.2 contains a function to print the structure (`print_component()`). To pass a structure into a function the data type of parameter passed must be defined. For the purpose the `typedef` keyword is used to define a new data type; in this case, it is named `CompType`. The program also uses braces to initialize the fields within the structure.

⬡ C
Program 10.2

```
#include <stdio.h>
#include <string.h>

typedef struct
{
    float cost;
    int     code;
    char    name[BUFSIZ];
} CompType;

void    print_component(CompType Comp);

int     main(void)
{
CompType Component={30.1,32201,"Resistor 1"};

    print_component(Component);
    return(0);
}

void    print_component(CompType Comp)
{
    printf("Name %s Code %d Cost %f\n",
            Comp.name,Comp.code,Comp.cost);
}
```

Test run 10.5 shows a sample run.

💻 **Test run 10.5**
```
Name Resistor 1 Code 32201 Cost 30.100000
```

C Program 10.3 uses a function to get data into the structure (get_component()). The parameter passed into this function will be a pointer to the base address of the structure. For this purpose, an ampersand is inserted before the structure name. The structure pointer operator (->) is used with the structure pointer to access a member of a field.

⬡ C **Program 10.3**
```c
#include <stdio.h>
#include <string.h>

typedef struct
{
   float cost;
   int   code;
   char  name[BUFSIZ];
} CompType;

void      get_component(CompType *Comp);
void      print_component(CompType Comp);

int       main(void)
{
CompType Component;

   get_component(&Component);
   print_component(Component);
   return(0);

}

void      get_component(CompType *Comp)
{

   Comp->cost=30.1;
   Comp->code=32201;
   strcpy(Comp->name,"Resistor 1");

}

void      print_component(CompType Comp)
{

   printf("Name %s Code %d Cost %f\n",
                    Comp.name,Comp.code,Comp.cost);

}
```

Test run 10.6 shows that the results are identical to the previous run.

💻 **Test run 10.6**
```
Name Resistor 1 Code 32201 Cost 30.100000
```

10.4 Array of structures

An array of structures can be set up in a way similar to normal array indexing. When an array is declared the compiler assigns enough memory to hold all its elements. Program 10.4 is similar to Program 10.3, but uses an array of structures to store up to 5 electrical components. Figure 10.1 shows a structure chart of this program. It uses `get_float()` and `get_int()` to filter any invalid inputs for the cost and code of a component. These functions were developed in Chapter 9 and have been reused as they have been well tested and are easily ported into any program. The cost of the electrical components is now limited between 0 and 1,000 and the component code from 0 to 32,767. A `get_string()` function has also been added to get the name of the component. Test run 10.7 shows a sample run.

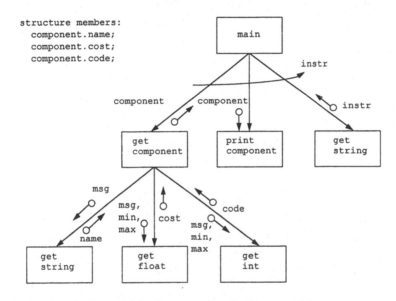

Figure 10.1   Structure chart for Program 10.4

Program 10.4
```c
#include        <string.h>
#include        <stdio.h>

#define    TRUE            1
#define    FALSE           0
#define    MAXCOMPONENTS   50

typedef struct
{
   float    cost;
   int      code;
   char  name[BUFSIZ];
} CompType;

void       get_component(int *n,CompType Comp[]);
void       print_component(int n,CompType Comp[]);
```

```
void      get_int(char msg[],int min,int max,int *val);
void      get_float(char msg[],float min,float max,float *val);
void      get_string(char msg[],char ins[]);

int   main(void)
{
int        NumComponents=0;
char       instr[BUFSIZ];
CompType   Component[MAXCOMPONENTS];

   do
   {
      get_component(&NumComponents,Component);
      print_component(NumComponents,Component);
      get_string("Do you wish to continue (type 'exit' to leave)",
                 instr);
   } while (strcmp(instr,"exit")!=0);
   return(0);
}

void      get_component(int *n,CompType Comp[])
{
int        i;

   i=*n;
   get_string("Enter name of component >> ",Comp[i].name);
   get_float("Enter cost of component >> ",0.0,1000.0,
                 &Comp[i].cost);
   get_int("Enter component code >> ",0,32767,&Comp[i].code);
   (*n)++;
}

void      print_component(int n,CompType Comp[])
{
int        i;

   for (i=0;i<n;i++)
      printf("Name %12s Code %8d Cost %8.2f\n",
                 Comp[i].name,Comp[i].code,Comp[i].cost);
}
void      get_string(char msg[],char ins[])
{
   printf("%s",msg);
   gets(ins);
}

void      get_float(char msg[],float min,float max,float *val)
{
char       instr[BUFSIZ];
int        rtn,okay;

   do
   {
      printf("%s",msg);
      gets(instr);
      rtn=sscanf(instr,"%f",val);
      if ((rtn!=1) || (*val<min) || (*val>max))
      {
         okay=FALSE;
         printf("Invalid input <%s>\n",instr);
      }
      else okay=TRUE;

   } while (!okay);
}

void      get_int(char msg[],int min,int max,int *val)
{
```

```
char     instr[BUFSIZ];
int      rtn,okay;
   /* get an integer value between min and max  */
   do
   {
      printf("%s",msg);
      gets(instr);
      rtn=sscanf(instr,"%d",val);
      if ((rtn!=1) || (*val<min) || (*val>max))
      {
         okay=FALSE;
         printf("Invalid input <%s>\n",instr);
      }
      else okay=TRUE;

   } while (!okay);
}
```

💻 Test run 10.7

```
Enter name of component >> resistor 1
Enter cost of component >> 1.23
Enter component code >> 32455
Name    resistor 1 Code    32455 Cost     1.23
Do you wish to continue (type 'exit' to leave)y
Enter name of component >> capacitor 1
Enter cost of component >> 2.68
Enter component code >> 32456
Name    resistor 1 Code    32455 Cost    1.23
Name    capacitor 1 Code   32456 Cost    2.68
Do you wish to continue (type 'exit' to leave)y
Enter name of component >> inductor 2
Enter cost of component >> 6.32
Enter component code >> 32457
Name    resistor 1 Code    32455 Cost    1.23
Name    capacitor 1 Code   32456 Cost    2.68
Name    inductor 2 Code    32457 Cost    6.32
Do you wish to continue (type 'exit' to leave)exit
```

(Pascal) Program 10.4

```
const MAXCOMPONENTS=50;

type
   comptype = record
      cost:real;
      code:integer;
      name:string;
   end;

   comp_arr=array[1..MAXCOMPONENTS] of comptype;



var   Component:comp_arr;
      NumComponents:integer;
      instr:string;

procedure print_component(n:integer;comp:comp_arr);
var        i:integer;
begin
   for i:=1 to n do
      writeln('Name ',Comp[i].name,' Code ',Comp[i].code,
         ' Cost ',Comp[i].cost);
end;
```

```
procedure get_string(msg:string;var ins:string);
begin
   write(msg);
   readln(ins);
end;

procedure get_real(msg:string; min, max:real; var value:real);
var instr:string;
    code:integer;
    okay:boolean;
begin
   repeat
      write(msg);
      readln(instr);
      val(instr,value,code);
      if (code<>0) then
      begin
         okay:=FALSE;
         writeln('Invalid input <',instr,'>');
      end
      else okay:=TRUE;

   until (okay=TRUE);
end;

procedure get_int(msg:string; min, max:integer; var value:integer);
var instr:string;
    code:integer;
    okay:boolean;
begin
   repeat
      write(msg);
      readln(instr);
      val(instr,value,code);
      if (code<>0) then
      begin
         okay:=FALSE;
         writeln('Invalid input <',instr,'>');
      end
      else okay:=TRUE;

   until (okay=TRUE);
end;

procedure get_component(var n:integer; var Comp:Comp_arr);
begin
   n:=n+1;
   get_string('Enter name of component >> ',Comp[n].name);
   get_real('Enter cost of component >> ',0.0,1000.0,Comp[n].cost);
   get_int('Enter component code >> ',0,1000,Comp[n].code);
end;

begin
   NumComponents:=0;
   repeat
      get_component(NumComponents,Component);
      print_component(NumComponents,Component);
      get_string('Do you wish to continue (type "exit" to leave)',instr);
   until (instr='exit');
end.
```

10.4.1 Complex arithmetic

Program 10.5 uses a structure (or record) to multiply to complex numbers. If these complex numbers are $a+jb$ and $c+jd$ and the result is z then:

$$z = (a + jb)(c + jd)$$
$$= ac + jad + jbc - bd$$
$$= (ac - bd) + j(ad + bc)$$
$$\text{Re}(z) = ac - bd$$
$$\text{Im}(z) = ad + bc$$

This operation is implemented in the function (procedure) multi_complex(). Test run 10.8 gives a test run for the result of *3+j4* and *6+j2*.

Program 10.5

```c
#include <stdio.h>
typedef  struct
{
   float x,y;
} complex;

void  multi_complex(complex a,complex b,complex *c);
void  show_complex(complex a);

int    main(void)
{
complex  z1,z2,z3;

   z1.x=3;  z1.y=4;
   z2.x=6;  z2.y=2;
   multi_complex(z1,z2,&z3);
   show_complex(z3);
   return(0);
}

void  multi_complex(complex a, complex b, complex *c)
{
   c->x=a.x*b.x-a.y*b.y;
   c->y=a.x*b.y+a.y*b.x;
}

void  show_complex(complex a)
{
   printf("%8.2f + j %8.2f\n",a.x,a.y);
}
```

Test run 10.8

```
10 + j 30
```

Program 10.5

```pascal
type
   rect=record
               x,y:real;
   end;

var z1,z2,z3:rect;

procedure multi_rect(a,b:rect;var c:rect);
```

```
begin
    c.x:=a.x*b.x-a.y*b.y;
    c.y:=a.x*b.y+a.y*b.x;
end;

procedure show_rect(a:rect);
begin
    writeln(a.x:8:2,'+j ',a.y:8:2);
end;

begin
    z1.x:=3; z1.y:=4;
    z2.x:=6; z2.y:=-2,
    multi_rect(z1,z2,z3);
    show_rect(z3);
end.
```

10.5 Exercises

10.5.1   Write a database program, based on Program 10.4, which gives a menu choice as to whether the user wishes to enter input a new electric component, to list all the components already in the database, or to exit the program. A sample run is given in Test run 10.9.

🖳   **Test run 10.9**

```
Do you wish to
(1)    Input a component
(2)    List all components
(3)    Exit from program
Enter option >>>
```

10.5.2   Write a program which converts from rectangular notation to polar form. In rectangular notation:

$$z = x + iy$$

in polar form this is:

$$Z = |z| \langle z \rangle$$

where $|z| = \sqrt{x^2 + y^2}$ and $\langle z \rangle = \tan^{-1}\left(\dfrac{y}{x}\right)$

10.5.3   Write a program using structures (or records) with complex impedance values that will determine the parallel impedance of two impedances Z_1 and Z_2. The formula for the impedances in parallel is given below:

$$Z_{EQ} = \frac{Z_1 Z_2}{Z_1 + Z_2} \; \Omega$$

10.5.4   Write a program which contains the following complex number manipulation routines:

(i)   complex number add.
(ii)   complex number subtract.
(iii)   complex number multiply.
(iv)   complex number divide.

10.5.5   Modify the database program in the text so that the user can select an option which will order the component names in alphabetic order.

PART B

C/Pascal
C++
Assembly Language
Visual Basic
HTML/Java
DOS
Windows 3.x
Windows 95
UNIX

Introduction to C++

11.1 Introduction

C is an excellent software development language for many general purpose applications. Its approach is that data and associated functions are distinct, data is declared and the functions are then implemented. Object-oriented programming languages allow the encapsulation of a set of data types and associated functions into objects. These objects are integrated entities.

C++ is by far the most popular object-oriented language. It was developed by Bjane Stoustrup at AT&T Bell Laboratories. A new programming language, based on C++, called Java, has since been developed by Sun Microsystems. C++ is well supported and there are many different development systems. Figure 11.1 shows Microsoft Visual C++ Version 4.0. Microsoft Visual Version 5.0 is similar but has enhanced support for WWW applications. The default file extension for a C++ program is CPP.

Figure 11.1   Microsoft Visual C++

11.2 Enhancements to C++

C++ has many enhancements over C. These include:

I/O stream support. This facility allows data to be directed to an input and/or an output stream. Section 11.3 outlines this.

Objects. An object incorporates data definitions and the declaration and definitions of functions which operate on that data.

Classes. These are used to implement objects and can be initialized and discarded with constructors and destructors, respectively. Section 11.6 outlines this.

Complex numbers. C++ supports the usage of complex numbers and their mathematical operations. Section 11.7 outlines these.

Data hiding. This allows certain data to be hidden from parts of a program which are not allowed access to it. Section 11.6 outlines this.

Overloading. This allows more than one definition and implementation of a function. Section 11.5 outlines this.

Virtual functions. This allows any one of a number of multiple defined functions to be called at run-time.

Template classes. This allows the same class to be used by different data.

11.3 I/O stream

In C, output is sent to the standard output using the `printf()` function. In C++, standard input and output are taken from streams. The standard output stream is `cout` (normally to the monitor) and the standard input stream is taken from `cin` (normally from the keyboard). To be able to use these streams, the `iostream.h` header file must be included in the program. The redirection operator (`<<`) is used to direct the data to the input or output. Program 11.1 shows an example program and Test run 11.1 shows a sample run.

Program 11.1
```
#include <iostream.h>

int     main(void)
{
int     a=10,b=12;

    cout << "Value of a is " << a << " and b is " << b;
    return(0);
}
```

Test run 11.1
```
The value of a is 10 and the value of b is 12
```

Program 11.2 shows another example and Test run 11.2 show a sample run.

⬡ C++

Program 11.2

```
#include <iostream.h>

#define  PI 3.14157

int   main(void)
{
float radius,area;

    cout << "Enter the radius > ";
    cin >> radius;
    area=PI*radius*radius;
    cout<< "Area is " << area;
    return(0);
}
```

🖥 **Test run 11.2**

```
Enter radius > 32
Area is 3619.967773
```

11.4 Comments

C++ supports single line comment with a double slash (//) or multiline comment which are defined between a (* and a *). The single line comment makes everything on the line after the double slash a comment. A few examples are:

```
PI=3.14157;      // Sets the value of pi
area=PI*r*r      // Area is PI times the square of the radius
```

11.5 Function overloading

Often the programmer requires to call a function in a number of ways but still wants the same name for the different implementations. C++ allows this with function overloading. With this the programmer writes a number of functions with the same number but which are called with a different argument list or return type. The compiler then automatically decides which one should be called. For example, in C++ Program 11.3 the programmer has defined three square functions named sqr(). The data type of the argument passed is of a different type for each of the functions, that is, either an int, a float or a double. The return type is also different. In this program, three variables, of different types, have been declared in the main() function. The compiler will then call the correct format function for each of the calls. Test run 11.3 shows a sample run. It can be seen that the implementation of the square function with a float data type produces some rounding errors.

⬡ C++

Program 11.3
```
#include <iostream.h>

int        sqr(int x);
float      sqr(float x);
double     sqr(double x);

int    main(void)
{
int        val1=4;
float      val2=4.1;
double     val3=4.1;

    cout << "Square of 4 is " << sqr(val1)<< "\n";
    cout << "Square of 4.1 is " << sqr(val2)<< "\n";
    cout << "Square of 4.1 is " << sqr(val3)<< "\n";
    return(0);
}

int    sqr(int x)
{
    return(x*x);
}

float    sqr(float x)
{
    return(x*x);
}

double   sqr(double x)
{
    return(x*x);
}
```

🖳 **Test run 11.3**
```
Square of 4 is 16
Square of 4.1 is 16.809999
Square of 4.1 is 16.81
```

The argument list of the overloaded function does not have to have the same number of arguments for each of the overloaded functions. C++ Program 11.4 shows an example of an overloaded function which has a different number of arguments for each of the function calls. In this case, the show_results() function can either be called with a message, followed by three values, or the three values, or with a single value.

⬡ C++

Program 11.4
```
#include <iostream.h>

void  get_values(char msg[],float *R1,float *R2);
float calc_parallel(float R1,float R2);
void  show_results(char msg[],float R1, float R2,float Requ);
void  show_results(float R1, float R2,float Requ);
void  show_results(float Requ);

int    main(void)
```

```
{
float r1,r2,requ;

    get_values("Enter R1 and R2 and press RETURN ",&r1,&r2);
    requ=calc_parallel(r1,r2);
    show_results("R1, R2 and Req are ",r1,r2,requ);

    return(0);

}

float calc_parallel(float R1,float R2)
{
    return((R1*R2)/(R1+R2));

}

void  get_values(char msg[],float *R1,float *R2)
{
    cout << msg;
    cin >> *R1 >> *R2;

}

void  show_results(char msg[],float R1, float R2,float Requ)
{
    cout << msg;
    cout << " " << R1 << " " << R2 << " " << Requ << "\n";

}
void  show_results(float R1, float R2,float Requ)
{
    cout << " " << R1 << " " << R2 << " " << Requ << "\n";
}

void  show_results(float Requ)
{
    cout << Requ << "\n";
}
```

11.6 Classes

Classes are a general form of structures (struct). In C, a struct gathers
together data members and in C++ the struct may also have function members. The main difference between a class and a struct is that in a
struct all members of the structure are public, while in a class the
members are, by default private (restricted). In a class, members are made
public with the public keyword, and private with the private keyword.

C++ Program 11.3 shows an example of a class. It can be seen that it is
similar to a structure, but that some parts of the class are private while
others are public. These are defined as follows:

- A public member of a class can be accessed by external code that is not
  part of the class. Often the public member is a function.
- A private class member can only be used by code defined in a member
  function of the same class.

It is obvious that all classes must have a public content so that they can be accessed by external functions. In C++ Program 11.5 the circuit `class` has a private part, in this case, the variable `rtemp`. The public parts are the functions `parallel()` and `serial()`, which determine the parallel and series resistance of two resistors. In the main function the class is declared to c1. The series calculation is called with `c1.series`(*r1,r2*) and the parallel calculation is called with `c1.parallel`(*r1,r2*).

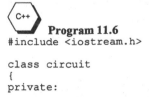

Program 11.5

```
#include <iostream.h>

class circuit
{
private:
   float rtemp;
public:
   float parallel(float r1, float r2)
   {
      return((r1*r2)/(r1+r2));
   }
   float series(float r1, float r2)
   {
      return(r1+r2);
   }
};

int    main(void)
{
   circuit c1;
   float res;

   res=c1.series(2000,1000);
   cout << "Series resistance is " << res << "ohms\n";

   res=c1.parallel(1000,1000);
   cout << "Parallel resistance is " << res << "ohms\n";

   return(0);
}
```

Test run 11.4
```
Series resistance is 2000 ohms
Parallel resistance is 500 ohms
```

The class would get cumbersome if the functions within it where to be defined in it. C++ Program 11.6 shows how the functions can be defined in another part of the program. In this case the class name is defined followed by a double colon (::) and then the function name.

Program 11.6

```
#include <iostream.h>

class circuit
{
private:
```

```
        float rtemp;
public:
        float series(float, float);
        float parallel(float,float);
};

int    main(void)
{
        circuit c1;
        float res;

        res=c1.series(2000,1000);
        cout << "Series resistance is " << res << " ohms\n";

        res=c1.parallel(1000,1000);
        cout << "Parallel resistance is " << res << " ohms\n";

        return(0);
}

float circuit::parallel(float r1, float r2)
{
        return((r1*r2)/(r1+r2));
}

float circuit::series(float r1, float r2)
{
        return(r1+r2);
}
```

The private part of a class is typically used to store variables which are local only to the class. An example is shown in C++ Program 11.7 where the variables local to the circuit class are r1 and r2. The get_res() function is then used to set these values and the other two functions then use them.

Program 11.7

```
#include <iostream.h>

class circuit
{
private:
        float r1,r2;
public:
        void  get_res(void);
        float series(void);
        float parallel(void);
};

int    main(void)
{
        circuit  c1;
        float    res;

        c1.get_res();

        res=c1.series();
        cout << "Series resistance is " << res << " ohms\n";

        res=c1.parallel();
        cout << "Parallel resistance is " << res << " ohms\n";
        return(0);
}
```

```
void  circuit::get_res(void)
{
    cout << "Enter r1 >> ";
    cin >> r1;
    cout << "Enter r2 >> ";
    cin >> r2;
}

float circuit::parallel(void)
{
    return((r1*r2)/(r1+r2));
}

float circuit::series(void)
{
    return(r1+r2);
}
```

🖥 **Test run 11.5**

```
Enter r1 >> 1000
Enter r2 >> 1000
Series resistance is 2000 ohms
Parallel resistance is 500 ohms
```

In many cases the use of passing parameters by pointers can be eliminated by using classes. This is shown in C++ Program 11.8 where the get_res() function is private to the circuit class. Then within the series() and parallel() functions the get_res() function is called. Note that no other function can call this function as it is private to circuit.

⬡ C++
Program 11.8

```
#include <iostream.h>

class circuit
{
private:
    float r1,r2;
    void  get_res(void);
public:
    float series(void);
    float parallel(void);
};

int   main(void)
{
    circuit  c1;
    float    res;
    res=c1.series();
    cout << "Series resistance is " << res << " ohms\n";

    res=c1.parallel();
    cout << "Parallel resistance is " << res << " ohms\n";

    return(0);
}

void  circuit::get_res(void)
{
```

```
   cout << "Enter r1 >> ";
   cin >> r1;
   cout << "Enter r2 >> ";
   cin >> r2;
}

float circuit::parallel(void)
{
   get_res();
   return((r1*r2)/(r1+r2));
}

float circuit::series(void)
{
   get_res();
   return(r1+r2);
}
```

11.7 Complex numbers

C++ contains a range of functions which support complex numbers. These are prototyped in complex.h and are outlined in Table 11.1.

The data type complex is used to declare a complex variable. In C++ Program 11.9 the variable z has been declared a complex variable. The complex() function is used to convert the x and y variables into a complex number. The functions real(), imag(), conj(), abs() and pow() are used and Test run 11.6 displays a sample run.

Program 11.9
```
#include <iostream.h>
#include <complex.h>

int main(void)
{
complex  z;
double   x = 3, y = 4;

   z = complex(x,y);
   cout << "z = "<< z << "\n";
   cout << "The real part is " << real(z) << "\n";
   cout << "The imaginary part is " << imag(z) << "\n";
   cout << "Complex conjugate is " << conj(z) << "\n";
   cout << "Magnitude is " << abs(z) << "\n";
   cout << "Squared is " << pow(z,2.0) << "\n";
   return (0);
}
```

Test run 11.6
```
z= (3,4)

The real part is 3
The imaginary part is 4
Complex conjugate is (3,4)
Magnitude is 5
Squared is (-7,24)
```

Table 11.1 Complex functions prototyped in `complex.h`

Function	Description	Calculation
`complex cos(complex z)`	Cosine of a complex number	$\cos(z) = (\exp(i * z) + \exp(-i * z)) / 2$
`complex sin(complex z)`	Sine of a complex number	$\sin(z) = (\exp(i * z) - \exp(-i * z)) / (2i)$
`complex tan(complex x)`	Cosine of a complex number	$\tan(z) = \sin(z) / \cos(z)$
`complex sqrt(complex x)`	For complex numbers x, sqrt(x) gives the complex root whose arg is arg(x) / 2.	$\mathrm{sqrt}(z) = \mathrm{sqrt}(\mathrm{abs}(z)) * (\cos(\arg(z)/2) + i * \sin(\arg(z)/2))$
`double abs(complex z);`	Complex absolute	$\mathrm{sqrt}(z.x * z.x + z.y * z.y)$
`complex acos(complex z);`	Inverse cosine of a complex number	$-i * \log(z + i*\mathrm{sqrt}(1 - z^{**}2))$
`complex asin(complex z)`	Inverse sine of a complex number	$-i * \log(i*z + \mathrm{sqrt}(1 - z^{**}2))$
`complex atan(complex z)`	Inverse tan of a complex number	$-0.5 * i*\log((1 + i*z)/(1 - i*z))$
`complex exp(complex z)`	Exponential of a complex number	$\exp(x + iy) = \exp(x) * (\cos(y) + i \sin(y))$
`double imag(complex x);`	The imaginary part of the complex number.	
`complex log(complex x)`	Logarithm of a complex number	$\log(z) = \log(\mathrm{abs}(z)) + i \arg(z)$
`complex log10(complex x)`	Logarithm using base 10 of a complex number	$\log10(z) = \log(z) / \log(10)$
`complex pow(complex x, complex y)` `complex pow(complex x, double y)` `complex pow(double x, double y)`	Power of a complex number	
`double real(complex x);`	Real part of complex number	
`complex polar(double mag, double angle)`	Calculates the complex number with the given magnitude and angle.	polar(mag, angle); complex(mag * cos(angle), mag * sin(angle))
`double conj(complex z)`	Conjugate of a complex number	Same as complex(real(z), −imag(z))

11.8 Exercises

11.8.1 Write an overloaded function which determines the gradient of a straight line. The parameters passed are either two (x, y) co-ordinates $(x1, y1, x2, y2)$ or if the change in x and the change in y (dx, dy). The gradient will be calculated as follows:

$$m = \frac{y_2 - y_1}{x_2 - x_1} \quad \text{or} \quad m = \frac{dy}{dx}$$

11.8.2 Write an overloaded function which determines the angle of a straight line. The parameters passed are either two co-ordinates $(x1, y1, x2, y2)$ or if the change in x and the change in y (dx, dy). The gradient will be calculated as follows:

$$\theta = \tan^{-1}\left(\frac{y_2 - y_1}{x_2 - x_1}\right) \quad \text{or} \quad \theta = \tan^{-1}\left(\frac{dy}{dx}\right)$$

11.8.3 The following C++ class definition contains date functions. The `get_date()` function should fill the variables `day`, `month` and `year`. The `valid_data()` function should then test if the entered date is valid.

```
class datefunction
{
private:
    day,month,year;

public:
    void   get_date(void);    // get day, month and year
    int    valid_date(void); // check valid date
};
```

Write a C++ program which contains this class definition.

11.8.4 C++ Program 11.10 has a `mathfunction` class which contains multiply and divide functions. Write a program which adds several other mathematical functions, to give the following:

```
class mathfunction
{

public:
    float mult(float a, float b);
    float div(float a, float b);
    float add(float a, float b);
    float sub(float a, float b);
    float sqrt(float a);
    float sin(float a);
    float cos(float a);
    float mod(int a, int b);
};
```

Program 11.10

```
#include <iostream.h>

class mathfunction
{

public:
   float mult(float a, float b);
   float div(float a, float b);
};

int    main(void)
{
   mathfunction m;
    float val;

   val=m.mult(5,6);

   cout << "5 times 6 is " << val << " ohms\n";

   val=m.div(5,6);

   cout << "5 divided by 6 is " << val << " ohms\n";

   return(0);
}


float mathfunction::mult(float a,float b)
{
   return(a*b);
}

float mathfunction::div(float a, float b)
{
   return(a/b);
}
```

11.8.5   Modify Exercise 11.8.4 so that the values are entered from the keyboard.

11.8.6   The following C++ class definition contains an initialization, increment, decrement and displaying functions. The initialization function should initialize the samples variable to 0; the increment function should add one onto the samples variable; the decrement should take one away from samples and the show_samples display the samples value. Complete the class declaration and write a C++ which calls it. The increment function should be called when the 'j' key is pressed (followed by the RETURN key) and the decrement function should be called when the 'l' key (followed by the RETURN key) is pressed. The value of samples should be displayed when the RETURN key is pressed without either the 'j' or 'l' keys.

```
class operation
{
private:
```

```
        samples;

    public:
        void  init_samples() {samples=0} ;  // initialize samples
        int   increment() { to be completed } ;  // increment samples
        int   decrement() { to be completed } ;  // decrement samples
        void  show_samples() { to be completed } ;  // show samples
    };
```

11.8.7   Using the complex number functions determine the following, for the complex number $a=4+i5$ and $b=6+i10$:

(a)   a+b

(b)   a×b

(c)   a−b

(d)   a÷b

(e) a^2

(f) b^3

(g) $|a|$

(h) e^a

(i) $\sqrt{a}$

Note that in electrical engineering a complex number is typically represented as $a+jb$.

12

More C++

12.1 Constructors and destructors

A constructor allows for the initialization of a class and a destructor allows the class function to be removed from memory. They are defined as follows:

- A constructor is a special initialization function that is automatically called whenever a class is declared. The constructor always has the same name as the class name, and no data types are defined for the argument list or the return type. Normally a constructor is used to initialize a class.
- A destructor is a member of a function which is automatically called when the class is destroyed. It has the same name as the class name but is preceded by a tilde (~). Normally a destructor is used to clean-up when the class is destroyed.

C++ Program 12.1 has a class which is named class_ex. The constructor for this class is class_ex() and the destructor is ~class_ex(). Test run 12.1 shows a sample run. It can be seen that initially when the program is run the message Constructing is displayed. This is because the class is initially declared to c1. Then when the function test() is called the Constructing message is again displayed as a new class is defined (c2). When the program leaves this function the destructor is called and thus the message Destructor is displayed. Finally, when the program quits the destructor is again called.

Program 12.1

```
#include <iostream.h>

void   test(void);

class class_ex
{
   private:

   public:
   class_ex() { cout << "Constructing\n"; } // Constructor
```

```
    ~class_ex() { cout << "Destructing\n"; }    // Destructor
    void show_msg(void) { cout << "Hello\n"; }

};

int   main(void)
{
class_ex c1;

    cout << "Test point 1\n";
    c1.show_msg();
    cout << "Test point 2\n";
    test();
    cout << "Test point 3\n";

    return(0);
}

void  test(void)
{
class_ex c2;

    cout << "In test function\n";
}
```

🖥 **Test run 12.1**
```
Constructing
Test point 1
Hello
Test point 2
Constructing
In test function
Destructing
Test point 3
Destructing
```

12.2 Function templates

Function templates allow functions to be capable of operating on arguments of any data type. To define a function template the following is done:

- Prefix the function with the `template` keyword.
- Define one or more identifiers that define parameterized types within angled brackets.

In C++ Program 12.2 the template is created for a `maximum()` function. The template definition:

```
template <class number>
```

defines that `number` can be of any data type. Thus:

```
number maximum(number val1, number val2);
```

could be matched to any one of the following:

```
int maximum(int val1,int val2);
float maximum(float val1,float val2);
double maximum(double val1,double val2);
```

The definition of the function is given as:

```
template <class number>
number maximum(number val1,number val2)
{
    if (val1>val2) return(val1);
    else return(val2);
}
```

⬡ C++

Program 12.2

```
#include <iostream.h>

template <class number>
number maximum(number val1,number val2);

int   main(void)
{
float value1=10.0, value2=100.0, value3;

    value3=maximum(value1,value2);

    cout << "Largest value is " << value3;

    return(0);

}

template <class number>
number maximum(number val1,number val2)
{
    if (val1>val2) return(val1);
    else return(val2);
}
```

12.3 Class templates

The class template is a generalization of the function template. These allow collections of objects of any type using the same class template. Program 12.3 shows an example of a class `calc_current`. In this case the template is defined as:

```
template <class number>
```

which defines that number can be of any data type. When defining the class variable the data type is defined. For example, to define the data type of number as an `int`, then:

```
calc_current<int> i1;
```

or as a float:

```
calc_current<float> i2;
```

In the case of C++ Program 12.3 the class which uses an `int` will perform integer operations and the float declaration will perform `float` operations.

⬡ C++
     Program 12.3

```
// Program to determine the current through a resistor with 10 V
// across the resistor
#include <iostream.h>

template <class number>
class calc_current
{
private:
     number v,res;
public:
     calc_current()     // constructor, set v to 10 Volts
     {
          v=10;
     }
     void get_res() { cout<< "Enter resistance "; cin >> res;}
     void print_current() { cout << v/res << "\n";}
};

int   main(void)
{
calc_current <int> I1;
calc_current <float> i2;

   i1.get_res();
   i1.print_current();

   i2.get_res();
   i2.print_current();

   return(0);
}
```

12.4 I/O streams

The standard C++ library provides an expanded set of functions, templates, string-handling classes and many other enhancements to C. One of the main enhancements is the stream I/O library.

The stream I/O library gets rids of a particular problem in C where functions such as `printf()` and `scanf()`, gives run-time errors when the data they are handling is in the incorrect data form. For example:

```
scanf("%d",&i);
```

would create an error if `i` was declared as a floating-point value. The stream I/O function for input from the standard input is:

```
cin >> i
```

This allows for automatic data type conversion. To support the stream I/O the `stdio.h` header file has been replaced by the header files `iostream.h`, `fstream.h` and `iomanip.h`, and classes which are automatically defined are:

cout   for standard output.
cin   for standard input.
cerr   for standard error.

The stream I/O class hierarchy is given in Figure 12.1. The `ios` (I/O state) class is declared in `iostream.h`, and it contains information about the state of the stream. It includes functions which are used to open and close streams, and it also has stream format flags.

The file I/O classes `ifstream` and `ofstream` are declared in the header file `fstream.h`. and the class `iostream` inherits both `istream` and `ostream`, and is declared in `iostream.h`. `ifstream` inherits all the standard input stream operations defined by `istream` and adds a few more, such as constructors and functions for opening files. `ofstream` similarly related to `ostream`.

Finally, `fstream`, declared in the header file `fstream.h`, inherits `iostream` and contains functions and constructors that allow files to be opened in input/output mode.

The classes can be summarized by:

Abstract Stream Base Class
ios   Stream base class.

Input Stream Classes
istream   General-purpose input stream class and base class
                         for other input streams.

ifstream   Input file stream class.

Output Stream Classes
ostream   General-purpose output stream class and base
                         class for other output streams.

ofstream   Output file stream class.

Input/Output Stream Classes
iostream   General-purpose input/output stream class and
                         base class for other input/output streams.

fstream   Input/output file stream class.

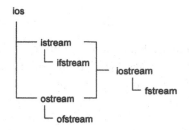

Figure 12.1   Block diagram of a simple computer system

12.4.1 Opening and closing a file

The functions used to open and close a stream are open() and close(), respectively.

open()

For a program to use a file stream (fstream) it must first be associated with a specific disk file with the open function. This function is used with an associated open_mode flag, which can be combined together with the bitwise OR (|) operator. Table 12.1 defines these flags.

Table 12.1   File open flags

Flag	Function
ios::app	Opens an output file for appending.
ios::ate	Opens an existing file and sets the file pointer to the end of the file.
ios::binary	Opens file in binary mode (Note that the default mode is text mode).
ios::in	Opens an input file. Use ios::in as an open_mode for an ofstream file to prevent truncating an existing file.
ios::nocreate	Open the file only if it already exists. Otherwise the operation fails.
ios::noreplace	Opens the file only if it does not exist. Otherwise the operation fails.
ios::out	Opens an output file. When you use ios::out for an ofstream object without ios::app, ios::ate, or ios::in, ios::trunc is implied.
ios::trunc	Opens a file and deletes the old file (if it already exists).

To declare a file pointer the class ofstream is used. For example, to declare a stream of myfile which has the file name of out.dat and is an output file:

```
#include <fstream.h>

int   main(void)
{
ofstream myfile("out.dat", ios::out );
   :     :
}
```

This type of declaration can be used when the stream has a fixed name and file_open flag. It is also possible to use the open function to open the file. The following achieves the same as the previous sample program.

```
#include <fstream.h>

int   main(void)
{
fstream myfile;

   myfile.open("c:\\out.dat",ios::out);
   :     :
}
```

The following gives an example of a binary file which is only for input:

```
myfile.open("c:\\out.dat",ios::binary | ios::in);
```

while the following gives an example of a binary file which is only for input and output:

```
myfile.open("c:\\out.dat",ios::binary | ios::out | ios::in);
```

close()

The close member function is used to close the file associated with an output file stream. The fstream destructor automatically closes the file when the class is destroyed. It is typically used when the stream needs to be associated with another file.

C++ Program 12.4 opens a file for output and writes the character 'H' to it and then closes the file.

Program 12.4

```
#include <fstream.h>

int   main(void)
{
fstream myfile;

   myfile.open("c:\\out.dat",ios::out); // open a file for output
```

```
    cout << "Writing to file";
    myfile.put('H');

    myfile.close();
    return(0);
}
```

12.4.2 Text I/O

The two main functions used for text I/O are put () and get ().

put()

The put member function writes a single character to the output stream. C++ Program 12.4 shows an example of the put () function.

get()

The unformatted get member function works like the >> operator with two exceptions:

- The get function includes white-space characters, whereas the extractor excludes white space.
- The get function is less likely to cause a tied output stream (cout, for example) to be flushed.

To extract a string:

stream.get(char *str, int *nCount*, char *delim* = '\n');
stream.get(unsigned char *str, int *nCount*, char *delim* = '\n');
stream.get(signed char *str, int *nCount*, char *delim* = '\n');

Or to extract a single character:

stream.get(char *ch);
stream.get(unsigned char *ch);
stream.get(signed char *ch);

where *nCount* is the maximum number of characters to store the line and *delim* is the delimiter character (which by default is a new-line).

getline()

The getline member function is similar to the get function, but they differ as the get function leaves the terminating character in the stream; whereas getline removes the terminating character.

C++ Program 12.5 gives an example of the getline () function.

Program 12.5
```
#include <string.h>

int   main(void)
{
char instr[100];

    cin.getline(instr,100); //get an input of a maximum of 100 chars

    cout.write(instr,strlen(instr));

    return(0);
}
```

12.4.3 Error processing functions

The I/O stream libraries also contain a number of member functions which can be used to test for errors while reading and writing to a stream. Table 12.2 lists these flags.

C++ Program 12.6 shows an example program which uses the `fail()` and `good()` functions. The `fail()` function allows the program to test if the file has been opened and the `good()` function allows the program to test if the end-of-file has been reached (or if an error has occurred).

Table 12.2   Error processing functions

Function	Return value
bad	Returns TRUE if there is an unrecoverable error.
fail	Returns TRUE if there is an unrecoverable error or an "expected" condition, such as a conversion error, or if the file is not found.
good	Returns TRUE if there is no error condition (unrecoverable or otherwise) and the end-of-file flag is not set.
eof	Returns TRUE on the end-of-file condition.
clear	Sets the internal error state. If called with the default arguments, it clears all error bits.
rdstate	Returns the current error state.

Program 12.6
```
#include <fstream.h>

int main(void)
{
char     ch;
ifstream infile;
```

```
infile.open( "c:\\in.dat", ios::in);
if( !infile.fail() )
{
   while ( infile.good() )    // EOF or failure stops the reading
   {
      infile.get( ch );
      if( !ch ) break;        // quit on null
      cout << ch;
   }
}
else
{
   cout << "ERROR: Cannot open out.dat." << "\n";
}
return(0);
}
```

Note that:

```
if( !infile.fail() )
```

can be replaced by:

```
if (infile)
```

as the ! operator is overloaded to perform the same function as the `fail()` function.

12.4.4 File pointers

The `seekg()` and `tellg()` are uses to set (`seekg`) or get (`tellg`) the current file point for an input stream. On the output stream the `seekp()` and `tellp()` are used. The `seekg` format is:

stream.`seekg(streampos` *pos*`);`
stream.`seekg(streamoff` *off*, `ios::seek_dir` *dir*`);`

where *pos* specifies the new position value, *off* is the new offset value, (streamoff and streampos are a typedef equivalent to a `long int`) and *dir* is the seek direction which must be one of the following:

- `ios::beg`   seek from the beginning of the stream.
- `ios::cur`   seek from the current position in the stream.
- `ios::end`   seek from the end of the stream.

C++ Program 12.7 gives an example of a file pointer being set to the beginning of a file with `myfile.seekp(0l,ios::beg))`, where `0l` represents a zero offset and `ios::beg` the flag for the beginning of the file.

The `tellg()` format is:

stream.tellg();

where the streampos type is a long int.

12.4.5 Binary I/O

read

The read member function is similar to the read function in C. It reads a specified number of bytes from a file into specified areas of memory. C++ Program 12.7 gives an example a single floating point value beginning written to a file and then read back.

Program 12.7
```
#include <fstream.h>

int    main(void)
{
fstream myfile;
float val1=10.1,val2;

    myfile.open("c:\\out.dat",ios::binary | ios::out | ios::in);

    myfile.write((char *) &val1,sizeof(float));

    myfile.seekp(0l,ios::beg);    // rewind to start of file

    myfile.read((char *)&val2,sizeof(float));

    cout << val2 << "\n";

    myfile.close();
    return(0);
}
```

write

The write member function is similar to the write() function in C. It writes a specified number of bytes from memory into a file. C++ Program 12.8 gives an example of writing a 10 element array to a file and then reading it back.

Program 12.8
```
#include <fstream.h>

int main(void)
{
int     arr[10]={1,2,3,4,5,6,7,8,9,10},val,i;
fstream  outfile,infile;

    outfile.open("c:\\out.dat",ios::binary | ios::out);

    cout << "Writing to file\n";
```

```
for (i=0;i<10;i++)
{
    outfile.write((char *)&arr[i],sizeof(int));
}
outfile.close();

infile.open("c:\\out.dat",ios::binary | ios::in);

cout << "Reading from file\n";

for (i=0;i<10;i++)
{
    infile.read((char *)&val,sizeof(int));
    cout << val << "\n";
}
infile.close();

return(0);
}
```

12.5 Exercises

12.5.1 Modify C++ Program 12.2 so that it implements a minimum function.

12.5.2 Using a function template write a program which will calculate the current in a circuit where the voltage and resistance can either be ints, floats or doubles. The current is given by:

$$I = \frac{V}{R} \quad A$$

12.5.3 Write a program which will read the text from a file in.dat and copy it, one character at a time, to the file out.dat.

12.5.4 Write a program in which the user enters 10 floating-point values from the keyboard. The program should then store these to a binary file (file1.dat). Finally the file should be read-back to check its contents.

PART C

C/Pascal
C++
Assembly Language
Visual Basic
HTML/Java
DOS
Windows 3.x
Windows 95
UNIX

Introduction

13.1 Introduction

High-level languages such as C, Pascal, Basic and FORTRAN are useful in representing operations in an algorithm-like manner. These high-level languages tend to hide much of the system operations away from the programmer. In most cases, the operation of the program within the computer is invisible and the location of the data or the program code in memory is also unimportant. This is important in most applications as the programmer can forget about much of the low-level operations and get on with high-level operations such as mathematical calculations, graphics, and so on.

There are three main reasons for learning Assembly Language, these are:

- It increases the understanding of the operation of the computer.
- Assembly Language codes generally can be used to finely tune the operation of the code. This fine tuning often speeds-up the operation of the program. This is because high-level compilers do not always optimize the generated code for speed.
- To gain access to certain hardware operations.

Many programs use a mixture of a high-level language and Assembly Language for machine or time critical operations.

13.2 Basic computer architecture

The main elements of a basic computer system are a central processing unit (or microprocessor), memory, and input/output (I/O) interfacing circuitry. These are connected by means of three main buses: the address bus; the control bus; and the data bus, as illustrated in Figure 13.1. External devices such as a keyboard, display, disk drives, and so on, can connect directly onto the data, address and control bus or can be connected via the I/O interfacing circuit. A bus is a collection of common electrical connections grouped by a single name.

Memory normally consists of RAM (random access memory) and ROM (read only memory). ROM stores permanent binary information, whereas

RAM is a non-permanent memory and will lose its contents when the power is taken away. Applications of this type of memory include running application programs and storing temporary information.

The main controller for the computer is the microprocessor. It fetches binary instructions from memory, decodes these instructions into a series of simple actions and carries out the actions in a sequence of steps which are synchronized by a system clock. To access a location in memory the microprocessor must put the address of the location on the address bus. The contents at this address are then placed on the data bus and the microprocessor reads the data on the data bus. To store data in memory the microprocessor places the data on the data bus. The address of the location in memory is put on the address bus and the data is then read from the data bus into the required memory address location.

The classification of a microprocessor relates to the maximum number of bits it can process at a time, that is their word length. The evolution has gone from 4-bit, 8-bit, 16-bit, 32-bit and to 64-bit architectures.

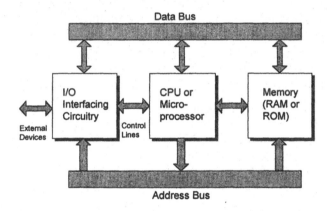

Figure 13.1 Block diagram of a simple computer system

13.3 Bits and bytes

A computer operates on binary digits named bits. These can either store a '1' or a '0' (ON/ OFF). A group of 4 bits is a nibble and a group of 8 bits a byte. These 8-bits provide 256 different combinations of ON/OFF, from 00000000 to 11111111. A 16-bit field is known as a word and a 32-bit field as a long word. Binary data is stored in memories which are either permanent or non-permanent. This data is arranged as bytes and each byte has a different memory address, as illustrated in Figure 13.2.

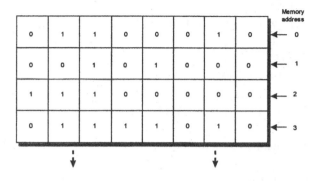

Figure 13.2   Memory storage (each address holds eight bits).

13.3.1 Binary numbers

A computer operates on binary digits which use a base-2 numbering system. To determine the decimal equivalent of a binary number each column is represented by 2 raised to the power of 0, 1, 2, and so on. For example, the decimal equivalents of 1000 0001 and 0101 0011 are:

2^7	2^6	2^5	2^4	2^3	2^2	2^1	2^0	
128	64	32	16	8	4	2	1	Decimal
1	0	0	0	0	0	0	1	129
0	1	0	1	0	0	1	1	83

Thus 01001111 gives:

$$(0\times128) + (1\times64) + (0\times32) + (0\times16) + (1\times8) + (1\times4) + (1\times2) + (1\times1) = 95$$

The number of decimal values that a binary number can represent relates to the number of bits. For example:

- 8 bits gives 0 to $2^8 - 1$ (255) different representations;
- 16 bits gives 0 to $2^{16} - 1$ (65 535) different representations;
- 32 bits gives 0 to $2^{32} - 1$ (4 294 967 295) different representations.

The most significant bit (msb) is at the left-hand side of the binary number and the least significant bit (lsb) on the right-hand side. To convert from decimal (base-10) to binary the decimal value is divided by 2 recursively and remainder noted. The first remainder gives the least significant digit (LSD) and the last the most significant digit (MSD). For example:

```
2 | 54
  | 27      r 0 <<< LSD
  | 13      r 1
  | 6       r 1
  | 3       r 0
  | 1       r 1
  | 0       r 1 <<< MSD
```

Thus 110110 in binary is 54 decimal.

13.4 Binary arithmetic

The basic binary addition operation is given next.

$$0 + 0 = 0 \qquad 1 + 0 = 1 \qquad 1 + 1 = 10 \quad 1+1+1 \;\; = 11$$

This is used when adding two binary numbers together. For example:

```
    0010001
    0001111
    0100000
     11111
```

13.5 Numbers and representations

Numbers are stored in several different ways. These can be:

- integers or floating point values.
- single precision or double precision numbers.
- signed or unsigned integers.

13.5.1 Negative numbers

Signed integers use a notation called 2s complement to represent negative values. In this representation the binary digits have a '1' in the most significant bit column if the number is negative, else it is a '0'. To convert a decimal value into 2s complement notation, the magnitude of the negative number is represented in binary form. Next, all the bits are inverted and a '1' is added. For example, to determine the 16-bit 2s complement of the value -65, the following steps are taken:

```
+65       00000000 01000001
invert    11111111 10111110
add 1     11111111 10111111
```

Thus, –65 is 11111111 1011111 in 16-bit 2s complement notation. Table 13.1 shows that with 16 bits the range of values that can be represented in 2s complement is from –32 767 to 32 768 (that is, 65 536 values).

Table 13.1   16-bit 2s complement notation

Decimal	2s complement
–32 768	10000000 00000000
–32 767	10000000 00000001
::::	::::
–2	11111111 11111110
–1	11111111 11111111
0	00000000 00000000
1	00000000 00000001
2	00000000 00000010
::::	::
32 766	01111111 11111110
32 767	01111111 11111111

When subtracting one value from another the value to be taken away is first converted into 2s complement format. This is then added to the other value and the result is in 2s complement. For example, to subtract 42 from 65, first 42 is converted into 2s complement (that is, –42) and added to the binary equivalent of 65. The result gives a carry into the sign bit and a carry-out.

```
  65            0100 0001
 -42            1101 0110
           (1)  0001 0111
```

For a 16-bit signed integer can vary from –32768 (1000 0000 0000 0000) to 32767 (0111 1111 1111 1111). A simple C program to convert from a 16-bit signed integer to 2s complement binary is given in Program 13.1.

Program 13.1
```
/* Program: int2bin.c                                        */
/* Program to convert from a 16-bit signed integer to 2s     */
/* complement binary                                         */
#include <stdio.h>
int   main(void)
{
unsigned  int  bit;
int            val;
   do
   {
      printf("Enter signed int(-32768 to 32767)>>");
      scanf("%d",&val);
      printf("Binary value is ");
      for (bit=0x8000;bit>0;bit>>=1)
      {
         if (bit & val) printf("1");
```

```
        else printf("0");
    }
    printf("\n");
} while (val!=0);
return(0);
}
```

A sample run is given in Test run 13.1.

🖥 **Test run 13.1**
```
Enter signed integer (-32768 to 32767)>>-1
Binary value is 1111111111111111
Enter signed integer (-32768 to 32767)>>-2
Binary value is 1111111111111110
Enter signed integer (-32768 to 32767)>>-3
Binary value is 1111111111111101
Enter signed integer (-32768 to 32767)>>1
Binary value is 0000000000000001
Enter signed integer (-32768 to 32767)>>2
Binary value is 0000000000000010
Enter signed integer (-32768 to 32767)>>100
Binary value is 0000000001100100
Enter signed integer (-32768 to 32767)>>-32767
Binary value is 1000000000000001
Enter signed integer (-32768 to 32767)>>-32768
Binary value is 1000000000000000
Enter signed integer (-32768 to 32767)>>32767
Binary value is 0111111111111111
Enter signed integer (-32768 to 32767)>>1
Binary value is 0000000000000001
Enter signed integer (-32768 to 32767)>>0
Binary value is 0000000000000000
```

13.5.2 Hexadecimal and octal numbers

In assembly language binary numbers are represented with a proceeding b, for example 010101111010b and 101111101010b are binary numbers. Binary digits are often commonly represented in hexadecimal (base 16) or octal (base 8) representation. Table 13.2 shows the basic conversion between decimal, binary, octal and hexadecimal numbers. In assembly language hexadecimal numbers have a proceeding h and octal number an o. For example, 43F1h is a hexadecimal value whereas 4310o is octal.

To represent a binary digit as a hexadecimal value the binary digits are split into groups of four bits (starting from the least significant bit). A hexadecimal equivalent value then replaces each of the binary groups. For example, to represent 0111010111000000b the bits are split into sections of 4 to give:

Binary	0111	0101	1100	0000
Hex	7	5	C	0

Table 13.2 Decimal, binary, octal and hexadecimal conversions

Decimal	Binary	Octal	Hex
0	0000	0	0
1	0001	1	1
2	0010	2	2
3	0011	3	3
4	0100	4	4
5	0101	5	5
6	0110	6	6
7	0111	7	7
8	1000	10	8
9	1001	11	9
10	1010	12	A
11	1011	13	B
12	1100	14	C
13	1101	15	D
14	1110	16	E
15	1111	17	F

Thus, 75C0h represents the binary number 0111010111000000b. To convert from decimal to hexadecimal the decimal value is divided by 16 recursively and each remainder noted. The first remainder gives the least significant digit and the final remainder the most significant digit. For example, the following shows the hexadecimal equivalent of the decimal number 1103:

```
16 | 1103
        68   r F  <<< LSD (least significant digit)
         4   r 4
         0   r 4  <<< MSD (most significant digit)
```

Thus the decimal value 1103 is equivalent to 044Fh.

In C, hexadecimal values are preceded by a 0 (zero) and the character 'x' (0x) and an octal number a preceding zero 0 (zero). In Pascal, a dollar sign is used to signify a hexadecimal value, for example $C4. Table 13.3 gives some example formats.

Table 13.3 Decimal, binary, octal and hexadecimal conversions

Assembly language representation	Pascal representation	C representation	Base of number
F2Ch	$F2C	0xF2C	hexadecimal
432o		0432	octal
321	321	321	decimal
1001001b			binary

13.6 Memory addressing size

The size of the address bus indicates the maximum addressable number of bytes. Table 13.4 shows the size of addressable memory for a given address bus size. For example:

- A 1-bit address bus can address up to two locations (that is 0 and 1).
- A 2-bit address bus can address 2^2 or 4 locations (that is 00, 01, 10 and 11).
- A 20-bit address bus can address up to 2^{20} addresses (1 MB).
- A 24-bit address bus can address up to 16 MB.
- A 32-bit address bus can address up to 4 GB.

Table 13.4 Addressable memory (in bytes) related to address bus size

Address bus size	Addressable memory (bytes)	Address bus size	Addressable memory (bytes)
1	2	15	32K
2	4	16	64K
3	8	17	128K
4	16	18	256K
5	32	19	512K
6	64	20	1M†
7	128	21	2M
8	256	22	4M
9	512	23	8M
10	1K*	24	16M
11	2K	25	32M
12	4K	26	64M
13	8K	32	4G‡
14	16K	64	16GG

* 1K represents 1024
† 1M represents 1 048 576 (1024 K)
‡ 1G represents 1 073 741 824 (1024 M)

13.7 Exercises

13.7.1 Complete the table below (assume 8-bit unsigned binary values):

Decimal	Binary (8-bit)
12	
235	
128	
255	
	1011 0110
	0110 1001

13.7.2 Determine the binary or decimal equivalents for the following values (assume 16-bit values and 2s complement notation):

Decimal	Binary
32	
354	
−34	
−128	
	1000 0110 1111 1111
	0000 0000 0000 0101
	1111 1111 1111 1000
−32651	
15481	
	1111 0000 0000 0000

13.7.3 Complete the table below (assume 16-bit signed integers and 2s complement):

Decimal	Hex	Octal	Binary
13			
95			
−432			
	f1		
	f2de		
	10bc		
		432	
		7776	
		120	
			1000 1111 0110 1010
			0000 1001 0011 0110

13.7.4 Determine the following subtractions using 2s complement notation (assume 8-bit values and the answer should be in 2s complement):

(i)   43–21
(ii) 12–46
(iii) 127–126
(iv) 0–72
(v)   32+75

14 Computer Architecture

14.1 History of the PC

In 1959, IBM built the first commercial transistorized computer named the IBM 7090/7094 series. It was so successful that it dominated the computer market for many years. Later, in 1965, they produced the famous IBM system 360 which was built with integrated circuits. Then in 1970 IBM introduced the 370 system, which included semiconductor memories. Unfortunately, these computers were extremely expensive to purchase and maintain.

Around the same time the electronics industry was producing cheap pocket calculators. The development of affordable computers happened when the Japanese company, Busicon, commissioned a small, at the time, company named Intel to produce a set of eight to twelve ICs for a calculator. Instead of designing a complete set of ICs, Intel produced a set of ICs which could be programmed to perform different tasks. These were the first ever microprocessors. Soon Intel (short for *Int*egrated *El*ectronics) produced a general-purpose 4-bit microprocessor, named the 4004, and a more powerful 8-bit version, named the 8080. Other companies, such as Motorola, MOS Technologies and Zilog were soon also making microprocessors.

IBM's virtual monopoly on computer systems soon started to slip as many companies developed computers based around the newly available 8-bit microprocessors, namely MOS Technologies 6502 and Zilog's Z-80. IBM's main contenders were Apple and Commodore who introduced a new type of computer – the personal computer (PC). The leading systems were the Apple I and the Commodore PET. These spawned many others, including the Sinclair ZX80/ZX81, the Sinclair Spectrum, the Commodore Vic-20 and the classic Apple II (all of which where based on or around the 6502 or Z-80).

IBM realized the potential of the microprocessor and used Intel's 16-bit 8086 microprocessor in their version of the PC. It was named the IBM PC and has since become the parent of all the PCs ever produced. IBM's main aim was to make a computer which could run business applications, such as word processors, spreadsheets and databases. To increase the production of this software they made information on the hardware freely available. This resulted in many software packages being developed and helped clone manufacturers to copy the original design. So the term 'IBM-compatible' was born and it quickly became an industry standard by sheer market dominance.

14.2 Intel microprocessors

Intel marketed the first microprocessor, named the 4004, and it caused a revolution in the electronics industry because previous electronic systems had a fixed functionality. With this processor the functionality could be programmed by software. It could handle just four bits of data at a time (a nibble), contained 2,000 transistors, operated with 46 instructions and allowed 4 KB of program code and 1 KB of data.

The second generation of Intel microprocessors began in 1974 with the 8-bit processors; these were named the 8008, 8080 and the 8085. As they could handle more bits at a time they were much more powerful than the previous 4-bit devices. They were typically used in early microcomputers and applications such as electronic instruments and printers. The 8008 had a 14-bit address bus and could thus address up to 16 KB of memory and the 8080 had a 16-bit address bus thus giving it a 64 KB limit.

The third generation of microprocessors began with the launch of the 16-bit processors. Intel released the 8086 microprocessor, which was mainly an extension to the original 8080 processor, and thus retained a degree of software compatibility. IBM's designers realized the power of the 8086 and used it in the original IBM PC and IBM XT (eXtended Technology). It had a 16-bit data bus and a 20-bit address bus, giving a maximum addressable capacity of 1 MB. It could also handle either 8 or 16 bits of data at a time (although in a messy way). The PC has evolved since using Intel processors.

A stripped-down 8-bit external data bus version called the 8088 was also available. This stripped down processor allowed designers to produce less complex (and cheaper) computer systems. An improved architecture version of the 8088/88, called the 80286, was launched in 1982, and was used in the IBM AT (Advanced Technology).

In 1985, Intel introduced its first 32-bit microprocessor, the 80386DX. This device was compatible with the previous 8088/8086/80286 (80X86) processors and gave excellent performance. It could handle 8, 16 or 32 bits at a time and had a full 32-bit data and address buses. This gave it a physical addressing capability of 4 GB. A stripped-down 16-bit external data bus and 24-bit address bus version called the 80386SX was released in 1988. Thus, because of its limited address bus size, it could only access up to 16 MB of physical memory.

In 1989, Intel introduced the 80486DX which was basically an improved 80386DX with a memory cache and math co-processor integrated onto the chip. It had an improved internal structure making it around 50% faster than a comparable 80386. The 80486SX was also introduced, which was merely a 80486DX with the link to the math co-processor broken. A major limiting factor on the speed of the processor became the speed of the system clock. For this reason clock doublers and treblers where released allowed the processor to use a higher clock speed than the system clock. Thus internal operations within the processors were much faster but the processor had to slow down to the system clock to communicate with external devices. Typically, systems with clock doubler processors operated around 75% faster than the comparable non-doubled processors (because much of the operation within the computer is done with the processor). Typical clock doubler processors are DX2-66 and DX2-50 which run from 33 MHz and 25 MHz clocks, respectively. Intel also produced a range of microprocessors that run at three or four times the system clock speed and are referred to as DX4 processors. These include the Intel DX4-100 (25 MHz clock) and Intel DX4-75 (25 MHz clock).

The Pentium (or P-5) is a 64-bit 'superscalar' processor. It can execute more than one instruction at a time and has a full 64-bit (8-byte) data bus and a 32-bit address bus, and can operate at speeds from 75 MHz to over 200 MHz (which runs from the 66 MHz system clock). In terms of performance, it operates almost twice as fast as the equivalent 80486. It also has improved floating-point operations (roughly three times faster) and is fully compatible with previous 80x86 processors. Figure 14.1 shows how Intel processors interface to external equipment.

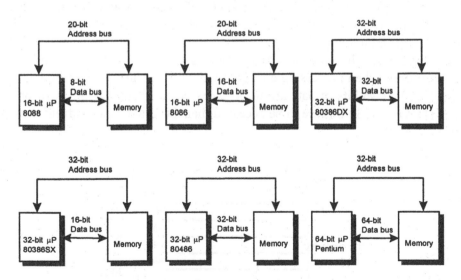

Figure 14.1   Intel microprocessors and their external interfacing

14.3 80386/ 80486 microprocessor

Figure 14.2 shows the main 80386/80486 processor connections. The Pentium processor connections are similar but it has a 64-bit data bus. There are

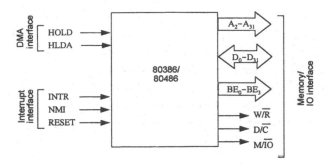

Figure 14.2   Some of the 80386/80486 signal connections.

three main interface connections: the memory/IO interface, interrupt interface and DMA (direct memory access) interface.

The write/read ($W / \overline{R}$) line determines whether data is written to (W) or read from ($\overline{R}$) memory. PCs can interface directly with memory or can interface to isolated memory. Signal line $M / \overline{IO}$ differentiates between the two types. If it is high then the direct memory is addressed, else if it is low then the isolated memory is accessed.

The 80386DX and 80486 have an external 32-bit data bus (D_0-D_{31}) and a 32-bit address bus ranging from A_2 to A_{31}. The two lower address lines, A_0 and A_1, are decoded to produce the byte enable signals $\overline{BE0}$, $\overline{BE1}$, $\overline{BE2}$ and $\overline{BE3}$. The $\overline{BE0}$ line activates when A_1A_0 is 00, $\overline{BE1}$ activates when A_1A_0 is 01, $\overline{BE2}$ activates when A_1A_0 and $\overline{BE3}$ activates when A_1A_0 is 11. Figure 14.3 illustrates this addressing.

The byte enable lines are also used to access either 8, 16, 24 or 32 bits of data at a time. When addressing a single byte, only the $\overline{BE0}$ line will be active (D_0-D_7); if 16 bits of data are to be accessed then $\overline{BE0}$ and $\overline{BE1}$ will be active (D_0-D_{15}); if 32 bits are to be accessed then $\overline{BE0}$, $\overline{BE1}$, $\overline{BE2}$ and $\overline{BE3}$ are active (D_0-D_{31}).

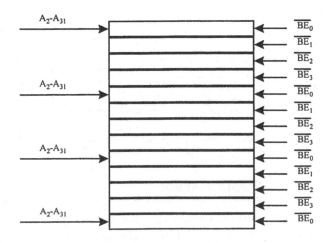

Figure 14.3   Memory addressing

The D/C̄ line differentiates between data and control signals. When it is high then data is read from or written to memory, else if it is low then a control operation is indicated, such as a shutdown command.

The interrupt lines are interrupt request (INTR), non-maskable interrupt request (NMI) and system reset (RESET), all of which are active high signals. The INTR line is activated when an external device, such as a hard disk or a serial port, wishes to communicate with the processor. This interrupt is maskable and the processor can ignore the interrupt if it wants. The NMI is a non-maskable interrupt and is always acted-on. When it becomes active the processor calls the non-maskable interrupt service routine. The RESET signal causes a hardware reset and is normally made active when the processor is powered-up.

14.4 Registers

All the PC-based Intel microprocessors are compatible with the original 8086 process and are normally backwardly compatible. Thus, for example, a Pentium can run 8086 and 80386 code. Microprocessors use registers to perform their operations. These registers are basically special memory locations in that they are given names. The 8086 has 14 registers which are grouped into four categories, as illustrated in Figure 14.4.

14.4.1 General purpose registers

There are four general purpose registers which are AX, BX, CX and DX. Each can be used to manipulate a whole 16-bit word or with two separate 8-bit bytes. These bytes are called the lower and upper order bytes. Each of these registers can be used as two 8-bit registers; for example, AL represents an 8-bit register which is the lower half of AX and AH represents the upper half of AX.

The AX register is the most general purpose of the four registers and is usually used for all types of operations. Each of the other registers has one or more implied extra functions. These are:

- AX, which is named the accumulator. It is used for all input/output operations and some arithmetic operations. For example, multiply, divide and translate instructions assume the use of AX.
- BX, which is named the base register. It can be used as an address register.
- CX, which is the count register. It is used by instructions which require to count. Typically it is used for controlling the number of times a loop is repeated and in bit shift operations.
- DX, which is the data register. It is used for some input/output and also when multiplying and dividing.

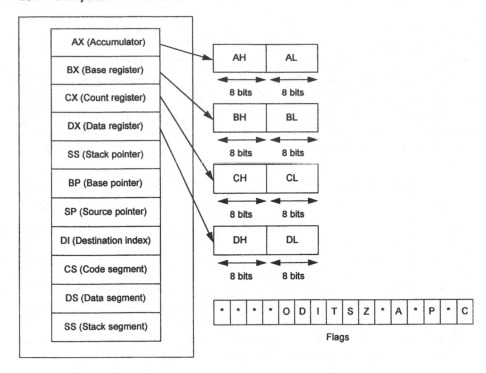

Figure 14.4 8086/88 registers

14.4.2 Addressing registers

The addressing registers are used in memory addressing operations, such as holding the source address of the memory and the destination address. These address registers are named BP, SP, SI and DI, which are:

- SI, which is the source index. This is used with extended addressing commands.
- DI, which is the destination index. The destination is used in some addressing modes.
- BP, which is the base pointer.
- SS, which is the stack pointer.

14.4.3 Status registers

Status registers are used to test for various conditions in an operation, such as 'is the result negative', 'is the result zero', and so on. The two status registers have 16 bits and are called the instruction pointer (IP) and the flag register (F):

- IP, which is the instruction pointer. The IP register contains the address of the next instruction of the program.

Table 14.1 Processor flags

Bit	Flag position	Name	Description
C	0	Set on carry	Contains the carry from the most significant bit (left hand bit) following a shift, rotate or arithmetic operation.
A	4	Set on 1/2 carry	
S	7	Set on negative result	Contains the sign of an arithmetic operation (0 for positive, 1 for negative).
Z	6	Set on zero result	Contains results of last arithmetic or compare result (0 for non-zero, 1 for zero).
O	11	Set on overflow	Indicates that an overflow has occurred in the most significant bit from an arithmetic operation.
P	2	Set on even parity	
D	10	Direction	
I	9	Interrupt enable	Indicates whether the interrupt has been disabled.
T	8	Trap	

- Flag register. The flag register holds a collection of 16 different conditions. Table 14.1 outlines the most used flags.

14.4.4 Segments registers

There are four areas of memory called segments, each of which have 16 bits and can thus address up to 64 KB (from 0000h to FFFFh). These segments are:

- Code segment (cs register). This defines the memory location where the program code (or instructions) is stored.
- Data segment (ds register). This defines where data from the program will be stored (ds stands for data segment register).
- Stack segment (ss register). This defines where the stack is stored.
- Extra segment (es).

All addresses are with reference to the segment registers.

The 8086 has a segmented memory, these registers are used to manipulate these segments. Each segment provides 64 KB of memory, this area of memory is known as the current segment. Segmented memory will be discussed in more detail in the next section.

14.4.5 Memory addressing

There are several methods of accessing memory locations, these are:

- Implied addressing which uses an instruction in which it is known which registers are used.
- Immediate (or literal) addressing uses a simple constant number to define the address location.
- Register addressing which uses the address registers for the addressing (such as AX, BX, and so on).
- Memory addressing which is used to read or write to a specified memory location.

14.5 Memory segmentation

The 80386, 80486 and Pentium processors run in one of two modes, either virtual or real. When using the virtual mode they act as a pseudo-8086 16-bit processor, known as the protected mode. In the real-mode they can use the full capabilities of their address and data bus. The mode and their addressing capabilities depend on the software and thus all DOS-based programs use the virtual mode.

The 8086 has a 20-bit address bus so that when the PC is running 8086-compatible code it can only address up to 1 MB of memory. It also has a segmented memory architecture and can only directly address 64 KB of data at a time. A chunk of memory is known as a segment and hence the phrase 'segmented memory architecture'.

Memory addresses are normally defined by their hexadecimal address. A 4-bit address bus can address 16 locations from 0000b to 1111b. This can be represented in hexadecimal as 0h to Fh. An 8-bit bus can address up to 256 locations from 00h to FFh.

Two important addressing capabilities for the PC relate to a 16- and a 20-bit address bus. A 16-bit address bus addresses up to 64 KB of memory from 0000h to FFFFh and a 20-bit address bus addresses a total of 1 MB from 00000h to FFFFFh. The 80386/80486/Pentium processors have a 32-bit address bus and can address from 00000000h to FFFFFFFFh.

A memory location is identified with a segment and an offset address and the standard notation is segment:offset. A segment address is a 4-digit hexadecimal address which points to the start of a 64 kB chunk of data. The offset is also a 4-digit hexadecimal address which defines the address offset from the segment base pointer. This is illustrated in Figure 14.5.

The segment:offset address is defined as the logical address, the actual physical address is calculated by shifting the segment address 4 bits to the left and adding the offset. The example given next shows that the actual address of 2F84:0532 is 2FD72h.

Segment (2F84):	0010	1111	1000	0100	0000
Offset (0532):		0000	0101	0011	0010
Actual address:	0010	1111	1101	0111	0010

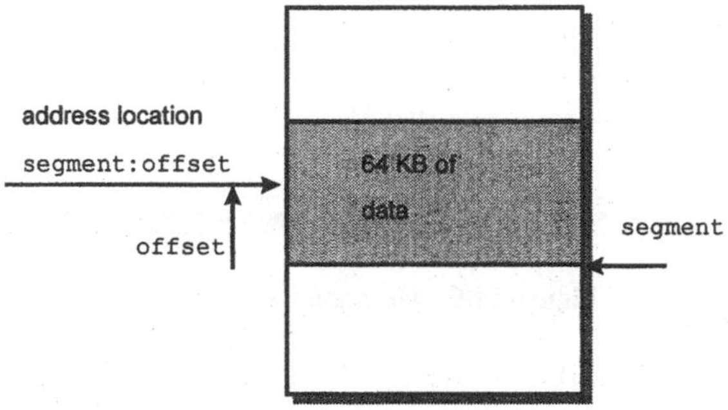

Figure 14.5   Memory addressing

14.5.1 Accessing memory using C and Pascal

In C the address 1234:9876h is specified as 0x12349876. Turbo Pascal accesses a memory location using the predefined array mem[] (to access a byte), memw[] (a word) or memw[] (a long integer). The general format is mem[segment:offset].

14.5.2 Near and far pointers

A near pointer is a 16-bit pointer which can only be used to address up to 64 KB of data whereas a far pointer is a 20-bit pointer which can address up to 1 MB of data. A far pointer can be declared using the far data type modifier, as shown next.

```
char    far *ptr;      /* declare a far pointer         */
ptr=(char far *) 0x1234567;/*initialize far pointer     */
```

In the program shown in Figure 14.6 a near pointer ptr1 and a far pointer ptr2 have been declared. In the bottom part of the screen the actual addresses stored in these pointers are displayed. In this case ptr1 is DS:1234h and ptr2 is 0000:1234h. Notice that the address notation of ptr1 is limited to a 4-digit hexadecimal address, whereas ptr2 has a segment:offset address. The address of ptr1 is in the form DS:XXXX where DS (the data segment) is a fixed address in memory and XXXX is the offset.

```
 File  Edit  Run  Compile  Project  Options  Debug  Break/watch
──────────────────────────────── Edit ────────────────────────────
      Line 13    Col 9    Insert Indent Tab Fill Unindent * C:NEW.C
#include <stdio.h>

int     main(void)
{
int     *ptr1;
int     far *ptr2;

        ptr1=(int *)0x1234;
        ptr2=(int far *)0x1234;

        printf("Pointer 1 is %p\n",ptr1);
        printf("Pointer 2 is %p\n",ptr2);

        return(0);
}

──────────────────────────────── Watch ───────────────────────────
•ptr2: 0000:1234
 ptr1: DS:1234
 F1-Help F5-Zoom F6-Switch F7-Trace F8-Step F9-Make F10-Menu
```

Figure 14.6   Near and far pointers

There are several modes in which the compiler operates. In the small model the compiler declares all memory addresses as near pointers and in the large model they are declared as far pointers. Figure 14.7 shows how the large memory model is selected in Borland C (Options→ Compiler→ Model→ Large). The large model allows a program to store up to 1 MB of data and code. Normally the small model is the default and only allows a maximum of 64 KB for data and 64 KB for code.

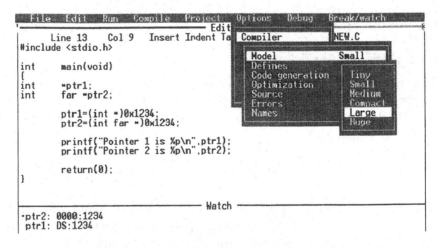

Figure 14.7   Compiling a program in the large model

14.6 View inside the processor

To be able to view the processor the user must use a debugging program.

Figure 14.8 shows an example of Turbo Debugger which is available with most of the Borland software development products and can be used to view the operation of a program. It can be seen that the machine code and equivalent assembly language macro appears in the top left hand window. A sample code line is:

```
cs:01FA→55            push    bp
```

which specifies that the memory location is 01FA in the code segment (cs:01FA). Machine code at this location is 55 (0101 0101) and the equivalent Assembly Language instruction is push bp. Note that the cs segment address in this case is 5757h, thus the actual physical address will be with reference to the address 57570h (see Section 14.5 for a fuller explanation).

```
+-[_]-CPU 80486----------------------------------Ð-------1-[□] [□]-+
¦                                              _   ax 0000    ¦c=0¦
¦   cs:01FA→55          push    bp             _   bx 062A    ¦z=1¦
¦   cs:01FB  8BEC       mov     bp,sp          _   cx 0009    ¦s=0¦
¦   cs:01FD  83EC08     sub     sp,0008        _   dx AB02    ¦o=0¦
¦   cs:0200  56         push    si             _   si 0145    ¦p=1¦
¦   cs:0201  57         push    di             _   di 060A    ¦a=0¦
¦                                              _   bp FFD2    ¦i=1¦
¦   cs:0202  B89401     mov     ax,0194        _   sp FFC8    ¦d=0¦
¦   cs:0205  50         push    ax             _   ds 58A0    ¦   ¦
¦   cs:0206  E8D40B     call    _puts          _   es 58A0    ¦   ¦
¦   cs:0209  59         pop     cx             _   ss 58A0    ¦   ¦
¦                                                  cs 5757    ¦   ¦
¦   cs:020A  B8B501     mov     ax,01B5            ip 01FA    ¦   ¦
Ã□                                          □¦              ¦   ¦
¦   ds:0000 00 00 00 00 54 75 72 62     Turb   ¦              ¦   ¦
¦   ds:0008 6F 2D 43 20 2D 20 43 6F o-C - Co   +--------------Ã
¦   ds:0010 70 79 72 69 67 68 74 20 pyright    ¦ ss:FFCA 0001  ¦
¦   ds:0018 28 63 29 20 31 39 38 38 (c) 1988   ¦ ss:FFC8→011D  ¦
+------------------------------------------¤--------------+
```

Figure 14.8 Example screen from Turbo Debugger

The contents of the flag register are shown on the right-hand side. In this case the flags are:

C=0, Z=1, S=0, O=0, P=1, A=0, I=1 and D=0.

The registers are shown to the left of the flag register. In this case the contents are:

AX=0000h,   BX=062Ah,   CX=0009h,   DX=AB02h,   SI=0145h,   DI=060Ah,
BP=FFD2h,   SP=FFC8h,   DS=58A0h,   ES=58A0h,   SS=58A0h,   CS=5757h,
IP=01FAh.

The data (in the data segment) is shown at the bottom left hand corner of the screen. The first line:

```
ds:0000 00 00 00 00 54 75 72 62     Turb
```

shows the first 8 bytes in memory (from DS:0000 to DS:0007). The first byte in memory is 00h (0000 0000) and the next is also 00h. After the 8 bytes are defined the 8 equivalent ASCII characters are shown. In this case, these are:

```
Turb
```

The ASCII equivalent character for 5A (1001 1010) is 'T' and for 75 (0111 0101) it is 'u'. Note that, in this case, the data segment register has 58A0h. Thus the location of the data will be referenced to the address 58A00h.

The bottom right-hand window shows the contents of the stack.

14.7 Machine code and assembly language

An important differentiation is between machine code and assembly language. The actual code which runs on the processor is machine code. These are made up to unique bit sequences which identifies the command and other values which these commands operate on. For example, for the debugger screen from Figure 14.8, the assembly language line to move a value into the AX register is:

```
mov     ax,0194
```

the equivalent machine code is:

```
B8  94  01
```

where the code B8h (1011 1000b) identifies the instruction to move a 16-bit value into the AX register and the value to be loaded is 0194h (0000 0001 1001 0100b). Note that the reason the 94h value is stored before the 01h value is that on the PC the least significant byte is stored in the first memory location and the most significant byte in the highest memory location. Figure 14.9 gives an example of storage within the code segment. In this case the two instructions are mov and push. In machine code these are B8h and 50h, respectively.

14.8 Exercises

14.8.1   For the debug screen given in Figure 14.10 determine the following:

      (i)   Contents of AX, BX, CX, DX, SI, DI.
      (ii)   Contents of AH, AL, BH and BL.
      (iii)   The first assembly language command.

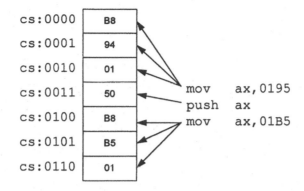

Figure 14.9 Example memory storage for code segment

(iv)   The memory address of the first line of code (Hint: the cs:02C2 and the value in the cs register need to be used).

(v)   The memory address of the data (Hint: the ds:0000 and the value in the ds register need to be used).

```
+-[_]-CPU 80486--------------------------------Ð-------1-[□] [□]-+
¦                                              -  ax 0100  ¦c=0¦
¦  cs:02C2>55          push   bp               _  bx 02CE  ¦z=1¦
¦  cs:02C3 8BEC        mov    bp,sp               cx 0001  ¦s=0¦
¦                                              _  dx 02CE  ¦o=0¦
¦  cs:02C5 B8AA00      mov    ax,00AA             si 02C8  ¦p=1¦
¦  cs:02C8 50          push   ax              _  di 02CE  ¦a=0¦
¦  cs:02C9 E8F609      call   _puts              bp 0000  ¦i=1¦
¦  cs:02CC 59          pop    cx              _  sp FFF8  ¦d=0¦
¦                                                ds 5846  ¦   ¦
¦  cs:02CD 33C0        xor    ax,ax           _  es 5846  ¦   ¦
¦  cs:02CF EB00        jmp    #PROG1_1#10  (_  ss 5846    ¦   ¦
¦                                              cs 5751    ¦   ¦
¦  cs:02D1 5D          pop    bp              _  ip 02C2  ¦   ¦
Ã□_____□¦         ¦   ¦
¦  ds:0000 00 00 00 00 42 6F 72 6C     Borl   ¦          ¦   Â
¦  ds:0008 61 6E 64 20 43 2B 2B 20  and C++   +----------------Â
¦  ds:0010 2D 20 43 6F 70 79 72 69 - Copyri   ¦  ss:FFFA 0000  ¦
¦  ds:0018 67 68 74 20 31 39 39 31  ght 1991  ¦  ss:FFF8 015B  ¦
+----------------------------------------------¤----------------+
```

Figure 14.10 Example screen from Turbo Debugger

15 ⌗ 8086/88 Instructions

15.1 Introduction

This chapter discusses 8086/88 assembly language. As previously mentioned all of the 80X86 and Pentium processors can run 8086 code. The 80386, 80486 and Pentium processors run in one of two modes, either virtual or real. When using the virtual mode they act as a pseudo-8086 16-bit processor, known as the protected mode. All DOS-based programs use the virtual mode.

15.2 Characters and numbers

Integers can be represented as binary, octal, decimal, or hexadecimal numbers; 8086 Assembly Language represents these with a preceding B, O, D or H, respectively. A decimal integer is assumed if there is no letter. Examples of numeric constants are:

```
01001100b   3eh   10b   ffffh   17o
```

Character constants are enclosed with single quotes when they have a fixed number of characters (such as 'b', 'fred', and so on), or if they have a variable number of characters they are enclosed with double quotes (such as "a", "fred", and so on).
   For example:

```
'c'                'Press ENTER'
"Input Value> "    "x"
```

15.3 Comments

Assembly Language programs probably need more comments than high-level language as some of the operations give little information on their purpose. The character used to signify a comment is the semi-colon (;) and all comments within a program are ignored by the assembler. For example, the following lines have comments:

```
                    ; This is a comment
    mov ax,1        ; move instruction will be discussed next
```

15.4 Move (MOV)

The move instruction (mov) moves either a byte (8 bits) or a word (16 bits) from one place to another. There are three possible methods:

• Moving data from a register to a memory location.
• Moving data from a memory location to a register.
• Moving data from one register to another.

Note that in 8086/88 it is not possible to move data directly from one memory location to another using a single instruction. To move data from one memory location to another then first the data is move from the memory location into a register, next it is moved from the register to the destination address.

Examples of moving a constant value into registers are:

```
mov   cx,20         ; moves decimal 20 into cx
mov   ax,10h        ; moves 10 hex into ax
mov   ax,01110110b  ; moves binary 01110110 into ax
```

An address location is identified within square bracket ([ ]). Then to move data into a specified address the address location must be loaded into a register. For example, to load the value of 50h (0101 0000) into address location 200h the following lines are used:

```
mov   bx,200h
mov   [bx],50h      ; load 50 hex into memory location 200h
```

The general format of the mov instruction is:

```
mov r/m , r/m/d       or
mov sr , r16/m16       or
mov r16/m16 , sr
```

Where r/m stands for register (such as AH, AL, BH, BL, CH, CL, DH, DL, AX, BX, CX, DX, BP, SI, DI) or memory location. And r/m/d stands for a register, memory or a constant value. The register sr stands for any of the segment registers (CS, DS, ES, SS) and r16/m16 stand for any 16-bit register (AX, BX, CX, SP, BP, SI, DI) and 16-bit memory address.

15.5 Addressing memory

An address location can be specified with either the BX, BP, SI or DI register. Examples are:

```
[BP]
[SI]
30 [BX]          ; which specifies the address BX+30
[BP+DI]
40 [BX+SI]       ; which specifies the address BX+SI+40
```

Program 15.1 gives an Assembly Language; it loads 1234h into address DS:0000h, 5678h into address DS:0002h and 22h into address DS:0005h.

Program 15.1

```
code segment
    mov bx,0
    mov [bx],1234h
    mov 2[bx],5678h
    mov 5[bx],1122h
    mov  ah,4ch
    int  21h
code ends
end
```

Figure 15.1 shows a sample run of Program 15.1. It can be seen that the mov [bx],1234 operation loads the value 34h into address location DS:0000h and 12h into address DS:0001h. This is because the processor loads the least significant byte into the lower address location.

```
+-[_]-CPU 80486-----------------------------------Ð-------1-[□] [□]-+
¦ cs:0000 BB0000       mov   bx,0000           -  ax 0000  ¦c=0¦
¦ cs:0003 C7073412     mov   word ptr [bx],1234 _  bx 0000  ¦z=0¦
¦ cs:0007 C747027856   mov   word ptr [bx+02],5678_ cx 0000  ¦s=0¦
¦ cs:000C C747052211   mov   word ptr [bx+05],1122_ dx 0000  ¦o=0¦
¦ cs:0011>B44C         mov   ah,4C             _  si 0000  ¦p=0¦
¦ cs:0013 CD21         int   21               _  di 0000  ¦a=0¦
¦ cs:0015 B253         mov   dl,53             _  bp 0000  ¦i=1¦
¦ cs:0017 018B7424     add   [bp+di+2474],cx   _  sp 0100  ¦d=0¦
¦ cs:001B 108B7C24     adc   [bp+di+247C],cl   _  ds 5707  ¦   ¦
¦ cs:001F 1483         adc   al,83             _  es 5707  ¦   ¦
¦ cs:0021 C404         les   ax,[si]           _  ss 5719  ¦   ¦
¦ cs:0023 56           push  si                _  cs 5717  ¦   ¦
¦ cs:0024 E85337       call  377A              _  ip 0011  ¦   ¦
Ã□_____□¦_____
¦ ds:0000 34 12 78 56 00 22 11 FE 4□xV "    _   +----------------Â
¦ ds:0008 1D F0 E0 01 21 24 AA 01 □-ó□!$¬□      ¦ ss:0104 0000    ¦
¦ ds:0010 21 24 89 02 7C 1E EE 0F !$ë□|-‾¤      ¦ ss:0102 3D08    ¦
¦ ds:0018 01 01 01 00 02 FF FF FF □□□ □         ¦ ss:0100>C483    ¦
+--------------------------------------------------¤----------------+
```

Figure 15.1 Sample run of Program 15.1

It can be seen from Figure 15.1 that the associated machine code for the instructions is:

```
BB 0000          mov   bx,0000
C707 3412        mov   word ptr [bx],1234
C747 02 7856     mov   word ptr [bx+02],5678
C747 05 2211     mov   word ptr [bx+05],1122
```

```
B4 4C              mov    ah,4C
```

Thus BBh is the machine code to load a value into the BX register, C707h loads as value into the address pointed to and C747h loads an offseted value into an address location.

15.6 Addition and subtraction (ADD and SUB)

As they imply, the ADD and SUB perform addition and subtraction of two words or bytes. The ADD and SUB instruction operate on two operands and put the result into the first operand. The source or destination can be a register or address. Examples are:

```
mov ax,100
add ax,20    ;adds 20 onto the contents of ax
sub ax,12    ;subtracts 12 from ax and puts result into ax

mov bx,10    ;move 10 into bx
add ax,bx    ;adds ax and bx and puts result into ax
```

The standard format of the add instruction is

```
add r/m , r/m/d
```

where r is any register, m is memory location and d is any constant value.

15.7 Compare (CMP)

The CMP instruction acts like the SUB instruction, but the result is discarded. It thus leaves both operands intact but sets the status flags, such as the O (overflow), C (carry), Z (zero) and S (sign flag). It is typically used to determine if two numbers are the same, or if one value is greater, or less than, another value. Examples are:

```
cmp 6,5      ;result is 1, this sets C=0 S=0; Z=0 O=0
cmp 10,10    ;result is 0, this sets Z=1 the rest as above
cmp 5,6      ;result is -1, this sets negative flag S=1
```

15.8 Unary operations (INC, DEC and NEG)

The unary operations operate on a single operand. An INC instruction increments the operand by 1, the DEC instruction decrements the operand by 1 and the NEG instruction makes the operand negative. Examples are:

```
mov al,10
inc al       ; adds 1 onto AL, AL will thus store 11
inc al       ; AL now stores 12
dec al       ; takes 1 away from AL (thus it will be equal to 11)
neg al       ; make AL negative, thus AL stores -11
```

15.9 Boolean bitwise instructions (AND, OR, XOR and NOT)

The Boolean bitwise instructions operate logically on individual bits. The XOR function yields a 1 when the bits in a given bit position differ; the AND function yields a 1 only when the given bit positions are both 1s. The OR operation gives a 1 when any one of the given bit positions are a 1. These operations are summarised in Table 15.1. For example:

```
        00110011              10101111            00011001
AND     11101110      OR      10111111     XOR    11011111
        00100010              10111111            11000110
```

Table 15.1 Bitwise operations

A	B	AND	OR	EX-OR
0	0	0	0	0
0	1	0	1	1
1	0	0	1	1
1	1	1	1	0

Examples of Assembly Language instructions which use bitwise operations are:

```
mov al,7dh     ;loads 01111101 into al
and al,03h     ; 01111101 AND 00000011 gives 00000001
               ;   al stores 00000001 or 1

mov ax,03f2h   ; loads 0000 0011 1111 0010 into AX
xor ax,ffffh   ; exclusive OR 1111 1111 1111 1111  with AX,
               ; AX now contain 1111 1100 0000 1101
```

15.10 Shift/rotate instructions (SHL, SAL, SHR, SAR, ROL, ROR, RCL and RCR)

The shift/rotate instructions are:

SHL –shift bits left   SHR–shift bits right
SAL –shift arithmetic left   SAR–shift arithmetic right
RCL –rotate through carry left RCR –rotate through carry right
ROL –rotate bits left   ROR –rotate bits right

The shift instructions move the bits, with or without the carry flag, and can either be an arithmetic shift or logical shift, whereas the rotate instructions are cyclic and may involve the carry flag. The SHL and SHR shift bits to the left and right, respectively. They shift the bits to the left or right where the bit shifted out is put into the carry flag and the bit shifted in is a 0. The rotate operations (ROL, ROR, RCL, RCR) are cyclic. Rotate with carry instructions (RCL and RCR) rotate the bits using the carry flag. Thus the bit shifted out is put into the carry flag and the bit shift in is taken from the carry flag. The rotate bits (ROL and ROR) rotate the bits without the carry flag. The SAL instruction is identical to SHL, but the SAR instruction differs from SHR in that the most significant bit is shifted to the right for each shift operation. This operation, and the others, are illustrated in Figure 15.2.

The number of shifts on the value is specified either as a unitary value (1) or the number of shift is stored in the counter register (CL). The standard format is

```
SAR r/m, 1/CL    SAL r/m, 1/CL    SHR r/m, 1/CL    SHL r/m, 1/CL
ROR r/m, 1/CL    ROL r/m, 1/CL    RCR r/m, 1/CL    RCL r/m, 1/CL
```

where r/m is for register or memory and 1 stands for one shift. If any more than one shift is required the CL register is used. These operations take a destination and a counter value stored in CL. For example, with bit pattern:

Initial conditions:

01101011 and carry flag 1

Result after:

SHR 00110101 CF 1 → SHL 11010110 CF 0 →
SAR 00110101 CF 1 → SAL 11010110 CF 0 →
ROR 10110101 CF 0 → RCR 10110101 CF 1 →
RCL 11010111 CF 0

The following is an example of the SAR instruction:

```
mov cl,03           ; Contents of AX
mov ax,10110111b    ; (10110111b)
sar ax,1            ; shifted one place to the right (0101 1011b)
sar ax,cl           ; shifted three places to the right (0000 0101b)
```

After sar ax,1 stores 005Bh (0000 0000 0101 1011b) then the sar ax,cl instruction moves the contents of AX by 3 bit positions to the right. The contents of AX after this operation will be 0005h (0000 0000 0000 0101b).

The following shows an example of the SHR instruction:

```
mov cl,03           ; Contents of AX
mov ax,10110111b    ; (10110111b)
shr ax,1            ; Shift right one (01011011)
shr ax,cl           ; shift right three (00001011)
```

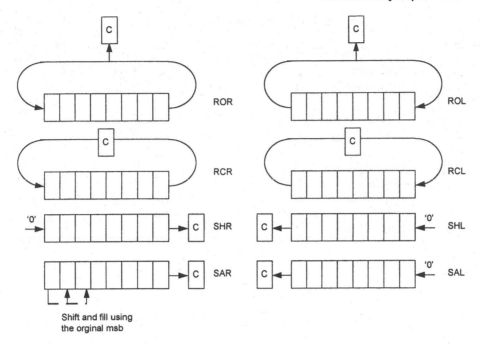

Figure 15.2 Rotate operations

And an example of the ROR instruction:

```
mov cl,03          ; Contents of BX are :
mov bl,10110111b   ; 10110111
ror bl,1           ; rotate one place to the right(1101 1011b)
ror bl,cl          ; rotate three places to the right(0111 1011b)
```

15.11 Unconditional jump (JMP)

The JMP instruction transfers program execution to another part of the program. It uses a label to identify the jump location; this is defined as a name followed by a colon. The JMP instruction is not conditional – the program will always jump. An example is given next:

```
        mov al,10101010b
        jmp nextst
          : :
          : :
nextst: mov bx,10
```

15.12 Conditional jumps

With the JMP the program always goes to the label, but the unconditional jumps will only branch if a certain condition is met, such as if the results is negative, or the result is zero, and so on. Table 15.2 outlines the condition jump instructions.

Table 15.2   Conditional jump instructions

Name	Description	Flag tests	Name	Description	Flag tests
JC	jump if carry	C=1	JZ	jump if zero	Z=1
JS	jump if sign	S=1	JNC	jump if not carry	C=0
JNS	jump is not sign	S=0	JL	jump if less	
JNZ	jump if not zero	Z=0	JLE	jump if less or equal	
JGE	jump if greater or equal	S=Ov	JG	jump if greater	Z=0 S=Ov
JA	jump if above	C=0 and Z=0	JNB	jump if not below	
JB	jump if below				

A few example are:

```
        mov   al,11
        cmp   al,10
        jle   fred  ; last operation was less than 10
                    ; then branch to label fred

fred2:  mov   ax,300
        sub   ax,1000 ; subtract 1000 from 300
        jg fred5      ; no jump since result was not greater

fred:   sub al,12    ;
        cmp al,0     ;
        jz fred2     ;no jump since not equal to zero
```

15.13 Subroutine calls (CALL and RET)

Subroutines allow a section of code to be called and for the program to return back to where it was called. The instructions are CALL and RET. An example is given next:

```
        call  fred ; Goto fred routine
        add   al,bl
          : :
          : :
fred:   mov   al,00h
        add   al,bl
        ret          ; return to place that called
```

15.14 Push and pop

The PUSH and POP instructions are typically used with subroutines. A PUSH instruction puts the operand onto a temporary storage called a stack (this will be covered in more detail later). The stack is a LIFO (last in, first out) where the last element to be loaded is the first to be taken off, and so on. The POP instruction is used to extract the last value which was put on the stack.

Typically they are used to preserve the contents of various registers so that their contents are recovered after a subroutine is called. For example, if a subroutine modifies the AX, BX and CX registers, then the registers are put on the stack with:

```
PUSH AX
PUSH BX
PUSH CX
```

Next the subroutine can use these registers for its own use. Finally, within the subroutine, the original registers are restored with:

```
POP CX
POP BX
POP AX
```

The order of the POP instructions must be the reverse of the PUSH instructions so that the contents are properly restored. For example:

```
        call sub1
        add   al,bl
          : :
          : :
sub1:   push ax
        push bx
        mov   ax,1111h
        mov   bx,1111h
        add   ax,bx
          : :
          : :
        pop bx
        pop ax
        ret             ; return to place that called
```

15.15 Moving around data in memory

Program 15.2 loads the memory locations from DS:0000h to DS:00FFh with values starting at 00h and ending at FFh. After the AL and BX registers have been initialized to 00h then the code runs round a loop until all the memory locations have been loaded. The BX register contains the address the value will be loaded to. This increments each time round the loop. The AL register stores the value to be loaded into the currently specified memory location. Figure 15.3 shows a sample run.

Program 15.2

```
code    segment
assume cs:code,ds:data

start:
   mov al,00h
```

```
    mov bx,00h
    loop1:
      mov [bx],al
      inc bx
      inc al
      cmp al,0ffh
      jne loop1

    mov  ah,4ch
    int  21h
code       ends
end        start
```

```
+-[_]-CPU 80486----------------------------------Đ-------1-[□] [□]-+
¦ cs:0000 B000          mov     al,00        -  ax 0020   ¦c=1¦
¦ cs:0002 BB0000        mov     bx,0000      _  bx 0020   ¦z=0¦
¦ cs:0005 8807          mov     [bx],al      _  cx 0000   ¦s=0¦
¦ cs:0007 43            inc     bx           _  dx 0000   ¦o=0¦
¦ cs:0008 FEC0          inc     al           _  si 0000   ¦p=0¦
¦ cs:000A>3CFF          cmp     al,FF        _  di 0000   ¦a=1¦
¦ cs:000C 75F7          jne     0005         _  bp 0000   ¦i=1¦
¦ cs:000E B44C          mov     ah,4C        _  sp 0100   ¦d=0¦
¦ cs:0010 CD21          int     21           _  ds 5705   ¦   ¦
¦ cs:0012 50            push    ax           _  es 5705   ¦   ¦
¦ cs:0013 6A01          push    0001         _  ss 5717   ¦   ¦
¦ cs:0015 9AB2467711    call    1177:46B2    _  cs 5715   ¦   ¦
¦ cs:001A FF76FE        push    word ptr [bp -  ip 000A   ¦   ¦
Ã□                                        □¦
¦ ds:0000 00 01 02 03 04 05 06 07  □□□□□□  ¦          ¦   ¦
¦ ds:0008 08 09 0A 0B 0C 0D 0E 0F       ¤  +---------------Â
¦ ds:0010 10 11 12 13 14 15 16 17 □□□□¶$□□  ¦ ss:0102 C033   ¦
¦ ds:0018 18 19 1A 1B 1C 1D 1E 1F □□□□□-   ¦ ss:0100□000C   ¦
+----------------------------------------¤---------------+
```

Figure 15.3   Sample debug screen

15.16 Assembler directives

There are various structure directives that allow the user to structure the program. These are defined in Table 15.3.

Table 15.3   Assembler directives

Directive	Name	Description
SEGMENT and ENDS	segment definition	The SEGMENT and ENDS directives mark the beginning and the end of a program segment. Its general format is: name segment : : : : main program : : name ends end
END	source file end	The END directive marks the end of a module. The assembler will ignore any statements after the end directive.

GROUP	segment group	
ASSUME	segment registers	The ASSUME directive specifies the default segment register name. For example:
		`assume cs:code`
		The general format is
		`assume segmentregister:segmentregistername`
ORG	segment origin	The ORG statement tells the assembler at which location the code should be located.
PROC and ENDP	procedure definition and end	These statement define the start and end of a procedure.

15.17 Data definition

Variables are declared in the data segment. To define a variable the DB (define byte) and DW (define word) macros are used. For example, to define (and initialize) a variable temp, which has the value 15 assigned to it, is declared as follows:

```
temp db 15
```

an uninitialized variable has a value which is a question mark, for example:

```
temp db ?
```

There are other definition types used, these are:

- dd (define doubleword – 2 times 16 bits which is 4 bytes).
- dq (define quadword, which is 8 bytes).
- dt (define 10 bytes).

The data definition is defined within the data segment. In Turbo Assembler (TASM) the data segment is defined after the .DATA directive (as shown in TASM Program 15.4). Microsoft Assembler (MASM) defines the data segment between the data segment and data ends, as shown in MASM Program 15.3.

Program 15.3 declares two variables named val1 and val2. The value val1 is loaded with the value 1234h and val2 is loaded with 5678h. Figure 15.4 shows an example screen after the three mov instructions have been executed. It can be seen val1 has been stored at DS:0000 and val2 at DS:0002.

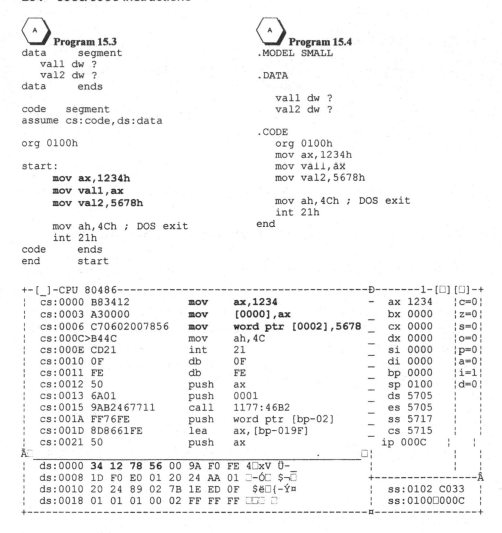

Figure 15.4   Sample debug screen

15.18 Equates (EQU)

To define a token to a certain value the **EQU** (equates) statement can be used. For example:

```
one      equ    1
outA     equ    1f1h
PI       equ    3.14159
prompt   equ    'Type Enter'
```

The general format is:

```
name EQU expression
```

The assembler simply replaces every occurrence of the token with the value given.

15.19 Exercises Part 1

In this tutorial the sample code should be inserted, by replacing the high-lighted code, into Program 15.5.

 **Program 15.5**
```
code segment
   mov bx,0
   mov [bx],1234h
   mov 2[bx],5678h
   mov 5[bx],1122h
   mov  ah,4ch
   int  21h
code ends
end
```

15.19.1 Enter the following code and run the debugger to determine the values given next.

```
mov al,54h
mov bl,36h
add al,bl
```

AL
Carry flag
Overflow flag
Sign flag
Zero flag

15.19.2 Enter the following code and run the debugger to determine the values given below.

```
mov al,54h
mov bl,36h
sub al,bl
```

AL
Carry flag
Overflow flag
Sign flag
Zero flag

15.19.3 Enter the following code and run the debugger to determine the values given below.

```
mov al,47h
```

```
mov bl,62h
sub al,bl
```

AL
Carry flag
Overflow flag
Sign flag
Zero flag

15.19.4 Enter the following code and run the debugger to determine the values given below.

```
mov al,54h
mov bl,36h
and al,bl
```

AL
Carry flag
Overflow flag
Sign flag
Zero flag

15.19.5 Enter the following code and run the debugger to determine the values given below.

```
mov al,73h
mov bl,36h
xor al,bl
```

AL
Carry flag
Overflow flag
Sign flag
Zero flag

15.19.6 Enter the following code and run the debugger to determine the values given below.

```
mov al,54h
not al
```

AL
Carry flag
Overflow flag
Sign flag
Zero flag

15.19.7 Enter the following code and run the debugger to determine the values given below.

```
mov ax,1f54h
mov bx,5a36h
add al,bl
```

AL
Carry flag
Overflow flag
Sign flag
Zero flag

15.19.8 Enter the following code and run the debugger to determine the values given below.

```
mov ax,3a54h
mov bx,0236h
mov cl,3
shr ax,1
shl bx,cl
```

AX
BX
Carry flag
Overflow flag
Zero flag

15.19.9 Enter the following code and run the debugger to determine the values given below.

```
mov ax,3a54h
mov bx,0236h
mov cl,3
sar bx,cl
```

AX
BX
Carry flag
Overflow flag
Zero flag

15.19.10 Enter the following code and run the debugger to determine the values given below.

```
mov ax,3a54h
mov bx,0236h
mov cl,3
ror ax,cl
rcl bx,cl
```

AX
BX
Carry flag
Overflow flag
Zero flag

15.19.11 Enter the following code and run the debugger to determine the values given below.

```
mov al,54h
mov bl,66h
cmp al,bl
```

AL
BL
Carry flag
Overflow flag
Zero flag

15.19.12 Enter the following code and run the debugger to determine the values given below.

```
mov al,32
mov ah,53
mov bx,236
xor ax,bx
```

AX
BX
Carry flag
Overflow flag
Zero flag

15.19.13 Enter the following code and run the debugger to determine the values given below.

```
mov ax,3a54h
mov bx,100h
mov [bx],ax
```

AX
Contents of address 100h
Contents of address 101h
Overflow flag
Zero flag

15.19.14 Enter the following code and run the debugger to determine the values given below.

```
mov ax,3a54h
mov bx,120h
mov 20[bx],ax
```

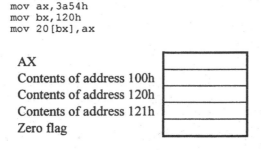

AX	
Contents of address 100h	
Contents of address 120h	
Contents of address 121h	
Zero flag	

15.19.15 Enter the following code and run the debugger to determine the values given below.

```
mov al,'a'
mov ah,'b'
```

AL	
BL	
Carry flag	
Overflow flag	
Zero flag	

15.20 Exercises Part 2

15.20.1 Write an Assembly Language program which contains a function which adds the contents of the AX and BX registers and puts the result into the CX register.

15.20.2 Write a program which loads the values 00h, 01h, 02h,...FEh, FFh into the memory locations starting from address DS:0008h. A basic layout is shown below.

```
loop:
    mov al,00h
    mov bx,08h
     :   :
     :   :
    jne loop
```

15.20.3 Write a program which will load the values FFh, FEh, FDh,...01h, 00h into the memory locations starting from address DS:0000h.

15.20.4 Write a program which moves a block of memory from DS:0020h to 0100h to addresses which start at address 0200h.

15.20.5 Write a program which determines the largest byte in the memory locations 0000h to 0050h.

8086 Interfacing and Timing

16.1 Introduction

There are two main methods of communicating external equipment, either they are mapped into the physical memory and given a real address on the address bus (memory mapped I/O) or they are mapped into a special area of input/output memory (isolated I/O). Figure 16.1 shows the two methods. Devices mapped into memory are accessed by reading or writing to the physical address. Isolated I/O provides ports which are gateways between the interface device and the processor. They are isolated from the system using a buffering system and are accessed by four machine code instructions. The IN instruction inputs a byte, or a word, and the OUT instruction outputs a byte, or a word. C and Pascal compilers interpret the equivalent high-level functions and produce machine code which uses these instructions.

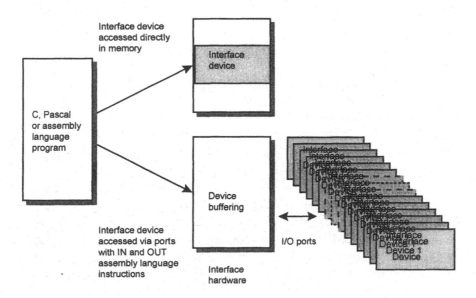

Figure 16.1   Memory mapping or isolated interfacing

16.2 Interfacing with memory

The 80X86 processor interfaces with memory through a bus controller, as shown in Figure 16.2. This device interprets the microprocessor signals and generates the required memory signals. Two main output lines differentiate between a read or a write operation (R/$\overline{W}$) and between direct and isolated memory access (M/$\overline{IO}$). The R/$\overline{W}$ line is low when data is being written to memory and high when data is being read. When M/$\overline{IO}$ is high, direct memory access is selected and when low, the isolated memory is selected.

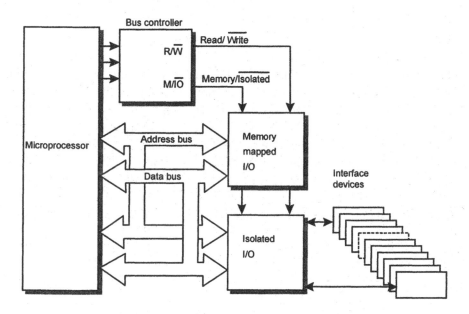

Figure 16.2   Access memory mapped and isolated I/O

16.3 Memory mapped I/O

Interface devices can map directly onto the system address and data bus. In a PC-compatible system the address bus is 20 bits wide, from address 00000h to FFFFFh (1 MB). If the PC is being used in an enhanced mode (such as with Microsoft Windows) it can access the area of memory above the 1 MB. If it uses 16-bit software (such as Microsoft Windows 3.1) then it can address up to 16 MB of physical memory, from 000000h to FFFFFFh. If it uses 32-bit software (such as Microsoft Windows 95) then the software can address up to 4 GB of physical memory, from 00000000h to FFFFFFFFh. Table 16.1 and Figure 16.3 gives a typical memory allocation.

Table 16.1 Memory allocation for a PC

Address	Device
00000h-00FFFh	Interrupt vectors
00400h-0047Fh	ROM BIOS RAM
00600h-9FFFFh	Program memory
A0000h-AFFFFh	EGA/VGA graphics
B0000h-BFFFFh	EGA/VGA graphics
C0000h-C7FFFh	EGA/VGA graphics

16.4 Isolated I/O

Devices are not normally connected directly onto the address and data bus of the computer because they may use part of the memory that a program uses or they could cause a hardware fault. On modern PCs only the graphics adaptor is mapped directly into memory, the rest communicate through a specially reserved area of memory, known as isolated I/O memory.

Isolated I/O uses 16-bit addressing from 0000h to FFFFh, thus up to 64 KB of memory can be mapped. Microsoft Windows 95 can display the isolated I/O memory map by selecting Control Panel → System → Device Manager, then selecting Properties. From the computer properties window the Input/output (I/O) option is selected. Figure 16.4 shows an example for a computer in the range from 0000h to 0064h and Figure 16.5 shows from 0378h to 03FFh.

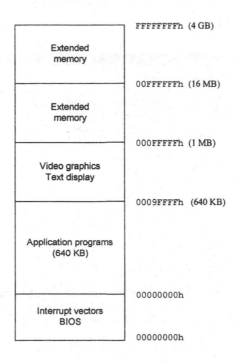

Figure 16.3 Typical PC memory map

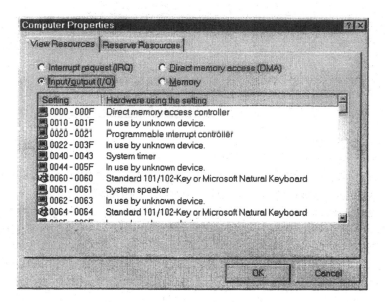

Figure 16.4   Example I/O memory map from 0000h to 0064h

It can be seen from Figure 16.5 that the keyboard maps into address 0060h and 0064h, the speaker maps to address 0061h and the system timer between 0040h and 0043h. Table 16.2 shows the typical uses of the isolated memory area.

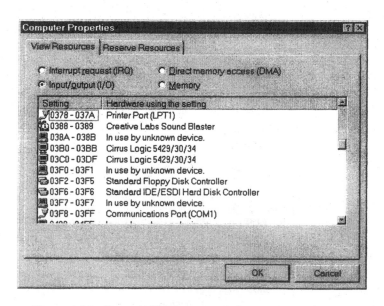

Figure 16.5   Example I/O memory map from 0378h to 03FFh

Table 16.2 Typical isolated I/O memory map

Address	Device
000h–01Fh	DMA controller
020h–021h	Programmable interrupt controller
040h–05Fh	Counter/Timer
060h–07Fh	Digital I/O
080h–09Fh	DMA controller
0A0h–0BFh	NMI reset
0C0h–0DFh	DMA controller
0E0h–0FFh	Math coprocessor
170h–178h	Hard-disk (Secondary IDE drive or CD-ROM drive)
1F0h–1F8h	Hard-disk (Primary IDE drive)
200h–20Fh	Game I/O adapter
210h–217h	Expansion unit
278h–27Fh	Second parallel port (LPT2:)
2F8h–2FFh	Second serial port (COM2:)
300h–31Fh	Prototype card
378h–37Fh	Primary parallel port (LPT1:)
380h–38Ch	SDLC interface
3A0h–3AFh	Primary binary synchronous port
3B0h–3BFh	Graphics adapter
3C0h–3DFh	Graphics adapter
3F0h–3F7h	Floppy disk controller
3F8h–3FFh	Primary serial port (COM1:)

16.4.1 Inputting a byte from an I/O port

The assembly language command to input a byte is:

```
IN AL,DX
```

where DX is the Data Register which contains the address of the input port. The 8-bit value loaded from this address is put into the register AL.

For Turbo/Borland C the equivalent function is inportb(). Its general syntax is as follows:

```
value=inportb(PORTADDRESS);
```

where PORTADDRESS is the address of the input port and value is loaded with the 8-bit value from this address. This function is proto-typed in the header file dos.h.

For Turbo Pascal the equivalent is accessed via the port[] array. Its general syntax is as follows:

```
value:=port[PORTADDRESS];
```

where PORTADDRESS is the address of the input port and value the 8-bit value at this address. To gain access to this function the statement uses dos requires to be placed near the top of the program.

Microsoft C++ uses the equivalent inp() function (which is prototyped in conio.h).

16.4.2 Inputting a word from a port

The assembly language command to input a word is:

```
IN AX,DX
```

where DX is the Data Register which contains the address of the input port. The 16-value loaded from this address is put into the register AX.

For Turbo/Borland C the equivalent function is inport(). Its general syntax is as follows:

```
value=inport(PORTADDRESS);
```

where PORTADDRESS is the address of the input port and value is loaded with the 16-bit value at this address. This function is prototyped in the header file dos.h.

For Turbo Pascal the equivalent is accessed via the portw[] array. Its general syntax is as follows:

```
value:=portw[PORTADDRESS];
```

where PORTADDRESS is the address of the input port and value is the 16-bit value at this address. To gain access to this function the statement uses dos requires to be placed near the top of the program.

Microsoft C++ uses the equivalent inpw() function (which is prototyped in conio.h).

16.4.3 Outputting a byte to an I/O port

The assembly language command to output a byte is:

```
OUT DX,AL
```

where DX is the Data Register which contains the address of the output port. The 8-bit value sent to this address is stored in register AL.

For Turbo/Borland C the equivalent function is `outportb()`. Its general syntax is as follows:

```
outportb(PORTADDRESS,value);
```
where PORTADDRESS is the address of the output port and `value` is the 8-bit value to be sent to this address. This function is prototyped in the header file `dos.h`.

For Turbo Pascal the equivalent is accessed via the `port[]` array. Its general syntax is as follows:

```
port[PORTADDRESS]:=value;
```

where PORTADDRESS is the address of the output port and `value` is the 8-bit value to be sent to that address. To gain access to this function the statement `uses dos` requires to be placed near the top of the program.

Microsoft C++ uses the equivalent `outp()` function (which is prototyped in `conio.h`).

16.4.4 Outputting a word

The assembly language command to input a byte is:

```
OUT  DX,AX
```

where DX is the Data Register which contains the address of the output port. The 16-bit value sent to this address is stored in register AX.

For Turbo/Borland C the equivalent function is `outport()`. Its general syntax is as follows:

```
outport(PORTADDRESS,value);
```

where PORTADDRESS is the address of the output port and `value` is the 16-bit value to be sent to that address. This function is prototyped in the header file `dos.h`.

For Turbo Pascal the equivalent is accessed via the `port[]` array. Its general syntax is as follows:

```
portw[PORTADDRESS]:=value;
```

> where PORTADDRESS is the address of the output port and value
> is the 16-bit value to be sent to that address. To gain access to this
> function the statement uses dos requires to be placed near the top
> of the program.

Microsoft C++ uses the equivalent outp() function (which is prototyped
in conio.h).

16.5 Digital I/O using the 8255

Each 8255 IC has 24 digital input/output lines. These are grouped into three
groups of 8 bits and are named Port A, Port B and Port C. A single 8-bit
register, known as the control register, programs the functionality of these
ports. Port C can be split into two halves to give Port C (upper) and Port C
(lower). The ports and the control register map into the input/output memory
with an assigned base address. The arrangement of the port addresses with
respect to the base address is given in Table 16.3.

Figure 16.6 shows the functional layout of the 8255. The control register
programs each of the ports to be an input or an output and also their mode of
operation. There are four main parts which are programmed: Port A, Port B,
Port C (lower) and Port C (upper).

Figure 16.7 shows the definition of the Control Register bits. The msb
(most significant bit) D7 either makes the device active or inactive. If it is
set to a 0 it is inactive, else it will be active. The input/output status of Port
A is set by D4. If it is a 0 then Port A is an output, else it will be an input.
The status of Port B is set by D1, Port C (lower) by D0 and Port C (upper)
by D3.

Port A can operate in one of three modes – 0, 1 and 2. These are set by
bits D5 and D6. If they are set to 00 then Mode 0 is selected, 01 to Mode 1
and 10 to Mode 2. Port B can be used in two modes (Mode 0 and 1) and is
set by bit D2. Examples of bit definitions and the mode of operation are
given in Table 16.4.

Table 16.3 PPI addresses

Port address	Function
BASE_ADDRESS	Port A
BASE_ADDRESS+1	Port B
BASE_ADDRESS+2	Port C
BASE_ADDRESS+3	Control register

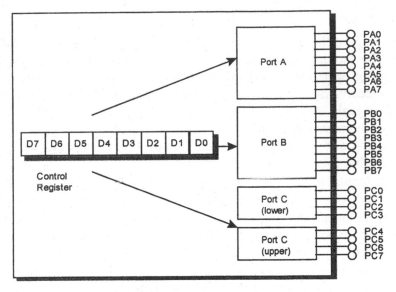

Figure 16.6   Layout of PPI

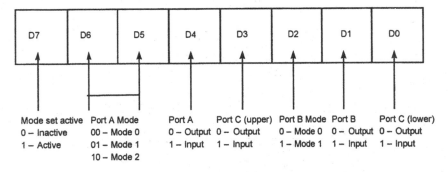

Figure 16.7   PPI Control Register Bit Definitions

16.5.1 Mode 0

Mode 0 is the simplest mode with no handshaking. In this mode the bits on Port C can be programmed as inputs or outputs.

Table 16.4   Example bit patterns for control register

Bit pattern	Mode of operation
01101000	Device is inactive as D7 set to 0
10011000	Mode 0 Port A input, Port C (upper) input, Mode 0 Port B output, Port C (lower) output
10101000	Mode 1 Port A output, Port C (upper) input, Mode 0 Port B output, Port C (lower) output

16.5.2 Mode 1

This mode gives handshaking for the synchronization of data. Handshaking is normally required when one device is faster than another. In a typical handshaking operation the originator of the data asks the recipient if it is ready to receive data. If it is not then the recipient sends back a 'not ready for data' signal. When it is ready it sends a 'ready for data' signal and the originator then sends the data.

If Ports A and B are inputs then the bits on Port C have the definitions given in Table 16.5.

When inputting data, the $\overline{\text{STB}}$ going low (active) writes data into the port. After this data is written into the port, the IBF line automatically goes high. This automatically remains high until the data is read from the port.

If any of the ports are outputs, then the bit definitions of Port C are given in Table 16.6.

In this mode, when data is written to the port the $\overline{\text{OBF}}$ line goes low, which indicates that data is ready to be read from the port. The $\overline{\text{OBF}}$ line will not go high until the $\overline{\text{ACK}}$ is pulled low.

Table 16.5 Mode 1 handshaking lines for inputting data

Signal	Port A	Port B
Strobe ($\overline{\text{STB}}$)	PC4	PC2
Input Buffer full (IBF)	PC5	PC1

Table 16.6 Mode 1 handshaking lines for outputting data

Signal	Port C	Port B
Output Buffer Full ($\overline{\text{OBF}}$)	PC7	PC1
Acknowledge ($\overline{\text{ACK}}$)	PC6	PC2

16.5.3 Mode 2

This mode allows bi-directional I/O. The signal lines are given in Table 16.7.

Table 16.7 Mode 2 operation for bi-directional I/O

Signal	Port A
$\overline{\text{OBF}}$	PC7
$\overline{\text{ACK}}$	PC6
$\overline{\text{STB}}$	PC4
IBF	PC5

16.6 Digital I/O programs

Program 16.1 outputs the binary code for 0 to 255 to Port B with a one-second delay between changes. The program exits when the output reaches 255. A delay routine has been added which uses the system timer. This will be discussed in more detail in Section 17.10. Figure 16.8 shows a typical set-up to test the program where Port B has been connected to eight light-emitting diodes (LEDs).

In 8086 Assembly Language a macro is defined using the equ statement. Program 16.1 uses these to define each of the port addresses. This helps to document the program and makes it easier to make global changes. For example, a different base address is relatively easy to set up, as a single change to BASE_ADDRESS automatically updates all port defines in the program. In this case the base address is 1F0h. This address should be changed to the required base address of the DIO card.

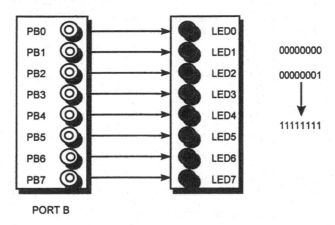

Figure 16.8   Possible system set-up

The statements:

```
        mov dx,CNTRLREG     ; set up PPI with
        mov al,90h          ; Port B as Output
        out dx,al
```

output the value 90h (1001 0000b) to the control register (CNTRLREG). The statements:

```
        mov ax,00h
loop1:
        mov dx,PORTB
        out dx,al           ; output to Port B

        call delay
        inc ax
        cmp ax,100h
        jnz loop1           ; repeat until all 1s
```

initially set the AL register to 00h. The next two statements (mov dx, PORTB and out dx,al) output the value of AL to PORT B. Next the delay routine is called (with call delay). This routine delays for a period of 1 second. Next the AL register is incremented (inc al). After this the AL register value is compared with 100h (0001 0000 0000b). If the result of the compare statement is not equal to zero then the program loops back to the loop1: label.

Program 16.1

```
code            SEGMENT
                ASSUME cs:code
BASEADDRESS EQU     01F0H ; change this as required
PORTA       EQU     BASEADDRESS
PORTB       EQU     BASEADDRESS+1
CNTRLREG    EQU     BASEADDRESS+3
; program to output to Port B counts in binary until from 00000000
; to 11111111 with approximately 1 second delay between changes
start:
        mov dx,CNTRLREG      ; set up PPI with
        mov al,90h           ; Port B as Output
        out dx,al

        mov ax,00h
loop1:
        mov dx,PORTB
        out dx,al            ; output to Port B

        call delay
        inc ax
        cmp ax,100h
        jnz loop1            ; repeat until all 1s

        mov ah,4cH           ; program exit
        int 21h              ;

        ; ROUTINE TO GIVE 1 SECOND DELAY USING THE PC TIMER
DELAY:  push ax
        push bx
        mov ax,18            ; 18.2 clock ticks per second;
        ; Address of system timer on the PC is 0000:046C (low word)
        ; and 0000:046E (high word)
        mov bx,0
        mov es,bx
        ; Add the number of required ticks of the clock (ie 18)
        add ax,es:[46CH]
        mov bx,es:[46EH]

loop2:  ; Compare current timer count with AX and BX
        ;  if they are equal 1 second has passed
        cmp bx,es:[46EH]
        ja loop2
        jb over
        cmp ax,es:[46CH]
        jg loop2

over:   pop bx
        pop ax
        ret

code    ENDS
        END     start
```

Program 16.2 reads the binary input from Port A and sends it to Port B. It will stop only when all the input bits on port A are 1s. It shows how a byte can be read from a port and then outputted to another port. Port A is used, in this example, as the input and Port B as the output. Figure 16.9 shows how Port A could be connected to input switches and Port B to the light-emitting diodes (LEDs). Loading the bit pattern 90h into the control register initializes the correct set-up for Ports A and B.

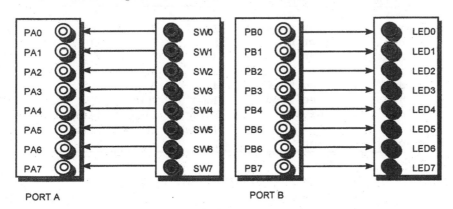

Figure 16.9   Typical system set-up

Program 16.2

```
code        SEGMENT
            ASSUME cs:code
; program to read from Port A and send to
; Port B. Program stops with all 1's
start:
        mov dx,1f3h
        mov al,90h
        out dx,al
loop:
        mov dx,1f0h
        in al,dx            ; read from Port A

        mov dx,1f1h
        out dx,al           ; output to Port B

        cmp al,ffh
        jnz loop:           ; repeat until all 1s
        mov ah,4cH          ; program exit
        int 21h             ;
code    ENDS
        END     start
```

16.7 Timing

Each instruction takes a finite time to complete. The speed of operation is determined by the processor clock speed. To find out how long a certain instruction will take determine the number of clock cycles to execute it and multiply this by the clock period. For example, if the clock rate is 8 MHz the clock period is 0.125 μs. The following gives the number of clock cycles for various instructions. Note that different processors take differing numbers of clock cycles to execute a command. Notice also that the 80386 processor is around twice as fast as the 8086 for many of the commands. This is due to improved architecture.

Table 16.8   Instruction timings for different processors

Command	Example	8086	80286	80386
mov	mov ax,1234	4	2	2
mov	mov dx,ax	2	2	2
out	out dx,al	8	3	11
inc	inc ax	3	2	2
dec	dec bx	3	2	2
and	and ax,0b6h	4	3	2
jne (nj)	jne fred	16(4)	7(3)	7(3)
div	div cx	80 (b)	14 (b)	14 (b)
		144 (w)	22 (w)	22 (w)
nop	nop	2	2	2

where (b) – byte divide, (w) – word divide, (nj) – no jump.

For example, the mov ax,1234 statement takes 0.5 µs assuming a 8 MHz clock. The following program will output an increment value every two seconds. Note the great improvement in the 80286/386 over the 8086 on dealing with mathematics operations. In the 8086 it takes 144 clock cycles to perform a word divide while the 80386 only takes 22 clock cycles (nearly seven times faster).

```
loop:    mov    dx,1f3h      ; control register address
         mov    al,90h       ; set port A input, port B output
         out    dx,al

         mov    dx,1f0h
         in     al,dx        ;read byte from port A

         mov    dx,1f1h
         out    dx,al        ; send byte to port B

         call   delay
         jmp    loop


    ; two second delay loop for 8 MHz clock
delay:   mov    ax,13
outer:   mov    bx,64777                 ;;; ) outer
inner:   dec    bx          ;;;;; ) inner
         jnz inner          ;;;;; ) loop
         dec ax
         jnz outer                     ;;; ) loop
         ret
```

Within the second (inner) loop:

```
    inner: dec bx
    jnz inner
```

will be executed 64 777 times. The number of cycles to do a dec and a jnz is 3+16 cycles. Thus it takes 19 cycles to complete this loop. The total time to complete this inner loop is thus:

Number of cycles × clock period = 19 × 0.125 μs = 2.375 μs.

Total time to complete this loop is 64 777 × 2.375 μs = 0.1538 s.

This inner loop is executed 7 times, thus the total delay time is 13 × 0.1538 = 2 s.

In general, for a general purpose loop with A and B as the variables in AX and BX, then:

```
delay:    mov    ax,A        ; 4 clock cycles
outer:    mov    bx,B        ; 4
inner:    dec    bx          ; 3
          jnz inner          ; 16
          dec ax             ; 3
          jnz outer          ; 16
          ret
```

First the inner loop:

```
inner:    dec    bx          ; 3
          jnz inner          ; 16
```

Then the number of cycles for inner loop will be B×19:

```
inner:    dec    bx          ; 3
          jnz inner          ; 16
          dec ax             ; 3
          jnz outer          ; 16
```

$$\text{Number of cycles} = A \times ((B \times 19) + 16 + 3)$$

If 19 × B is much greater than 19 then the following approximation can be made:

$$\text{Number of cycles} = 19 \times A \times B$$

Thus:

$$\text{Time taken} = \frac{\text{Number of clock cycles}}{\text{Clock frequency}}$$
$$= \frac{19 \times A \times B}{\text{Clock frequency}}$$

For example, in the last example (assuming a 4 MHz clock) the value of A is 13 and B is 64 777, thus:

$$\text{Time taken} = \frac{19 \times 13 \times 64\,777}{8 \times 10^6} = 2\,\text{s}$$

Typical processor clocks are:

8086	4.77 MHz, 8 MHz.
386	16 MHz, 25 MHz, 33 MHz.
486	33 MHz, 50 MHz, 66 MHz, 100 MHz.
Pentium	60 MHz, 90 MHz, 120 MHz, 200 MHz.

16.8 Exercises

16.8.1   Write a program to input a byte from Port A.

16.8.2   Write a program which will send to Port B all 1's.

16.8.3   Write a program which will read a byte from Port A. This byte is then sent to Port B.

16.8.4   Write a program that sends a 'walking-ones' code to Port B. The delay between changes should be one second. A 'walking-ones' code is as follows:

```
00000001
00000010
00000100
00001000
  :  :
10000000
00000001
00000010
and so on.
```

Hint: Use the shift left operator, that is << or shl.

16.8.5   Write separate programs which output the patterns in (a) and (b). The sequences are as follows:

```
(a)   00000001    (b)   10000001
      00000010          01000010
      00000100          00100100
      00001000          00011000
      00010000          00100100
      00100000          01000010
      01000000          10000001
      10000000          01000010
```

```
01000000          00100100
00100000          00011000
00010000          00100100
   ::           and so on.
00000001
00000010
and so on.
```

16.8.6 Write separate programs which output the following sequences:

(a)
```
1010 1010
0101 0101
1010 1010
0101 0101
and so on.
```

(b)
```
1111 1111
0000 0000
1111 1111
0000 0000
and so on.
```

(c)
```
0000 0001
0000 0011
0000 1111
0001 1111
0011 1111
0111 1111
1111 1111
0000 0001
0000 0011
0000 0111
0000 1111
0001 1111
and so on.
```

(d)
```
0000 0001
0000 0011
0000 0111
0000 1111
0001 1111
0011 1111
0111 1111
1111 1111
0111 1111
0011 1111
0001 1111
0000 1111
and so on.
```

(e)   The inverse of (d) above.

16.8.7 Write a program that reads a byte from Port A and sends the 1s complement representation to Port B. Note that 1s complement is all bits inverted.

16.8.8 Change the program in Exercise 16.8.9 so that it gives the 2s complement value on Port B. *Hint*: Either complement all the bits of the value and add 1 or send the negated value.

16.8.9 Write a program which will count from 00h to ffh with 1 s delay between each count. The output should go to Port B.

16.8.10 Write a program which will sample Port A every 1s then sends it to Port B.

16.8.11 Write a program which will simulate the following logic functions.

```
NOT    PB0 = not (PA0)
AND    PB0 = PA0 and PA1
```

OR   PB0 = PA0 or PA1

where PA0 is bit 0 of Port A, PA1 is bit 1 of Port A and PB0 is bit 0 of Port B.

16.8.12   Write a program which will simulate a traffic light sequence. The delay between changes should be approximately 1 second.

PB0 is RED
PB1 is AMBER
PB2 is GREEN

and the sequence is:

RED
AMBER
GREEN
AMBER
RED
AMBER
GREEN
and so on.

16.8.13   Modify the program in 16.8.12 so that the sequence is:

RED
RED and AMBER
GREEN
AMBER
RED
RED and AMBER
GREEN
and so on.

16.8.14   Write a program which will input a value from Port A. This value is sent to Port B and the bits are rotated with a delay of 1 second.

16.8.15   Write a program which will sample port A when bit 0 of Port C is changed from a 0 to a 1. Values are then entered via Port A by switching PC0 from a 0 to a 1. These values are put into memory starting from address 100h. The end of the input session is given by PC1 being set (ie PC1 is equal to a 1). When this is set all the input values are sent to Port B with a 2 s interval.

17 8086 Interrupts

17.1 Interrupts (INT)

The interrupt function (INT) interrupts the processor. It can be used to gain access to either DOS or BIOS functions. BIOS functions are typically used to gain access to the hardware, whereas DOS functions are used to quit from programs, read a character from the keyboard and write a character to the screen.

17.2 Interrupt 21h: DOS services

Programs access DOS functions using interrupt 21h. The functionality of the call is set by the contents of the AH register. Other registers are used either to pass extra information to the function or to return values back. For example, to determine the system time the AH is loaded with the value 2AH. Next, the processor is interrupted with interrupt 21H. Finally, when the program returns from this interrupt the CX register will contain the year, DH the month, DL the day and AL the day of the week.

Table 17.1 is only a small section of all the DOS related interrupts. For example, function 2Fh contains many functions that control the printer. Note that for the *Get free disk space* function the total free space on a drive is AX×BX×CX and total space on disk, in bytes, is AX×CX×DX.

Table 17.1 DOS interrupts

Description	Input registers	Output registers
Read character from keyboard with echo	AH=01h	AL=character returned
Write character to output	AH=02h DL=character to write	
Write character to printer	AH=05h DL=character to print	
Read character with no echo	AH=07h	AL=character read from keyboard

Get k/b status	AH=0Bh	AL=0 no characters AL=FFh characters
Get system date	AH=2Ah	CX=year, DH=month DL=day, AL=day of week (0–Sunday, and so on)
Set system date	AH=2Bh, CX=year, DH-month, DL=day	
Get system time	AH=2Ch	CL=hour CL=minute DH=second DL=1/100 seconds
Set system time	AH=2Dh, CH=hour CL=minute, DH=second DL=1/100 second	
Get DOS version	AH=30h	AL=major version number AH=minor version number
Terminate and stay resident	AH=32h DL=driver (0–default, 1– A: ,and so on)	
Get boot drive	AX=3305h	DL=boot drive (1=A, and so on)
Get free disk space	AL=36h DL=drive number (0=A, etc.)	AX=sectors per cluster BX=number of free clusters CX=bytes per sector DX=total clusters on driver
DOS exit	AH=4ch	

In Program 17.1, function 02h (write character to the output) is used to display the character 'A'. In this case, the function number 02h is loaded into AH and the character to be displayed is loaded into DL.

Program 17.1

```
code    segment
assume cs:code,ds:data

start:
    mov ah,2
    mov dl,'A'
    int 21h

    mov ah,4Ch ; DOS exit
    int 21h
code    ends
end     start
```

Program 17.2 uses the function 01h to get a character from the keyboard and then the function 02h to display it.

Program 17.2

```
code    segment
assume cs:code,ds:data

start:
    mov ah,1
    int 21h        ; read character from keyboard

    mov dl,al      ; put character read-in into dl
    mov ah,2       ; DOS interrupt to print to screen
    int 21h        ; show character to the screen

    mov ah,4Ch ; DOS exit
    int 21h

code    ends
end     start
```

Program 17.3 displays the default boot drive. In most cases the default boot drive will be C.

Program 17.3

```
code    segment
assume cs:code,ds:data

start:
    mov ax,3305h
    int 21h
    add dl,'A'-1
    mov ah,2
    int 21h

    mov ah,4Ch ; DOS exit
    int 21h

code    ends
end     start
```

17.3 Interrupt 10h: BIOS video mode

Interrupt 10h allows access to the video display. Table 17.2 outlines typical interrupt calls. Program 17.4 uses the BIOS video interrupt to display a border around the screen which changes colour each second from black to light blue. These colours are set-up with an `enum` data type definition. In this case, BLACK is defined as 0, BLUE as 1, and so on.

To display a border the AH register is loaded with 0Bh, BH with 00h and BL with the border colour. Next, the interrupt 10h is called with these parameters. Figure 17.1 shows the bit definition for the colours.

Table 17.2   BIOS video interrupt

Description	Input registers	Output registers
Set video mode	AH=00h	AL = video mode flag 0 (Text:40×25 B/W) 1 (Text:40×25 B/W) 2 (Text:80×25 Colour) 3 (Text:60×25 Colour) 4 (Graphics:320×200 Colour) 5 (Graphics:320×200 B/W) 6 (Graphics: 640×200 Colour)
Set cursor position	AH = 02h BH = 00h DH = row (00h is top) DL = column (00h is left)	
Read cursor position	AH = 02h BH = 00h	DH = row (00h is top) DL = column (00h is left)
Write character and attribute at cursor position	AH = 09h AL = character to display BH = 00h BL = attribute (text mode) or colour (graphics mode) if bit 7 set in graphics mode, character is Exclusive-ORed on the screen CX = number of times to write characters	
Read character and attribute at cursor position	AH = 08h BH = 00h	AH = attribute (see Figure 17.1) AL = character
Set background/ border colour	AH = 0Bh BH = 00h BL = background/border colour (border only in text modes)	

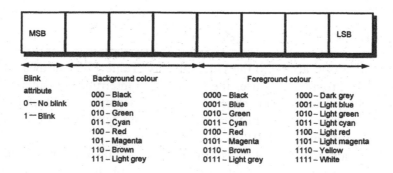

Figure 17.1   Character attribute

⬡ **Program 17.4**

```
code    segment
assume cs:code,ds:data

start:
     mov ah,0bh
     mov bh,0
     mov cl,0
loop1:
     mov bl,cl
     int 10h

     inc cl
     call delay
     cmp cl,7
     jne loop1
     mov ah,4Ch ; DOS exit
     int 21h

DELAY:   push ax
         push bx
         mov ax,18          ; 18.2 clock ticks per second;
         ; Address of system timer on the PC is 0000:046C (low word)
         ; and 0000:046E (high word)
         mov bx,0
         mov es,bx
         ; Add the number of required ticks of the clock (ie 18)
         add ax,es:[46CH]
         mov bx,es:[46EH]

loop2:   ; Compare current timer count with AX and BX
         ;  if they are equal 1 second has passed
         cmp bx,es:[46EH]
         ja loop2
         jb over
         cmp ax,es:[46CH]
         jg loop2

over:    pop bx
         pop ax
         ret

code     ends
end      start
```

17.4 Interrupt 11h: BIOS equipment check

Interrupt 11h returns a word which gives a basic indication of the types of equipment connected. It is useful in determining if there is a math co-processor present and the number of parallel and serial ports connected. Table 17.3 shows the format of the call.

Table 17.3 BIOS equipment check interrupt

Description	Input registers	Output registers
Get equipment list		AX = BIOS equipment list word

17.5 Interrupt 13h: BIOS disk access

Interrupt 13h allows access to many disk operations. Table 17.4 lists two typical interrupt calls.

Table 17.4   BIOS disk access interrupt

Description	Input registers	Output registers
Reset disk system	AH = 00h DL = drive (if bit 7 is set both hard disks and floppy disks are reset)	Return: AH = status
Get status of last operation	AH = 01h DL = drive (bit 7 set for hard disk)	Return: AH = status

17.6 Interrupt 14h: BIOS serial communications

BIOS interrupt 14h can be used to transmit and receive characters using RS-232 and also to determine the status of the serial port. Table 17.5 lists the main interrupt calls. Program 17.5 initializes COM2: with 4800 baud, even parity, 1 stop bit and 7 data bits.

Table 17.5   BIOS serial communications interrupts

Description	Input registers	Output registers
Initialize serial port	AH = 00h AL = port parameters (see Figure 17.2) DX = port number (00h–03h)	AH = line status (see get status) AL = modem status (see get status)
Write character to port	AH = 01h AL = character to write DX = port number (00h–03h)	AH bit 7 clear if successful AH bit 7 set on error AH bits 6–0 = port status (see get status)
Read character from port	AH = 02h DX = port number (00h–03h)	Return: AH = line status (see get status) AL = received character if AH bit 7 clear
Get port status	AH = 03h DX = port number (00h–03h)	AH = line status bit 7: timeout 6: transmit shift register empty 5: transmit holding register empty 4: break detected 3: framing error 2: parity error 1: overrun error 0: receive data ready

17.7 Interrupt 17h: BIOS printer

The BIOS printer interrupt allows a program either to get the status of the printer or to write a character to it. Table 17.7 outlines the interrupt calls.

Program 17.5

```
code    segment
assume cs:code,ds:data

start:
     mov ah,02h
     mov al,0d2h ; 1101 0010
     mov dx,01h
     int 14h

     mov ah,4Ch ; DOS exit
     int 21h
code    ends
end     start
```

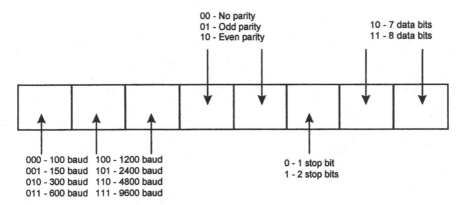

Figure 17.2 Bit definitions for serial port initialization

17.8 Interrupt 16h: BIOS keyboard

Interrupt 16h allows access to the keyboard. Program 17.6 uses the BIOS keyboard interrupt to display characters, entered from the keyboard, to the screen. Initially, the code:

```
start:
     mov ah,1    ;
     int 16h     ; test for keystroke repeat loop
     jz start    ; until the zero flag is not set
```

tests to see if a key has been pressed. It repeats until the zero flag is unset (this happens when a key has been pressed. The check for keystroke interrupt call sets the zero flag (ZF) if there are no characters in the buffer, otherwise it will be a 0. Next the code:

```
     mov ah,0    ; get keystroke
     int 16h
```

gets the key that has been pressed and puts the result into the DL register. Finally, the DOS interrupt is used to display the character to the screen.

Table 17.6   BIOSkeyboard interrupt

Description	Input registers	Output registers
Get keystroke	AH = 00h	AH = scan code
		AL = ASCII character
Check for keystroke	AH = 01h	Return: ZF set if no keystroke available
		ZF clear if keystroke available
		AH = scan code
		AL = ASCII character
Get shift flags	AH = 02h	AL = shift flags
		bit 7: Insert active
		6: CapsLock active
		5: NumLock active
		4: ScrollLock active
		3: Alt key pressed
		2: Ctrl key pressed
		1: left shift key pressed
		0: right shift key pressed

Program 17.6

```
code    segment
assume cs:code,ds:data

start:
    mov ah,1     ;
    int 16h      ; test for keystroke repeat loop
    jz start     ; until the zero flag is not set

    mov ah,0     ; get keystroke
    int 16h

    mov dl,al    ;
    mov ah,2     ; display keystroke
    int 21h

    mov ah,4Ch   ; DOS exit
    int 21h

code    ends
end     start
```

17.9 Interrupt 19h: BIOS reboot

Interrupt 19h reboots the system without clearing memory or restoring interrupt vectors. For a warm boot, equivalent to Ctrl-Alt-Del, then `1234h` should be stored at `0040h:0072h`. For a cold boot, equivalent to a reset, then `0000h` is stored at `0040h:0072h`. Care should be taken with this interrupt as it may cause the PC to 'hang'.

17.10 Interrupt 1Ah: BIOS system time

The BIOS system time interrupt allows a program to get access to the system timer. Table 17.8 outlines the interrupt calls.

Table 17.7 BIOS printer interrupt

Description	Input registers	Output registers
Initialize printer port	AH = 01h DX = printer number (00h-02h)	AH = printer status bit 7: not busy 6: acknowledge 5: out of paper 4: selected 3: I/O error 2: unused 1: unused 0: timeout
Write character to printer	AH = 00h AL = character to write DX = printer number (00h-02h)	AH = printer status
Get printer status	AH = 02h DX = printer number (00h-02h)	AH = printer status

Table 17.8 BIOS system time interrupt

Description	Input registers	Output registers
Get system time	AH = 00h	CX:DX = number of clock ticks since midnight AL = midnight flag, non-zero if midnight passed since time last read
Set system time	AH = 01h	CX:DX = number of clock ticks since midnight
Set real-time clock time	AH = 03h CH = hour (BCD) CL = minutes (BCD) DH = seconds (BCD) DL = daylight savings flag (00h standard time, 01h daylight time)	
Get real-time clock time	AH = 02h	Return: CF clear if successful CH = hour (BCD) CL = minutes (BCD) DH = seconds (BCD) DL = daylight savings flag (00h standard time, 01h daylight time) CF set on error

17.11 C and Pascal interrupts

See Section D.3.

17.12 Exercises

17.11.1   Using BIOS video interrupt 10h write programs which perform the following:

   (a)   fill a complete screen with the character 'A' of a text colour of red with a background of blue;

   (b)   repeat (a), but the character displayed should cycle from 'A' to 'Z' with a one-second delay between outputs;

   (c)   repeat (a), but the foreground colour should cycle through all available colours with a one-second delay between outputs;

   (d)   repeat (a) so that the background colour cycles through all available colours with a one-second delay between outputs.

17.11.2   Using BIOS keyboard interrupt 16h write a program, in C or Pascal, that displays the status of the Shift, Caps lock, Cntrl, Scroll and Num keys.

PART D

C/Pascal
C++
Assembly Language
Visual Basic
HTML/Java
DOS
Windows 3.x
Windows 95
UNIX

Introduction

18.1 Introduction

Microsoft Windows has become the de-facto PC operating system. All versions up to, and including, Windows 3.11 used DOS as the core operating system. New versions of Windows, such as Windows NT and Windows 95 do not use DOS and can thus use the full capabilities of memory and of the processor. The most popular programming languages for Windows programming are:

- Microsoft Visual Basic.
- Microsoft Visual C++ and Borland C++.
- Delphi (which is available from Borland).

Visual Basic has the advantage over the other language in that it is relatively easy to use and to program with, although the development packages which are used with C++ and Delphi make constructing the user interface relatively easy. Visual Basic Version 4 is shown in Figure 18.1 and Version 5 is shown in Figure 18.2.

18.2 Event-driven programming

Traditional methods of programming involve writing a program which flows from one part to the next in a linear manner. Most programs are designed using a top-down structured design, where the task is split into a number of submodules, these are then called when they are required. This means that it is relatively difficult to interrupt the operation of a certain part of a program to do another activity, such as updating the graphics display.

Visual Basic in general is:

- **Object-oriented**. Where the program is designed around a number of ready-made objects.
- **Event-driven**. Where the execution of a program is not predefined and its execution is triggered by events, such as a mouse click, a keyboard press, and so on.

- **Designed from the user interface outwards**. The program is typically designed by first developing the user interface and then coded to respond to events within the interface.

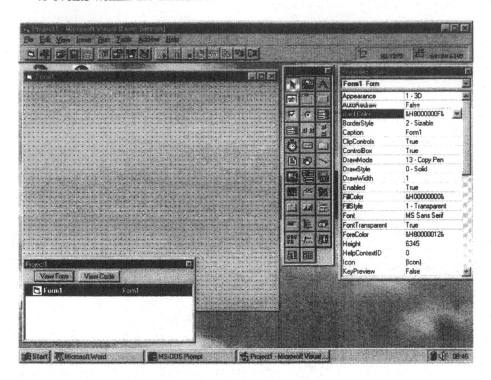

Figure 18.1 Visual Basic 4 user interface

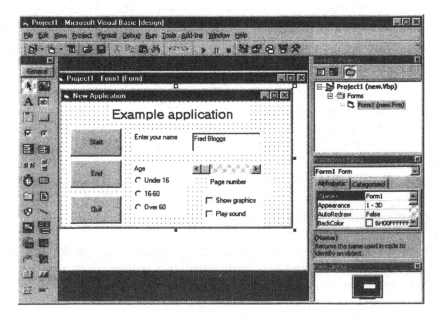

Figure 18.2 Visual Basic 5 user interface

18.3 Visual Basic files

A listing of a sample directory which contains Visual Basic files is:

```
DDE       BAS         143   12/01/96   0:00  DDE.BAS
DDE       VBP         338   12/01/96   0:00  DDE.VBP
EXECUTE   FRM       2,431   12/01/96   0:00  EXECUTE.FRM
MAIN      FRM      18,468   12/01/96   0:00  MAIN.FRM
```

The files are:

- **Project files**. Projects bind together the individual elements of a Visual Basic program. Initially the user creates a project file for the program and this is loaded whenever the program is being developed. The default extension for a project is either .MAK or .VBP extension. Projects generally make it easier to control the various elements of a program.
- **Modules**. Code that is attached to a form is accessible from anywhere on that form, but a program may have more than one form. It will sometimes be necessary to have program code that can be reached from any form, and in this case the code would be written on a module. Modules disappear from view when the program runs – only forms have an on-screen existence. There may be several modules in one program, and each is saved as a separate file. These are marked by a .BAS extension.
- **Forms**. A form forms the anchor for all parts of a Visual Basic program. Initially it is a blank window and the user pastes controls onto it to create the required user interface. Code is then associated with events on the form, such as responding to a button press or a slider control, although some control elements do not have associated code. A program can have one or more forms, each of which displays and handles data in different ways. To make forms shareable with other programs then each is saved separately with a .FRM extension.
- **Icons**. These are, normally, small graphics images and have a .ICO extension.
- **Graphics images**. These are normally either BMP (bitmapped) files or WMF (windows metafile) files with the .BMP and .WMF extensions, respectively.
- **Others**. Other files also exist, such as VBX which is Visual Basic eXtension.

18.4 Other terms

Visual Basic uses a number of other terms to describe design procedure, these are:

- **Controls**. The VB interface contains a window with control objects which are pasted onto a form. These controls can be simple text, menus, spreadsheet grids, radio buttons, and so on. Each control has a set of properties that defines their operation, such as their colour, the font size, whether it can be resized, and so on. Some controls, such as command buttons, menus, and so on, normally have code attached to them, but simple controls, such as text and a graphics image can simply exist on a form with no associated code.

- **Procedures**. As with C and Pascal, Visual Basic uses procedures, or sub-routines, to structure code. Most of these are associated with an event that occurs from a control and some will be free-alone with no associated event.

18.5 Main screen

Figure 18.3 shows the Visual Basic 4 desktop (Visual Basic 5 is similar, but the windows do not float on the display). It contains a menu form, controls, main form, project windows and properties window.

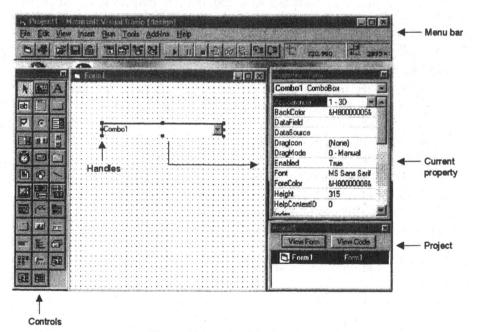

Figure 18.3   Visual Basic 4 desktop

18.5.1 Menu bar and toolbar

The menu bar and tool bar appear in a single, floating window, as shown in Figure 18.4. The menu bar contains options for file manipulation (File), editing (Edit), viewing (View), running (Run), testing the program (Debug), manipulating windows (Window) and getting help (Help). These can either be selected with the mouse, using the function key F10 and then selecting the option with the arrow keys and pressing return or use the hot-key. The hot key is Alt and the underlined character, thus Alt-F selects the File menu, Alt-E selects the Edit menu, and so on.

Figure 18.4 Visual Basic 4 menu bar and toolbar

The toolbar contains shortcut buttons for commonly used menu items. To the right of the toolbar there are two indicators; these display the position and size of a selected form or control. This area of the window is also used in the compilation phase to display the status of the compilation.

The toolbar buttons are:

Creates a new form.

Creates a new standard module.

Opens an existing project.

Saves the currently active project.

Locks/unlocks the controls on the active form.

Displays the Menu Editor.

Displays the Properties window.

Displays the Object Browser.

Displays the Project window.

Runs the application.

Pauses program execution.

Stops execution.

Toggles a breakpoint (breakpoint on or off).

Displays the value of the current selection in the Code window.

Displays the structure of active calls.

Traces through each line of code and steps into procedures.

Executes code one procedure or statement at a time in the Code window.

18.5.2 Project window

The Project window displays all the forms and modules used in the currently active project; an example is shown in Figure 18.5. A new project is opened by selecting New Project from the File menu, whereas to open an existing project the Open Project option is selected from the File menu, else the open existing project option is selected from the toolbar.

Only one project can be opened at a time, but that project can have any number of forms. In the Project window the user can do the following:

- Open a Form window for an existing form by selecting the form name and clicking the View Form button.
- Open the Code window for an existing form by selecting the module name and clicking the View Code button.
- Remove a file from a project by selecting the file in the Project window, and then from the File menu choose Remove File.

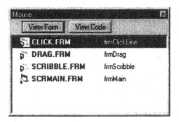

Figure 18.5 Visual Basic Project window

18.5.3 Form window

The Form window, as shown in Figure 18.6, creates application windows and dialog boxes. A new form is created by selecting the Form from the Insert menu (or use the toolbar shortcut) and an existing form is opened by selecting the form name and then clicking the View Form button. An existing form is added to a project by selecting the Add File command from the File menu. Any associated code on a form can be viewed by clicking on the View Code button.

Each form has a Control-menu box, Minimize and Maximize buttons, and a title bar, and can be moved and resized. Table 18.1 shows the key combinations in the Form window.

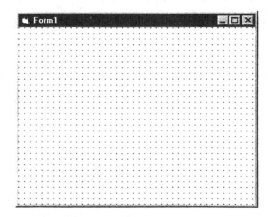

Figure 18.6 Visual Basic Form window

Table 18.1 Key combination in Form window

Key combination	Operation
alpha	Enter a value in the Properties window for the selected property.
CLICK–Drag	Select multiple controls.
CTRL+CLICK+ DRAG	Add or remove controls from the current selection.
CTRL–C	Copy the selected controls to the Clipboard.
CTRL–CLICK	Add or remove a control from the selection.
CTRL–E	Display the Menu Editor.
CTRL–J	Bring control to front (if controls are overlapping).
CTRL–K	Send control to back (if controls are overlapping).
CTRL–V	Paste from Clipboard onto the form.
CTRL–X	Cut the selected controls to the Clipboard.
CTRL–Z	Undo a deleted control.
DEL	Delete the selected controls.
F4	Display the Properties window
F7	Open the Code window for the selected object.
SHIFT–CTRL– *alpha*	Select a property in the Property list of the Properties window.
SHIFT–TAB	Cycle backward through controls in tab order.
TAB	Cycle forward through controls in tab order.

18.5.4 Toolbox

The Toolbox contains the icons for controls. These are standard Visual Basic controls and any custom controls and insertable objects, as shown in Figure 18.7. The Toolbox is displayed, if it is not already in view, with Toolbox from the View menu and it is closed by double-clicking the Control-menu box.

Figure 18.6 shows some typical controls, these include:

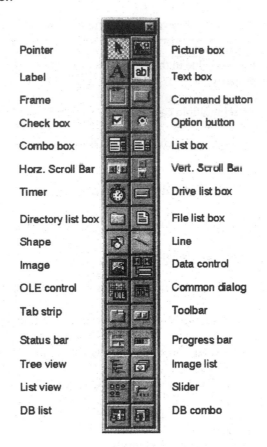

Pointer		Picture box
Label		Text box
Frame		Command button
Check box		Option button
Combo box		List box
Horz. Scroll Bar		Vert. Scroll Bar
Timer		Drive list box
Directory list box		File list box
Shape		Line
Image		Data control
OLE control		Common dialog
Tab strip		Toolbar
Status bar		Progress bar
Tree view		Image list
List view		Slider
DB list		DB combo

Figure 18.7   Toolbox controls

- **Pointer**. The pointer does not draw any control objects and is used to resize or move a control once it has been drawn on a form. When a control is added to a form then the pointer is automatically selected.

- **PictureBox**. Displays graphical images (BMP, WMF, ICO or DIB).

- **Label**. Used to display text that cannot be changed by the user.

- **TextBox**. Allows the user to either enter or change text.

- **Frame**. Used to create a graphical or functional grouping for controls. These are grouped by first drawing a Frame around them and then drawing controls inside the frame.

- **CommandButton**. Used to carry out a command.

- **CheckBox**. Used to create a check box, where the user can indicate if something is on or off (true or false), or, when there is more than one option, a multiple of choices.

- **OptionButton**. Used to display a number of options but only one can be chosen (this differs from the check box which only allows one option to be chosen).

- **ComboBox**. Used to give a combination of a list box and test box, where the user can either enter a value in a text box or choose an item from the list.

- **ListBox**. Used to display a list of items and the user is allowed to choose one of them. This list has a scroll button to allow the list to be scrolled.

- **HScrollBar** (horizontal scroll bar). Used to scroll up and down through a list of text or graphical information. It can also be used to indicate the current position on a scale or by the user to indicate a given strength of value.

- **VScrollBar** (vertical scroll bar). Used to scroll across a list of text or graphical information. It can also be used to indicate the current position on a scale or by the user to indicate a given strength of value.

- **Timer**. Used to generate timed events at given intervals.

- **DriveListBox**. Used to display currently connected disk drives.

- **DirListBox** (directory list box). Used to display directories and paths.

- **FileListBox**. Used to display a list of files.

- **Shape**. Used to draw shapes, such as rectangles, rounded rectangles, squares, rounded squares, ovals or circles.

- **Line**. Used to draw a variety of line styles on your form at design time (transparent, solid, dash, dot, dash-dot and dash-dot-dot).

- **Image**. Used to display a graphical image, such as a bitmap (BMP), icon (ICO), or metafile (WMF). These images can only be used to display and image and do not have the same control functions as Picture-Box.

- **Data**. Used to provide access to data in databases.

- **OLE Controller**. Used to link and embed objects from other applications (such as Word Documents, Excell Spreadsheets, and so on). OLE stands for Object Linking and Embedding.

- **CommonDialog**. Used to create customized dialog boxes for operations such as printing files, opening and saving files, setting fonts and help functions.

- **DBList** (data-bound list box). Used as an enhanced ListBox which can be customized to display a list of items from which the user can choose one. The list can be scrolled if it has more items than can be displayed at one time.

- **DBCombo** (data-bound combo box). Used as an enhanced Combo which can be customized to display a list of items from which the user can choose one. Use to draw a combination list box and text box. The user can either choose an item from the list or enter a value in the text box.

- **DBGrid** (data-bound grid). Use to display a series of rows and columns and to manipulate the data in its cells. DBGrid is a custom control and has increased data access capabilities that the standard Grid does not have.

18.6 Properties window

The Properties window displays the properties of the currently selected form, control or menu. They allow properties such as the colour, font type and size of text, background colour of a form, type of graphic image, and so on.

First, the item to be changed is selected and then the Properties option is chosen from the View menu, else the function key F4. The Property window is closed with a double-click on the Control-menu box.

The Properties window contains two main parts, these are:

- **The Object box**. This is found below the title bar and identifies the currently selected form or control on the form. In Figure 18.8 the command button has an associated Properties window. The Object box in this case is:

```
Command1 CommandButton
```

Where CommandButton is the control item and is named Command1, other control items are PictureBox, Label, TextBox, Frame, CheckBox, OptionButton, ComboBox, ListBox, HScrollBar, VScrollBar, Timer, DriveListBox, FileListBox, Shape, Line, Image, Data, and so on. Click the arrow at the right side of the Object box (▼) to get a list of the controls on the current form. From the list, choose the current form or a control on the form whose properties you want to change. An example is given in Figure 18.9. In this case there are three command buttons (named Command1, Command2 and Command3) on the form, a drive list box (named Drive1), and so on. The list also contains the currently active form (in this case it is named Form1). Names of controls are assigned consecutively, so that the first command button is Command1, the second is Command2, and so on.

- **The Properties list.** This is a two-column list that shows all properties associated with a form or control and their current settings. To change a properties setting then the properties name is selected and the new setting is either typed or selected from a menu. Properties that have predefined settings (such as a range of colours or true/false) display the list of settings by clicking the down arrow at the right of the settings box (▼), or they can be cycled through by double-clicking the property name in the left column. In Figure 18.8 the Default property has either a True or False setting. A ⚏ in the second column indicates either the selection of colours from a palette or the selection of picture files through a dialog box.

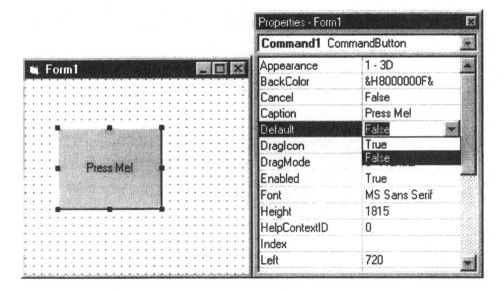

Figure 18.8 Properties window

Figure 18.10 shows an example of colour settings. Note that the colour appears as a 24-bit hexadecimal equivalent (with 8 other attribute bits), but when the user selects the colour it appears as a colour in the palette. This 24-bit colour is made up to red, green and blue (RGB). The standard format is:

`&HaaBBGGRR&`

The RR hexadecimal digits give the strength of the red from 00h to FFh (0 to 255), the GG hexadecimal digits give the strength of green and BB gives the strength of blue. Thus for the colour strength parts: white is &H*aa*FFFFFF&, black is &H*aa*000000&, red is &H*aa*0000FF&, yellow is &H*aa*00FFFF& and cyan is &H*aa*FFFF00&.

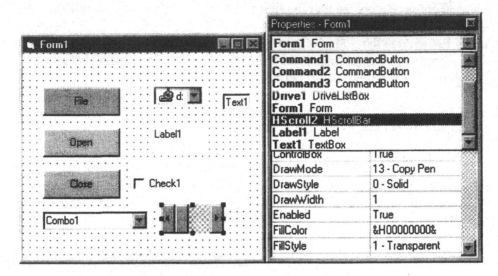

Figure 18.9   Example list of controls

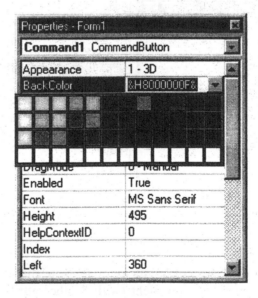

Figure 18.10   Setting colour

18.7 Controls and Event

Controls have associated properties and also a number of events. These events can be viewed by double clicking on a control or by selecting View Code from Project window. Figure 18.11 shows an example of the Code windows. It can be that the object can be selected by pulling down the menu of the left-hand side and the associated events in the right-hand menu. The associated code with that object and event is shown in the window below these menu options.

Figure 18.12 shows an example of the events that occur when the control is a command button (in this case the object is Command1). It can be seen that the associated events are: Click, DragDrop, DragOver, GotFocus, Key-Down, KeyPress and KeyUp. Each of these can have associated sections of code to react to the event. For example, the Click event is initiated when the user clicks the mouse button on the command button and KeyDown is initiated when a key has been pressed down.

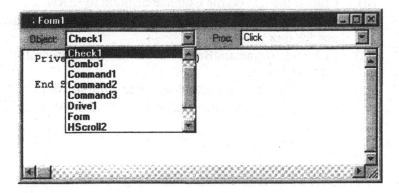

Figure 18.11   Selecting objects

Figure 18.12   Events when the control is a command button

18.8 Exercises

18.8.1   Place a CommandButton on a form and display its properties. Note all of the properties and, with the help of the help manual, identify of the function of each of the properties. Note that help on a property can be found by highlighting the property and pressing F1.

18.8.2 With a CommandButton identify the events that are associated with it.

18.8.3 Determine the actual colours of the following RGB colour values:

(i) &H0080FF80& (ii) &H00FF8080&

(ii) &H00C000C0& (iv) &H00E0E0E0&

18.8.4 Conduct the following:

(i) Add a command button to a form.
(ii) Change the text on the CommandButton to 'EXIT'.
(iii) Change the font on the CommandButton to 'Times Roman' and the font size to 16.
(iv) Resize the CommandButton so that the text fits comfortably into the button.
(iv) Change the background colour of the form to yellow.
(v) Change the Caption name of the form to 'My Application'.

18.8.5 Develop the form given in Figure 18.13.

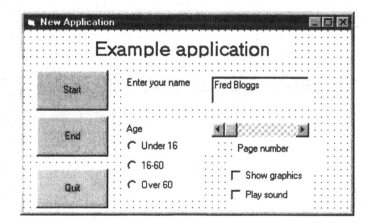

Figure 18.13 Exercise

18.8.6 Explain why, in the previous exercise, that a radio button is used for the age option and a check box is used to select the choices of Show Graphics and Play Sounds. Which of the following would be radio buttons or check boxes:

(i) Items on a shopping list.
(ii) Selection of a horse to win a race.
(iii) Selection of paint colour on a new car.
(iv) Selection of several modules on a course.

19 Visual Basic Language

19.1 Introduction

This chapter discusses the Visual Basic language and how a program is developed.

19.2 Programming language

Visual Basic has an excellent on-line manual in which the user can either search for the occurrence of keywords (with Help → Search For Help On) or view the contents of the manual (Help → Contents). The left-hand side of Figure 19.1 shows an example manual page after the user has selected Help → Contents → Visual Basic Help, and then leads to other parts of the manual, such as the Programming Language and Contents Topic. The left-hand side of Figure 19.1 shows an example of the Contents list.

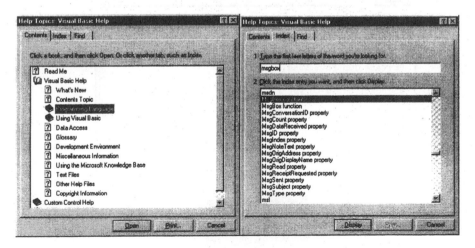

Figure 19.1 Visual Basic 4 on-line help manual

19.3 Entering a program

To start the development of a program with no controls on a form then the user selects the View Code from the Project window. Figure 19.2 shows the basic steps. After the View Code is selected then the user selects the Form object from within the form code window. Next the code is entered between the `Private Sub Load_From()` and `End Sub`. This code is automatically run when the form is run, as the procedure is Load. The code in Figure 19.2 simply displays the text "Hello to you" to a window.

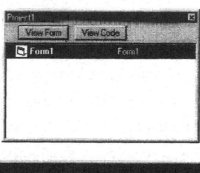

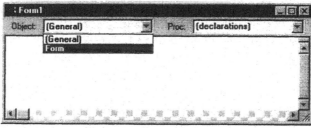

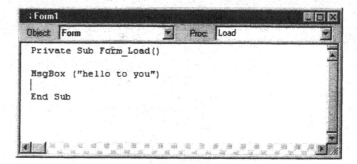

Figure 19.2 Steps taken to enter code

19.4 Language reference

This section contains a condensed reference to Visual Basic.

19.4.1 Data types and declaring variables

A variable is declared with the Dim keyword. These variables must conform to the following:

- Begin with an alphabet character.
- Cannot contain a dot, '$', '!', '@', '#' or '%'.
- Must be less than 256 characters long.

Examples are:

```
Dim val1 As Boolean
Dim x, y As Single
Dim i As Integer
Dim newdate As Date
Dim name As String * 30
```

which declares a Boolean variable called val1, two single precision floating-point variables: x and y, an integer named i, a date named newdate and a string of 30 characters named name.

Variables are assigned values with the assignment operation (=), such as:

```
val1 = True
x = 2.134
y = 10.1
newdate = Now
name = "Fred"
```

which sets the Boolean variable val1 to a True, the value of x to 2.134, y to 10.1, the date newdate is assigned the current date from the function Now and the string name is assigned the string "Fred". Note that strings of characters are defined between inverted commas (" ").

19.4.2 Operators

The basic operators in Visual Basic are similar to the ones used in Pascal. Table 19.1 shows the main operators. The operator precedence is:

- Arithmetic operators have the highest precedence, followed by comparison operators and finally logical operators.
- Arithmetic and logical operators are evaluated in the order given in Table 19.1.
- Comparison operators all have equal precedence.
- Multiplication and division have the same precedence, then the operation

is executed from left to right. The same occurs with addition and subtraction.

Table 19.1   Operator precedence

Arithmetic	Comparison	Logical
Exponentiation (^)	Equality (=)	Not
Negation (−)	Inequality (<>)	And
Multiplication and division (*,/)	Less than (<)	Or
Integer division (\)	Greater than (>)	Xor
Modulo arithmetic (Mod)	Less than or Equal to (<=)	
Addition and subtraction (+,−)	Greater than or Equal to (>=)	
String concatenation (&)		

19.4.3 Data types

As with C and Pascal, Visual Basic has a whole range of data types. Their range depends on their format (such as characters, integers and floating point values) and the number of bytes used to store them. Table 19.2 outlines the main pre-defined (intrinsic) data types. A user defined type can also be defined using the Type statement.

Table 19.2   Data types

Data type	Storage size	Range
Byte	1 byte	0 to 255
Boolean	2 bytes	True or False
Integer	2 bytes	−32,768 to 32,767
Long (long integer)	4 bytes	−2,147,483,648 to 2,147,483,647
Single (single-precision floating-point)	4 bytes	$\pm 3.402823 \times 10^{38}$ to $\pm 1.401298 \times 10^{-45}$
Double (double-precision floating-point)	8 bytes	$\pm 4.94065645841247 \times 10^{-324}$ to $\pm 1.79769313486232 \times 10^{308}$
Currency (scaled integer)	8 bytes	$\pm 922,337,203,685,477.5808$
Date	8 bytes	January 1, 100 to December 31, 9999.
String (variable-length)	10 bytes + string length	0 to approximately 2 billion
String (fixed-length)	Length of string	1 to approximately 65,400

The main data types are:

Boolean
Boolean variables can either be True or False and are stored as 16-bit (2-byte) values. Boolean variables are displayed as either True or False. Like C, when other numeric data types are converted to Boolean values then a 0 becomes False and any other values become True. When Boolean values are converted to other data types, False becomes 0 while True becomes −1.

Byte
Byte variables are stored as unsigned, 8-bit (1-byte) ranging from 0 to 255.

Currency
Currency data types are normally used in money calculations which have high accuracy. They are stored in 64-bit (8-byte) integer format which are scaled by 10,000 to give a fixed-point number with 19 significant digits and 4 decimal places. This gives a range of presentation providing a range of ±922,337,203,685,477.5808

Date
Date variables are stored as IEEE 64-bit (8-byte) floating-point numbers that represent dates ranging from 1 January 100 to 31 December 9999 and times from 0:00:00 to 23:59:59. Literal dates must be enclosed within number sign characters (#), for example, #January 1, 1993# or #1 Jan 93#. Any recognizable literal date values can be assigned to Date variables.

Double
Double (double-precision floating-point) variables are stored as IEEE 64-bit (8-byte) floating-point numbers ranging in value from $\pm 4.94065645841247 \times 10^{-324}$ to $\pm 1.79769313486232 \times 10^{308}$

Integer
Integer variables are stored as 16-bit (2-byte) numbers ranging in value from −32,768 to 32,767.

Long
Long (long integer) variables are stored as signed 32-bit (4-byte) numbers ranging in value from −2,147,483,648 to 2,147,483,647.

Single
Single (single-precision floating-point) variables are stored as IEEE 32-bit (4-byte) floating-point numbers, ranging in value from $\pm 3.402823 \times 10^{38}$ to $\pm 1.401298 \times 10^{-45}$.

String There are two kinds of strings:

Variable-length strings: which can contain up to 2,147,483,648 (2^{31}) characters (or 65,536 for Microsoft Windows version 3.1 and earlier).

Fixed-length strings: which can contain up to 65,536 character (2^{16}).

19.4.4 Convert between data types

Visual Basic has strong data type checking where the compiler generates an error when one data type is assigned directly to a variable with another data type. Thus different data types may need to be converted into another type so that they can be used. The basic conversion functions are:

Cbool(*expr*) Which converts an expression into Boolean. The expression argument is any valid numeric or string expression. If expression is zero then a False is returned, else a True is returned.

Cbyte(*expr*) Which converts an expression into Byte. The expression argument is any valid numeric or string expression. If expression lies outside the acceptable range for the Byte data type then an error occurs.

Ccur(*expr*) Which converts an expression into Currency. The expression argument is any valid numeric or string expression. If expression lies outside the acceptable range for the Currency data type then an error occurs.

Cdate(*expr*) Which converts an expression into Date. The date argument is any valid date expression.

Cdbl(*expr*) Which converts an expression into Double. The expression argument is any valid numeric or string expression.

Cint(*expr*) Converts an expression to an Integer. The expression argument is any valid numeric or string expression. Cint differs from the Fix and Int functions, which truncate, rather than round, the fractional part of a number. When the fractional part is exactly 0.5, the Cint function always rounds it to the nearest even number. For example, 0.5 rounds to 0, and 1.5 rounds to 2.

Clng(*expr*)   Which converts an expression into Long. The expression argument is any valid numeric or string expression. As with Cint the value is rounded to the nearest whole number.

Csng(*expr*)   Which converts an expression into Single. The expression argument is any valid numeric or string expression. If expression lies outside the acceptable range for the Single data type, an error occurs.

Cstr(*expr*)   Which converts an expression into String. The expression argument is any valid numeric or string expression. If the expression is Boolean then a string is returned with either True or False, else a numeric value returns a string containing the number.

Int(*expr*)
Fix(*expr*)   Returns the integer portion of a number. Int differs from Fix in that Int when the number is negative then it returns the first negative integer which is less than or equal to number, whereas Fix returns the first negative integer greater than or equal to number. For example, if the value is −12.3 then Int converts this to −13 while Fix converts it to −12.

A typical conversion is from a numeric or date variable into a string format. Program 19.1 shows an example of a Visual Basic program which contains the CStr function which is used to convert from two floating point values (x and y), an integer (i) and date (newdate) into a string format. This is then used to display the values to a window using the MsgBox function. The program also uses the Fix function to round-up the value of x. Figure 19.3 shows a sample run.

Program 19.1

```
Private Sub Form_Load()
Dim x, y As Double
Dim i As Integer
Dim newdate As Date

x = 43.2
y = 3.221
i = 100
newdate = #1/1/99#

MsgBox ("Values of x and y are: " + CStr(x) + "," + CStr(y))
MsgBox ("The whole part of x is : " + CStr(Fix(x)))
MsgBox ("Date is : " + CStr(newdate))
MsgBox ("Value of i is " + CStr(i))

End Sub
```

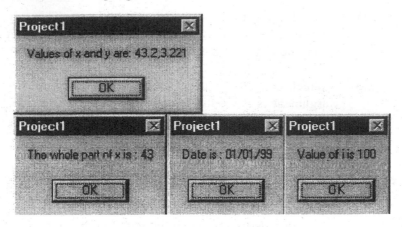

Figure 19.3   Sample run for Program 19.1

19.4.5 Input/output

The functions that can be used to input and output information are InputBox and MsgBox, respectively. Both these functions input and output information in the form of a string of characters. Thus when outputing non-string variables, such as integers and floating-point values, they must first be converted to a string using one of the string conversion functions. The same must be done for input, where the input string must be converted into the required data type, again using the data type conversion functions.

Output
The MsgBox function displays a message in a dialog box with specified buttons and then waits for the user to select a button. The value returned indicates the chosen button. The basic format is:

MsgBox(*prompt*[, *buttons*][, *title*][, *helpfile*, *context*])

where the parameters in brackets are optional. The parameters are:

prompt   String of be displayed in the dialog box.

buttons   Numeric value that is the sum of values that specifies the number, the type of buttons to display, the icon style and the default button. Table 19.3 outlines these values and if it is omitted then the default value for buttons is 0.

title   String which contains the title bar of the dialog box. If it is omitted then the application name is placed in the title bar.

helpfile   String that identifies the Help file to use to provide context-sensitive Help for the dialog box. If helpfile is provided then context must also be provided.

context   Numeric value that is the Help context number the Help author assigned to the appropriate Help topic. If context is provided, helpfile must also be provided. When both help-file and context are provided, the user can press F1 to view the Help topic corresponding to the context.

Table 19.1 defines the button settings. The values from 0 to 5 define the type of the button to be displayed. For example, a value of 5 will have two buttons, which are Retry and Cancel. The values 16, 32, 48 and 64 identify the icon to be displayed. For example, a value of 32 will display a question bubble. The 0, 256 and 512 define which button is the default. Each of these values can be added together to create the requires set of buttons, icon and default button. For example, to create a dialog box with the OK and Cancel buttons, a Critical icon and the Cancel button to be the default, then the setting would be:

```
setting = 1 + 16 + 256
```

which is 273. Note that to aid documentation in the program then the prede-fined constant values can be used, so for the previous example:

```
setting = vbOKCancel + vbCritical + vbDefaultButton2
```

Table 19.3   Button settings

Constant	Value	Description
vbOKOnly	0	Display OK button only
vbOKCancel	1	Display OK and Cancel buttons. See example 1 in Figure 19.4.
vbAbortRetryIgnore	2	Display Abort, Retry, and Ignore buttons. See ex-ample 2 in Figure 19.4.
vbYesNoCancel	3	Display Yes, No, and Cancel buttons. See example 3 in Figure 19.4.
vbYesNo	4	Display Yes and No buttons. See example 4 in Fig-ure 19.4.
vbRetryCancel	5	Display Retry and Cancel buttons. See example 5 in Figure 19.4.
vbCritical	16	Display Critical Message icon. See example 1 in Figure 19.5.

`vbQuestion`	32	Display Warning Query icon. See example 2 in Figure 19.5.
`vbExclamation`	48	Display Warning Message icon. See example 3 in Figure 19.5.
`vbInformation`	64	Display Information Message icon. See example 4 in Figure 19.5.
`vbDefaultButton1`	0	First button is default
`vbDefaultButton2`	256	Second button is default
`vbDefaultButton3`	512	Third button is default

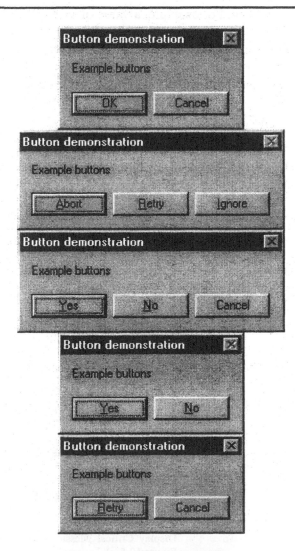

Figure 19.4   Buttons for MsgBox

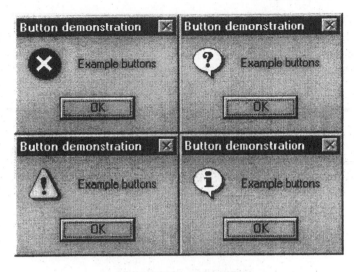

Figure 19.5 Icons for MsgBox

The MsgBox function returns a value depending on the button pressed; these return values are outlined in Table 19.4. For example, if the user presses the OK button then the return value will be 1. If the dialog box has a Cancel button then the user pressing ESC has the same effect as choosing Cancel. If the dialog box contains a Help button, context-sensitive Help is provided for the dialog box. However, no value is returned until one of the other buttons is chosen.

Table 19.4 MsgBox return values

Constant	Value	Button chosen
vbOK	1	OK
vbCancel	2	Cancel
vbAbort	3	Abort
vbRetry	4	Retry
vbIgnore	5	Ignore
vbYes	6	Yes
vbNo	7	No

Program 19.2 gives an example of a program which displays a dialog box with Yes and No buttons, and a question mark icon. The response will thus either be a 6 (if the Yes button is selected) or a 7 (if the No button is selected). Figure 19.6 shows a sample run.

Program 19.2

```
Private Sub Form_Load()
Dim msg, title As String
Dim style, response As Integer

msg = "Example buttons"
```

```
style = vbYesNo + vbQuestion
title = "Button demonstration"

response = MsgBox(msg, style, title)
MsgBox ("Response was " + CStr(response))


End Sub
```

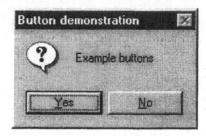

Figure 19.6 Example run

Input

The InputBox function prompts the user to input text, or choose a button. It then returns the contents of the text box. The basic format is:

InputBox(*prompt*[, *title*][, *default*][, *xpos*][, *ypos*][, *helpfile, context*])

where the parameters in brackets are optional. The parameters are:

prompt   String of be displayed in the dialog box.

title   String which contains the title bar of the dialog box. If it is omitted then the application name is placed in the title bar.

default   String which is displayed in the text box and is the default response if no other input is provided. If this field is omitted then the text box is initially empty.

xpos   Numeric value that specifies (in twips) the horizontal distance of the left edge of the dialog box from the left edge of the screen. If xpos is omitted then the dialog box is horizontally centred.

ypos   Numeric value that specifies (in twips) the vertical distance of the upper edge of the dialog box from the top of the screen. If ypos is omitted then the dialog box is vertically positioned approximately one-third of the way down the screen.

helpfile   String that identifies the Help file to use to provide context-sensitive Help for the dialog box. If helpfile is provided then context must also be provided.

context   Numeric value that is the Help context number the Help author assigned to the appropriate Help topic. If context is provided, helpfile must also be provided. When both help-file and context are provided, the user can press F1 to view the Help topic corresponding to the context.

Program 19.3 shows an example usage of the InputBox function. In this case the message for the title is 'Input demonstration', the default value is '10' and the value is return into the `inval` variable.

Program 19.3

```
Private Sub Form_Load()
Dim msg, title, default As String
Dim inval As Integer

msg = "Enter a value"
title = "Input demonstration"
default = "10"

inval = InputBox(msg, title, default)

MsgBox ("Value is " + CStr(inval))

End Sub
```

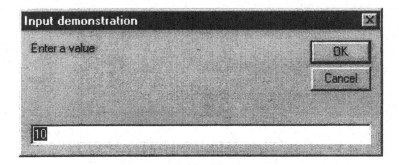

Figure 19.7   Example run

19.4.6 Decisions and loops

The decisions and loops in Visual Basic are similar to the ones used in C and Pascal.

If statement
The basic if statement format is:

```
if (expression) then
   statement block
end if
```

or, in general:

```
if (condition1) then
   statement block
elseif (condition2)
   statement block
 ::::
else
   statement block
end if
```

Where the condition can be a numeric or string expression that evaluates to True or False. The statement block contains one or more statements separated by colons.

As with C and Pascal, if the first condition is True then the first statement block is executed, else if the second condition is True then the second statement block is executed, and so on. If none of the conditions are True then the final else statement block is executed (if it exists).

Note that Else and ElseIf are both optional and there can be any number of ElseIf clauses but none of them can occur after the Else clause

Program 19.4 gives an example of a program in which the user enters a value from 0 to 2 and the program displays the equivalent resistor colour code colour.

Program 19.4
```
Private Sub Form_Load()
Dim msg, title, default As String
Dim inval As Integer

msg = "Enter color code value (0-2)"
title = "Color code"
default = "0"

inval = InputBox(msg, title, default)

If (inval = 0) Then
    MsgBox ("BLACK")
ElseIf (inval = 1) Then
    MsgBox ("BROWN")
ElseIf (inval = 2) Then
    MsgBox ("RED")
ElseIf (inval > 2 And inval < 10) Then
    MsgBox ("Colors not added")
Else
    MsgBox ("No color value")
End If
End Sub
```

Case

The Case statement is similar to the case and switch statements used in Pascal and C. Its general form is:

```
Select Case expression
Case expressionval1
    statement block1
Case expressionval2
    statement block2
Case Else
    else statement block1
End Select
```

The expression can be any numeric or string expression. A match of the expression to the expression value causes the corresponding statement block to be executed. If none of the block match then the Case Else statement block is executed. If testexpression matches any expressionlist expression associated with a Case clause, the statements following that Case clause are executed up to the next Case clause, or, for the last clause, up to the End Select. Control then passes to the statement following End Select. If testexpression matches an expressionlist expression in more than one Case clause, only the statements following the first match are executed.

Multiple expressions or ranges can be added to the Case cause, such as:

```
Case -1 To 3, 10 To 20, 51, 53
```

It is also possible to specify ranges and multiple expressions for strings. For example the following matches the string to 'apple' and everything, alphabetically between, 'banana' and 'carrot':

```
Case "apple", "banana" To "carrot"
```

Program 19.5 shows an example of a program which is similar to Program 19.4 but uses a case statement to select the resistor colour code colour.

Program 19.5
```
Private Sub Form_Load()
Dim msg, title, default As String
Dim inval As Integer

msg = "Enter color code value (0-9)"
title = "Color code"
default = "0"

inval = InputBox(msg, title, default)

Select Case inval
Case 0
    MsgBox ("BLACK")
Case 1
    MsgBox ("BROWN")
Case 2
    MsgBox ("RED")
Case 3, 4, 5, 6, 7, 8, 9
    MsgBox ("Colors not added")
Case Else
    MsgBox ("No Color")
End Select
End Sub
```

For loop

The For loop is similar to the for loop in Pascal. It repeats a group of statements a number of times. Its general form is:

```
For counter = start To end [Step step]
    statements
Next [counter]
```

or

```
For counter = start To end Step stepsize
    statements
Next [counter]
```

Where counter is a numeric variable which as used a loop counter, start is the Initial value of counter, end is the final value of counter and step is the amount by which the count is changed for each loop. This value can either be positive or negative. The default step size, if not specified is 1.

Program 19.6 uses a for loop to calculate the factorial value of an entered value.

Program 19.6
```
Private Sub Form_Load()

Dim fact As Long
Dim i, inval As Integer

inval = InputBox("Enter a value")

fact = 1

For i = 2 To inval
    fact = fact * i
Next i

MsgBox ("Factorial of " + CStr(i) + " is " + CStr(fact))

End Sub
```

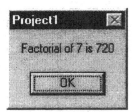

Figure 19.8   Example run

Do .. while loop

The Do..while loop is similar to the while() statements used in Pascal and C. Its general form is:

```
Do While condition
    statement block
Loop
```
or

```
Do Until condition
    statement block
Loop
```

or

```
Do
    statement block
Loop While condition
```

or

```
Do
    statement block
Loop Until condition
```

Program 19.7 uses a do..while loop to test to see if the user input is within a valid range. In this case the valid input is between 0 and 10 for the voltage input and greater than 0 and up to 10 for the current input. If the user enters a value which is outside this range then a MsgBox is displayed with an error message (INVALID: re-enter).

📄 **Program 19.7**

```
Private Sub Form_Load()
Dim voltage, current, resistance As Double

Do
    voltage = InputBox("Enter a voltage (0-10)")
    If (voltage < 0 Or voltage > 10) Then
        MsgBox ("INVALID: re-enter")
    End If
Loop While (voltage < 0 Or voltage > 10)

Do
    current = InputBox("Enter a current (0-10)")
    If (current <= 0 Or current > 10) Then
        MsgBox ("INVALID: re-enter")
    End If
Loop While (current <= 0 Or current > 10)

resistance = voltage / current

MsgBox ("Resistance is " + Cstr(resistance))

End Sub
```

19.5 Exercises

19.5.1 Complete Program 19.5 so that it implements the complete resistor colour code (see Table 6.1 on page 117).

19.5.2 Change the program in 19.5.1 so that it loops until the user enters a valid value (between 0 and 9).

19.5.3 Change the program in 19.5.2 so that after the result has been displayed the user is prompted as to whether to repeat the program (OK) or exit the program (Cancel).

19.5.4 Write a program which will continually display the current date and time. The sample code given below displays a single date and time.

```
Private Sub Form_Load()
Dim CurrDate, CurrTime As Date

CurrTime = Time ' System time
CurrDate = Date   ' System date
MsgBox (CurrTime + CurrDate)

End Sub
```

19.5.5 Modify the program in Exercise 19.5.4 so that the OK and Cancel buttons are shown. If the user selects the Cancel button then the program should exit, else the program should display the new date. The sample code given below displays a single date and time with the OK and Cancel buttons.

```
Private Sub Form_Load()
Dim CurrDate, CurrTime As Date
Dim response As Integer

CurrTime = Time ' System time
CurrDate = Date   ' System date
response = MsgBox(CurrDate +
    CurrTime, vbOKCancel, "Date")

End Sub
```

19.5.6 Complete Worksheet 2 on page 37 using Visual Basic. Note that for W2.2 the square root function in Visual Basic is `sqr`, and in W2.3 the inverse tangent function is `atn`.

 **Forms**

20.1 Introduction

This chapter discusses how forms are constructed and how code is associated with the form.

20.2 Setting properties

Each control object has a set of properties associated with it. For example, the TextBox control in Figure 20.1 has an object name of Text1. This object has a number of associated properties, such as Alignment, Appearance, BackColor, and so on. These properties can be changed within the program by using the dot notation. For example, to change the font to 'Courier New', the text displayed in the object to 'Hello' and the height of the window to 1000 the following can be used:

```
Text1.Font = "Courier New"
Text1.Text = "FALSE"
Text1.Height = 1000
```

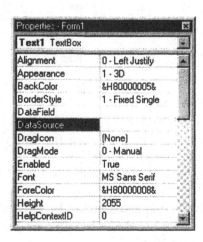

Figure 20.1   Object properties

20.3 Forms and code

Visual Basic programs are normally designed by first defining the user interface (the form) and writing the code which is associated with events and controls. The best way to illustrate the process is with an example.

20.3.1 Multiple choice example

In this example the user is to design a form with a simple question and three optional examples. It should display if the answer is correct (TRUE) or wrong (FALSE). The program should continue after each selection until the user selects an exit button.

Step 1: The label control is selected A. Then the text 'What is the capital of France' is entered the caption field, as shown in Figure 20.2.

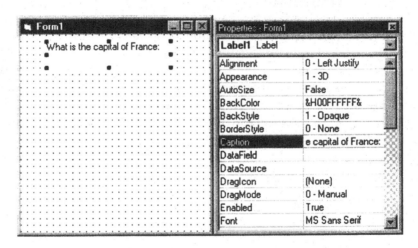

Figure 20.2   Step 1

Step 2:

Next the user add a command button by selecting the command control button.
The button is then added to the form and the Caption property is set to 'Edinburgh', as shown in Figure 20.3.

Step 3:
Next the user adds another command button by selecting the command control. The button is then added to the form and the Caption property is set to 'Paris', as shown in Figure 20.4.

Step 4:
Next the user adds another command button by selecting the command control. The button is then added to the form and the Caption property is set to 'Munich', as shown in Figure 20.5.

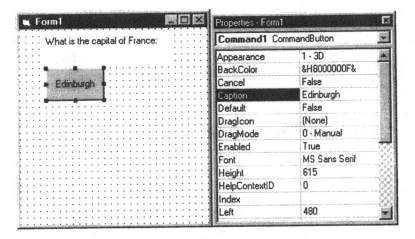

Figure 20.3   Step 2

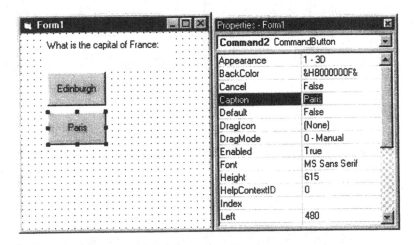

Figure 20.4   Step 3

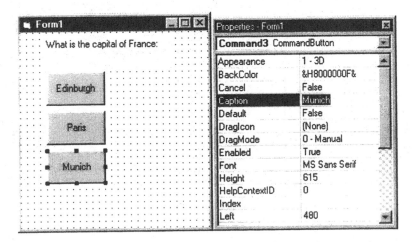

Figure 20.5   Step 4

Step 5:

Next the user adds a TextBox ![ab]. This is then added to the right-hand side of the form, as shown in Figure 20.6. The Text property is then changed to have an empty field. The TextBox will be used to display text from the program.

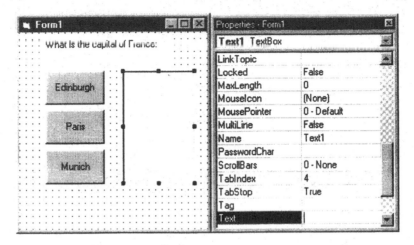

Figure 20.6   Step 5

Step 6:

Next the user adds another command button by selecting the command control. The button is then added to the form and the Caption property is set to 'Exit', as shown in Figure 20.7. A character in the name can be underlined by putting an & before it. Thus '&Exit' will be displayed as 'Exit'.

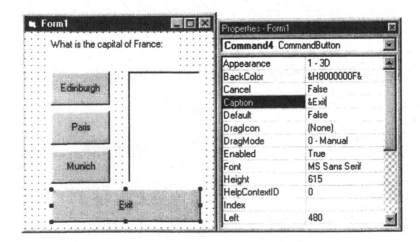

Figure 20.7   Step 6

Step 7:
Next the code can be attached to each of the command buttons. This is done by either double clicking on the command button or by selecting the button and pressing the F7 key. To display to the TextBox (the object named Text1) then the text property is set with:

```
Text1.Text = "FALSE"
```

which display the string "FALSE" to the text window. The associated code is shown in Figure 20.8.

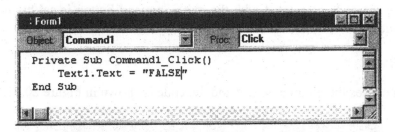

Figure 20.8   Visual Basic 4 on-line help manual

Step 8:
Next the code associated with the second command button is set, with:

```
Text1.Text = "TRUE"
```

as shown in Figure 20.9.

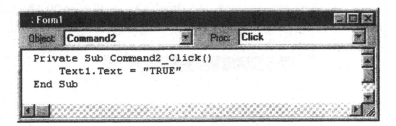

Figure 20.9   Visual Basic 4 on-line help manual

Step 9:
Next the code associated with the third command button is set, with:

```
Text1.Text = "FALSE"
```

as shown in Figure 20.10.

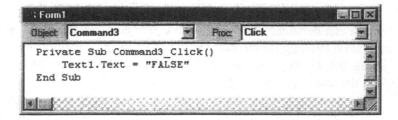

Figure 20.10   Visual Basic 4 on-line help manual

Step 10:
Finally the code associated with the exit command button is set by adding
the code:

```
End
```

which causes the program to end and the code is shown in Figure 20.11.

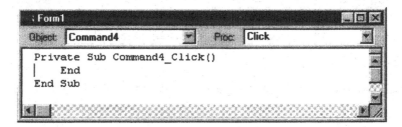

Figure 20.11   Visual Basic 4 on-line help manual

The program can then be executed with Run→Start. Figure 20.12 shows a
sample run.

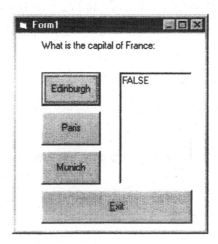

Figure 20.12   Sample test run

Next the form and the project are saved using the File→Save option. If the file has not been saved before then the user will be prompted to give the project and the form a new file name. In this case save the project and the form as VB03_01.VBP and VB03_01.FRM. These are listed in Program 20.1. It can be seen that the form file (VB03_01.FRM) contains the Visual Basic code along with the properties and definitions of the controls, whereas the project file (VB03_01.VBP) defines the user's environment, such as:

- The names of any forms.
- The control types.

📄 **Program 20.1: VB03_01.FRM**

```
VERSION 4.00
Begin VB.Form Form1
    BackColor       =   &H00FFFFFF&
    Caption         =   "Form1"
    ClientHeight    =   3630
    ClientLeft      =   2085
    ClientTop       =   3195
    ClientWidth     =   3495
    Height          =   4035
    Left            =   2025
    LinkTopic       =   "Form1"
    ScaleHeight     =   3630
    ScaleWidth      =   3495
    Top             =   2850
    Width           =   3615
    Begin VB.CommandButton Command4
        Caption     =   "&Exit"
        Height      =   615
        Left        =   480
        TabIndex    =   5
        Top         =   2880
        Width       =   2775
    End
    Begin VB.TextBox Text1
        Height      =   2055
        Left        =   1920
        TabIndex    =   4
        Top         =   720
        Width       =   1335
    End
    Begin VB.CommandButton Command3
        Caption     =   "Munich"
        Height      =   615
        Left        =   480
        TabIndex    =   3
        Top         =   2160
        Width       =   1095
    End
    Begin VB.CommandButton Command2
        Caption     =   "Paris"
        Height      =   615
        Left        =   480
        TabIndex    =   2
        Top         =   1440
        Width       =   1095
    End
    Begin VB.CommandButton Command1
        Caption     =   "Edinburgh"
```

```
            Height      =      615
            Left        =      480
            TabIndex    =      1
            Top         =      720
            Width       =      1095
         End
         Begin VB.Label Label1
            BackColor   =      &H00FFFFFF&
            Caption     =      "What is the capital of France:"
            Height      =      495
            Left        =      480
            TabIndex    =      0
            Top         =      120
            Width       =      2295
         End
      End
   End
   Attribute VB_Name = "Form1"
   Attribute VB_Creatable = False
   Attribute VB_Exposed = False


   Private Sub Command1_Click()
       Text1.Font = "Courier New"
       Text1.Text = "FALSE"
       Text1.Height = 1000
   End Sub


   Private Sub Command2_Click()
       Text1.Text = "TRUE"
   End Sub


   Private Sub Command3_Click()
       Text1.Text = "FALSE"
   End Sub

   Private Sub Label2_Click()

   End Sub


   Private Sub Command4_Click()
       End
   End Sub
```

📄 **Program 20.1: VB03_01.VBP**
```
Form=vb03_01.Frm
Object={F9043C88-F6F2-101A-A3C9-08002B2F49FB}#1.0#0; COMDLG32.OCX
Object={BDC217C8-ED16-11CD-956C-0000C04E4C0A}#1.0#0; TABCTL32.OCX
Object={3B7C8863-D78F-101B-B9B5-04021C009402}#1.0#0; RICHTX32.OCX
Object={6B7E6392-850A-101B-AFC0-4210102A8DA7}#1.0#0; COMCTL32.OCX
Object={FAEEE763-117E-101B-8933-08002B2F4F5A}#1.0#0; DBLIST32.OCX
Object={00028C01-0000-0000-0000-000000000046}#1.0#0; DBGRID32.OCX
Reference=*\G{BEF6E001-A874-101A-8BBA-
00AA00300CAB}#2.0#0#C:\WINDOWS\SYSTEM\OLEPRO32.DLL#Standard OLE Types
Reference=*\G{EE008642-64A8-11CE-920F-
08002B369A33}#1.0#0#c:\windows\system32\msrdo32.dll#Microsoft Remote
Data Object 1.0
ProjWinSize=56,29,273,181
ProjWinShow=0
Name="Project1"
HelpContextID="0"
StartMode=0
VersionCompatible32="0"
```

```
MajorVer=1
MinorVer=0
RevisionVer=0
AutoIncrementVer=0
ServerSupportFiles=0
```

20.4 Temperature conversion program

In this example the user will enter either a temperature in Centigrade or Fahrenheit and the program will convert to an equivalent Fahrenheit or Centigrade temperature. The steps taken, with reference to Figure 20.13, are:

1. Add a Label control and change its Caption property to 'Centigrade'.
2. Add a Label control and change its Caption property to 'Fahrenheit'.
3. Add a TextBox control and put it beside the Centigrade Label. Next change its Text property to '0'.
4. Add a TextBox control and put it beside the Fahrenheit Label. Next change its Text property to '32'.
5. Add a CommandButton control and put it below the text boxes. Next change its Caption property to 'C to F'. This command button will convert the value in the Centigrade to Fahrenheit and put the result in the Fahrenheit text box.
6. Add a CommandButton control and put it beside the other command button. Next change its Caption property to 'F to C'. This command button will convert the value in the Fahrenheit to Centigrade and put the result in the Centigrade text box.
7. Select the form and change the Caption property to 'Temperature Conversion'.

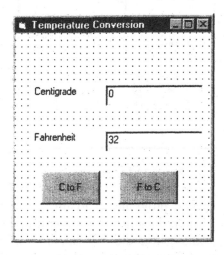

Figure 20.13 Temperature conversion form

Next the code associated with each control can be added, as follows:

1. First add code to the first command button (C to F) which will be used to convert the text from the Centigrade text box (Text1) and display it to the Fahrenheit text box (Text2). This is achieved with:

```
Private Sub Command1_Click()
Dim val As Double

    val = 9 / 5 * CDbl(Text1.Text) + 32
    Text2.Text = CStr(val)

End Sub
```

2. Next add code to the second command button (F to C) which will be used to convert the text from the Fahrenheit text box (Text2) and display it to the Centigrade text box (Text1). This is achieved with:

```
Private Sub Command2_Click()
Dim val As Double
    val = 5 / 9 * (CDbl(Text2.Text) - 32)
    Text1.Text = CStr(val)
End Sub
```

A test run of the program is given in Figure 20.14.

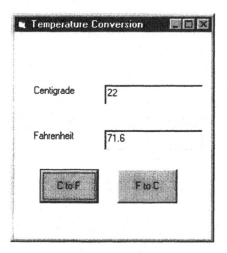

Figure 20.14 Sample run

The temperature conversion program up to this point has several weaknesses. One of these is that it does not have an exit option (this will be left as an exercise) and the other is that the user can enter a value which is not a valid temperature value and the program will accept it. For example, if the user enters a string of characters then the program stops and displays the message shown in Figure 20.15.

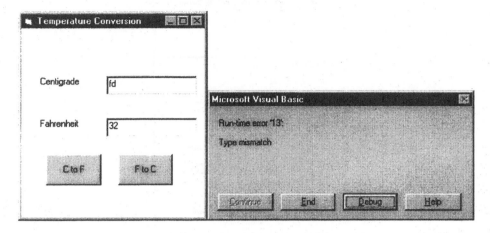

Figure 20.15 Sample run with invalid input

To overcome this problem the value that is entered is tested to see if it is a valid numeric value, using the IsNumeric() function. This returns a TRUE if the value can be converted to a numeric value, else a FALSE. The modified code for the two command buttons is given next and a sample run is shown in Figure 20.16.

```
Private Sub Command1_Click()
Dim val As Double

    If (IsNumeric(Text1.Text)) Then
        val = 9 / 5 * CDbl(Text1.Text) + 32
        Text2.Text = CStr(val)
    Else
        Text2.Text = "INVALID"
    End If

End Sub
```

```
Private Sub Command2_Click()
Dim val As Double

    If (IsNumeric(Text1.Text)) Then
        val = 5 / 9 * (CDbl(Text2.Text) - 32)
        Text2.Text = CStr(val)
    Else
        Text2.Text = "INVALID"
    End If

End Sub
```

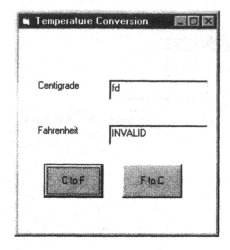

Figure 20.16   Sample run with invalid input

20.5 Quadratic roots program

In this example the program calculates the roots of a quadratic equation with a user entered values of a, b and c. The general form of a quadratic equation is:

$$ax^2 + bx + c = 0$$

the general solution is:

$$x_{1,2} = \frac{-b \pm \sqrt{b^2 - 4ac}}{2a}$$

This leads to three types of roots, these are:

if ($b^2 > 4ac$) then there are two real roots;
else if ($b^2 = 4ac$) then there is a single root of $-b/4a$;
else if ($b^2 < 4ac$) then these are two complex roots which are:

$$x_{1,2} = -\frac{b}{2a} \pm j\frac{\sqrt{4ac - b^2}}{2a}$$

The steps taken, with reference to Figure 20.17, are:

1.   Add a Label control and change its Caption property to 'a'.
2.   Add a Label control and change its Caption property to 'b'.
3.   Add a Label control and change its Caption property to 'c'.

4. Add a Label control and change its Caption property to 'x1'.
5. Add a Label control and change its Caption property to 'x2'.
6. Add a TextBox control and put it beside the a Label. Next change its Text property to '0' (this is the Text1 object).
7. Add a TextBox control and put it beside the b Label. Next change its Text property to '0' (this is the Text2 object).
8. Add a TextBox control and put it beside the c Label. Next change its Text property to '0' (this is the Text3 object).
9. Add a TextBox control and put it beside the x1 Label. Next change its Text property to '0' (this is the Text4 object).
10. Add a TextBox control and put it beside the x2 Label. Next change its Text property to '0' (this is the Text5 object).
11. Add a CommandButton control and put it below the text boxes. Next change its Caption property to 'Calculate'. This command button will be used to determine the roots of the equation.
12. Add a CommandButton control and put it beside the other command button. Next change its Caption property to 'Exit'. This command button will be used to exit the program.
13. Select the form and change the Caption property to 'Quadratic Equation'.

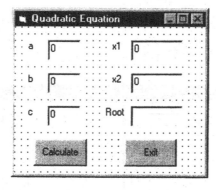

Figure 20.17 Quadratic equation form

Next the code associated with each control can be added, as follows:

1. First add code to the first command button (Calculate) which will be used to calculate the roots and display to the roots text box (Text4 and Text5). This is achieved with:

```
Private Sub Command1_Click()
Dim a, b, c As Double
Dim aval, bval, cval As String

    aval = Text1.Text
    bval = Text2.Text
    cval = Text3.Text
    a = CDbl(aval)
```

```
        b = CDbl(bval)
        c = CDbl(cval)

        If (Not (IsNumeric(aval)) Or _
            Not (IsNumeric(bval)) Or Not (IsNumeric(cval))) Then
            Text4.Text = ""
            Text5.Text = ""
            Text6.Text = "INVALID"
        ElseIf ((b * b) > (4 * a * c)) Then
            Text4.Text = CStr((-b + Sqr(b * b - 4 * a * c)) / (2 * a))
            Text5.Text = CStr((-b - Sqr(b * b - 4 * a * c)) / (2 * a))
            Text6.Text = "Real"
        ElseIf (b * b < 4 * a * c) Then
            Text4.Text = CStr(-b / (2 * a))
            Text5.Text = "j" + CStr(Sqr(4 * a * c - b * b) / (2 * a))
            Text6.Text = "Complex"
        Else
            Text4.Text = CStr(-b / (2 * a))
            Text5.Text = ""
            Text6.Text = "Singlar"
        End If
End Sub
```

Notice that the code includes the _ character which allows the programmer to continue a statement onto another line.

2. Next add code to the second command button (Exit):

```
Private Sub Command2_Click()
        End
End Sub
```

Figure 20.18 shows two sample runs.

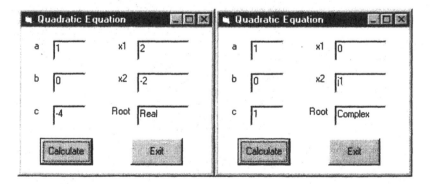

Figure 20.18   Sample runs

20.6 Resistance calculation with slider controls program

An excellent method of allowing the user to input a value within a fixed range is to use a slider control. These slider controls can either be vertical (VScroll) or horizontal (HScroll). The main properties, as shown in Figure 20.19, of a scroll bar are:

- Max. Which defines the maximum value of the scroll bar.
- Min. Which defines the minimum value of the scroll bar.
- Value. Which gives the current slider value.

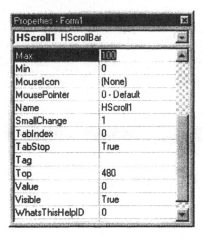

Figure 20.19   Scroll bar properties

As an example, a horizontal slider will be setup with a voltage range of 0 to 100. The value of the slider will be shown. The steps taken, with reference to Figure 20.20, are:

1. Add a Label control and change its Caption property to 'Voltage'.
2. Add a Label control and change its Caption property to '0'.
3. Add a HScrollBar control below the labels. Next change its Max property to '100' and its Min property to '0'.

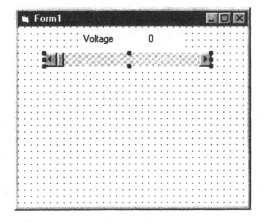

Figure 20.20   Voltage form

Next the code associated with each control can be added, as follows:

1. Add code to the horizontal scroll bar which will be used to display its value to the voltage value label (Label2). This is achieved with:

```
Private Sub HScroll1_Change()
    Label2.Caption = HScroll1.Value
End Sub
```

This will take the value from the scroll bar (HScroll1.Value) and display to the second label box (Label2). When the program is run then the user can move the scroll bar back and forward which causes a change in the displayed voltage value (from 0 to 100). A sample run is shown in Figure 20.21.

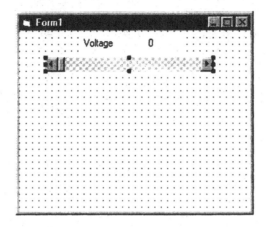

Figure 20.21 Sample run

This project can now be enhanced by adding another slider for current and displaying the equivalent resistance (which is voltage divided by current). The steps taken, with reference to Figure 20.22, are:

1. Add a Label control and change its Caption property to 'Voltage'.
2. Add a Label control and change its Caption property to '0'.
3. Add a HScrollBar control below the labels (HScroll1). Next change its Max property to '100' and its Min property to '0'.
4. Add a Label control and change its Caption property to 'Current'.
5. Add a Label control and change its Caption property to '0'.
6. Add a HScrollBar control below the labels (HScroll2). Next change its Max property to '100' and its Min property to '0'.
7. Add a Label control and change its Caption property to 'Resistance'.
8. Add a Label control and change its Caption property to '0'.

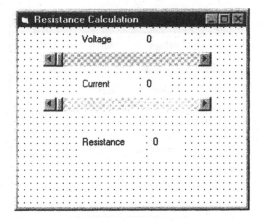

Figure 20.22   New form

Next the code associated with each control can be added, as follows:

1. Add code to the first horizontal scroll bar (HScroll1) which will be used to display its value to the voltage value label (Label2). The resistance label (Label6) is also updated with the result of the voltage divided by the current. This is achieved with:

```
Private Sub HScroll1_Change()
    Label2.Caption = HScroll1.Value
    If (HScroll2.Value <> 0) Then
        Label6.Caption = 100 * HScroll1.Value / HScroll2.Value
    End If
End Sub
```

2. Add code to the second horizontal scroll bar (HScroll2) which will be used to display its value to the current value label (Label4) with the value of the scroll bar divided by 100. The resistance label (Label6) is also updated with the result of the voltage divided by the current. This is achieved with:

```
Private Sub HScroll2_Change()
    Label4.Caption = HScroll2.Value / 100
    If (HScroll2.Value <> 0) Then
        Label6.Caption = 100 * HScroll1.Value / HScroll2.Value
    End If
End Sub
```

Figure 20.23 shows a sample run.

One of the problems of the design is that the controls and form have names which do not document their function. Visual Basic uses a naming convention which uses the type of control and consecutively adds a number, as shown in the design in Figure 20.24.

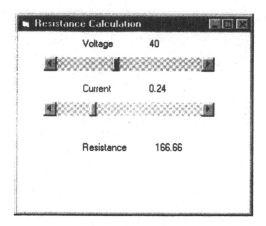

Figure 20.23 Sample run

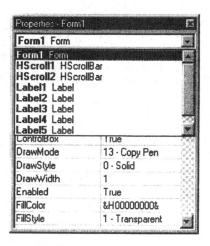

Figure 20.24 Object names

In this case an improved naming convention might be:

```
Label1        Voltage_Label
Label2        Voltage_Show
Label3        Current_Label
Label4        Current_Show
Label5        Resistance_Label
Label6        Resistance_Show
HScroll1      Voltage_Value
HScroll2      Current_Value
Form1         Resistance_Calc
```

These are set by selecting the properties of each of the objects and then changing the Name property to the required name. An example of changing the name of the form to Resistance_Calc is given in Figure 20.25.

Figure 20.25 Changing the name of the form

Figure 20.26 shows the list of objects after each of their names has been changed. Notice that it is now easier to locate the required object.

Figure 20.26 Changing the name of the form

Next the code must be modified so that the references are to the newly named objects. The code for the voltage scroll bar (Voltage_Value) is now:

```
Private Sub Voltage_Value_Change()
    Voltage_Display.Caption = Voltage_Value.Value
    If (Current_Value.Value <> 0) Then
     Resistance_Value.Caption = 100 * Voltage_Value.Value / Current_Value.Value
    End If
End Sub
```

and the code for the current scroll bar (Current_Value) is now:

```
Private Sub Current_Value_Change()
    Current_Display.Caption = Current_Value.Value / 100
```

```
    If (Current_Value.Value <> 0) Then
        Resistance_Value.Caption = 100 * Voltage_Value.Value /
                                   Current_Value.Value
    End If
End Sub
```

20.7 Exercises

20.7.1 Write a Visual Basic program in which the user enters either a value in either radians or degrees and the program converts to either degrees or radians, respectively. Figure 20.27 shows a sample run.

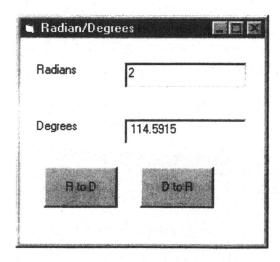

Figure 20.27 Radians to degrees conversion

20.7.2 Modify the program in Exercise 20.7.1 so that invalid entries are not accepted.

20.7.3 Modify the program in Exercise 20.7.2 so that the conversion value is automatically converted when the user enters a value (that is, there is no need for the command buttons).

20.7.4 Write separate Visual Basic programs with slider controls for the following formula:

(i) $F = ma$ range: m=0.01 to 1000 g, a=0.01 to 100 m.s^{-2}
(ii) $V = IR$ range: I=0.1 to 100 A, V=0.1 to 100 V

20.7.5 Write a Visual Basic program that calculates the values of m and c for a straight line. The values of (x_1, y_1) and (x_2, y_2) should be generated with slider controls (with a range of −100 to +100 for each of the values). Figure 20.28 shows a sample design.

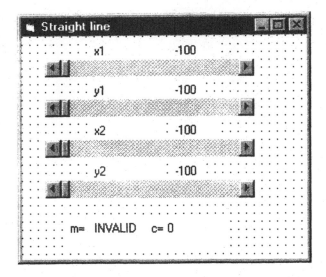

Figure 20.28 Straight line program design

20.7.6 Modify the program in Exercise 20.7.5 so that a divide by zero does not occur when the difference in the x values is zero. If this is so then the program should display 'INFINITY' for the gradient. If the two coordinates are the same then the program should display the message 'INVALID' for the gradient.

20.7.7 Write a Visual Basic program which has a multiple choice option question which is repeated. The program should keep a running tally of the number of correct answers and the number of incorrect answers.

20.7.8 Complete Worksheet 3 on page 39 using Visual Basic.

21 Menus and Dialog Boxes

21.1 Introduction

This chapter discusses how menus and dialog boxes are used.

21.2 Menu editor

Most Windows programs have menus in which the user selects from a range of defined pull-down menus with defined options. Visual Basic has an easy-to-use function called the Menu Editor which is used to create custom menus. The Menu Editor is started from the Tools Menu or from the toolbar shortcut ▣.

An example screen from the Menu Editor is shown in Figure 21.1. It includes:

- **Caption**. Which is a text box in which the name of the menu bar or menu option is entered. A hyphen (-) is entered as a caption if a menu separator bar is required. This bar helps to separate menu options. Often in menus the user can select a menu option by pressing the ALT key and an assigned key (hot key). To specify the ALT-hot key then an & is inserted before the letter of the menu option. When the program is run then this letter is underlined. For example Fi&le would be displayed as File and the assigned keys would be ALT-L. A double ampersand specifies the ampersand character.
- **Name**. Which is a text box in which the control name for the menu option is specified. This is used by the program code and is not displayed to the user when the program is run.
- **Index**. Which is a numeric value that can be used to specify the menu option. Typically it is used when calling a single function which services several menu items. For example, a File menu may have the options: New, Open and Save, then a single function could be created to service these requests and the index value would be used which option has been selected.
- **Shortcut**. Which is a pull-down menu that can be used to specify a

shortcut key (Cntrl-A to Cntrl-Z, F1 to F12, Cntrl-F1 to Cntrl-F12, Shift-F1 to Shift-F12, Shift-Cntrl-F1 to Shift-Cntrl-F12, Cntrl-Ins, Shift-Ins, Del, Shift-Del, Alt-Bkspace).

- **HelpContextID.** Which is a text box in which a unique numeric value is specified for the context ID. This value can be used to find the appropriate Help topic in the Help file identified by the HelpFile property.
- **NegotiatePosition.** Which is a pull-down list box which allows the user to specify the menu's NegotiatePosition property and determine how the menu appears in a form. Value options are 0 (None), 1 (Left) and 2 (Middle) and 3 (Right).
- **Checked.** Which is a check box which specifies if a check mark is to appear initially at the left of a menu item. It is generally used to specify if a toggled menu option is initially on or off.
- **Enabled.** Which is a check box which specifies if the menu item is to respond to events. If it is not enabled then the menu item appears dimmed.
- **Visible.** Which is a check box which specifies if the menu item is to appear in the menu.

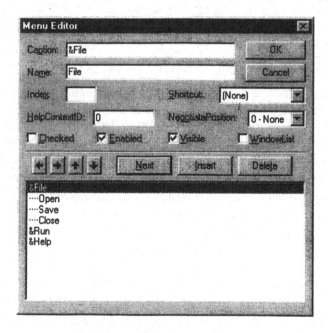

Figure 21.1   Menu editor

21.2.1 Creating a menu system

The user enters the caption and name and then uses the outlining buttons to either promote or demote the item to a higher or lower level (with the left and right arrow buttons) or up and down (with the up and down arrow buttons). With this up to 4 levels of submenus can be created.

The list box displays a hierarchical list of menu items with indented submenu items which indicate their hierarchical position. A menu option is inserted using the Insert button and deleted with the Delete button. The OK button closes the Menu Editor and saves the most recent changes.

The code associated with a menu item is defined by the menu item name. For example if a menu item has the caption of `File` and a name of `File-Option` then the associated code function will be `FileOption_Click`.

For example, to create a menu with:

```
File
   Open
   Save
   Close
Exit
```

1. Create a caption `&File` and add the name of `MenuFile`. Next press the Next button.
2. Create a caption `&Open` and add the name of `MenuFileItem`. Add an index value of 0. Then select the right arrow button to move the option to the next level, as shown in Figure 21.2. Next press the Next button.
3. Create a caption `&Save` and add the name of `MenuFileItem`. Add an index value of 1, as shown in Figure 21.3. Next press the Next button.
4. Create a caption `&Close` and add the name of `MenuFileItem`. Add an index value of 2, as shown in Figure 21.4. Next press the Next button.
5. Create a caption `&Exit` and add the name of `Exit`. Then select the left arrow button to move the option to the next level, as shown in Figure 21.5. Next press the OK button and the form shown in Figure 21.6 should be displayed.

The associated code modules which are related to these menus items are `MenuFile_Click()`, `MenuFileItem_Click()` and `Exit_Click()`. The `MenuFileItem_Click()` module has the `Index` parameter passed to it. This has a value of 0 when `Open` is select, a 1 when `Save` is selected and a 2 when `Close` is selected. The `MenuFileItem` is specified with:

```
Private Sub MenuFileItem_Click(Index As Integer)

End Sub
```

thus the value of Index will either be 0, 1 or 2.

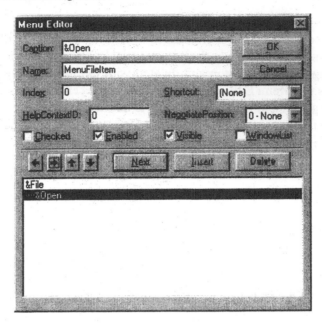

Figure 21.2 Adding a menu item

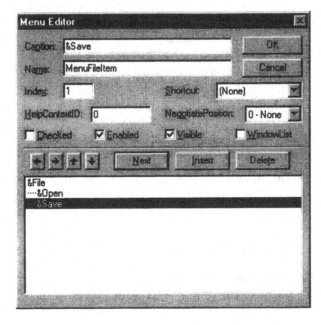

Figure 21.3 Adding a menu option

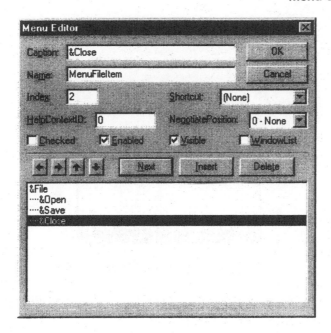

Figure 21.4 Adding a menu option

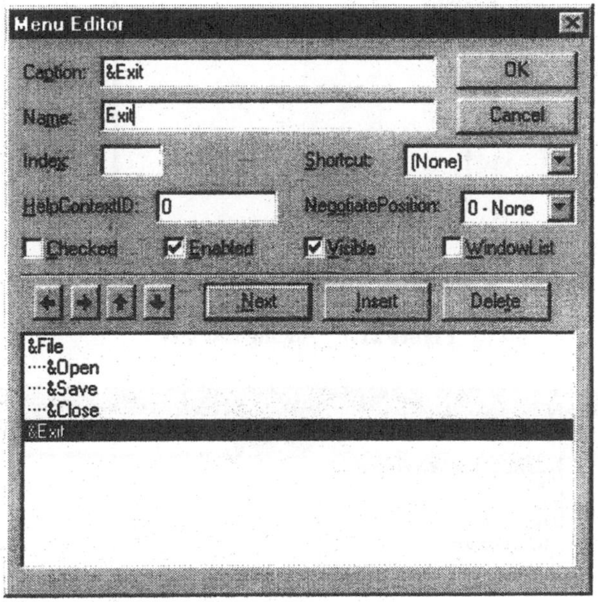

Figure 21.5 Adding a menu option

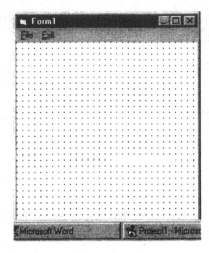

Figure 21.6 Menu editor

Next a text box is added to the form (Text1), as shown in Figure 21.7. The resulting objects are shown in Figure 21.8.

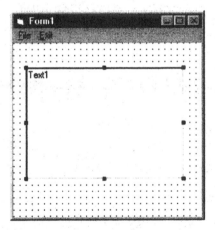

Figure 21.7 Adding text box

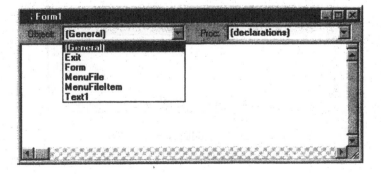

Figure 21.8 Program objects

Next the code can be added. The `MenuFileItem` object services the Open, Save and Close options and the Index parameter passed into it. In the code below the Case statement is used to test the Index parameter and, in this case, display some text to the text box.

📄 **Program 21.1**

```
Private Sub MenuFileItem_Click(Index As Integer)

    Select Case Index
    ' Check index value of selected menu item.
    Case 0
        Text1.Text = Text1.Text & " Open "
    Case 1
        Text1.Text = Text1.Text & " Save "
    Case 2
        Text1.Text = Text1.Text & " Close "
    End Select

End Sub
```

This code appends either 'Open', 'Save' or 'Close' to the text box (Text1). The `&` character concatenates two strings together, thus the code `Text1.Text & "Open"` simply adds the text 'Open' to the text already in the text window. The exit code is added as follows:

📄 **Program 21.2**

```
Private Sub Exit_Click()
    End
End Sub
```

Figure 21.9 shows a sample run of the developed program. The Exit menu option is used to quit the program and the File options (File, Save and Close) simply display the required text to the text box.

Figure 21.9   Example run

21.3 Common dialog control

The CommonDialog control allows for file operations such as opening, saving or printing files. It is basically a control between Visual Basic and the Microsoft Windows dynamic-link library COMMDLG.DLL. Thus this file must be in the Microsoft Windows SYSTEM directory for the common dialog control to work.

A dialog box is added to an application by first adding the CommonDialog control to a form and setting its properties. When developing the program the common dialog box is displayed as an icon on the form. A program calls the dialog with one of the following (assuming that the dialog box is named CommonDialog1):

CommonDialog1.Filter   A string which displays the filename filter. The | character is used to differentiate different For example the following filter enables the user to select text files or graphic files that include bitmaps and icons:

```
Text (*.txt)|*.txt|Pictures(*.bmp;*.ico)|*.bmp;*.ico
```

CommonDialog1.Filename   Which returns or sets the path and filename of a selected file.

CommonDialog1.FilterIndex   Which defines the default filter (with reference to the Filter).

CommonDialog1.ShowSave   Which returns or sets the path and filename of a selected file. Displays the CommonDialog control's Save As dialog box. The object placeholder represents an object expression that evaluates to an object in the Applies To list.

CommonDialog1.ShowOpen   Which displays the CommonDialog control's Open dialog box.

CommonDialog1.ShowPrinter Which displays the CommonDialog control's Printer dialog box.

CommonDialog1.ShowFont   Which displays the CommonDialog control's Font dialog box.

CommonDialog1.ShowHelp   Which runs WINHELP.EXE and displays the specified help file.

A dialog box is added to a form in any position, as shown in Figure 21.10.

This box can be placed anywhere as it will not be seen on the form when the program is run.

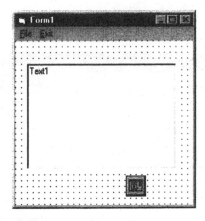

Figure 21.10 Example run

The common dialog box can be used to determine the filename of a file to be opened or saved. It is used in Program 21.3 (which is a modification of Program 21.1) in the Open and Save file menu options. The filter is set to:

```
"All Files (*.*)|*.*|Text Files (*.txt)|*.txt|Temp Files (*.tmp)|*.tmp"
```

This displays, in the Type of File field, the three options `"All Files (*.*),*.*,Text Files (*.txt),*.txt` and `Temp Files (*.tmp)|*.tmp"`. The default file type is set to the second option with the filter index setting of:

```
CommonDialog1.FilterIndex = 2
```

Figure 21.11 shows a sample run and the dialog box. It can be seen that the default Type of File is set to `"Text Files (*.txt)"` and Figure 21.12 shows an example message box.

Program 21.3
```
Private Sub MenuFileItem_Click(Index As Integer)
Dim filename As String

  Select Case Index
  Case 0
    CommonDialog1.Filter = _
    "All Files (*.*)|*.*|Text Files (*.txt)|*.txt|Temp Files (*.tmp)|*.tmp"
    CommonDialog1.FilterIndex = 2
    CommonDialog1.ShowOpen
    filename = CommonDialog1.filename
    MsgBox ("Open Filename: " & filename)
  Case 1
    CommonDialog1.Filter = _
    "All Files (*.*)|*.*|Text Files (*.txt)|*.txt|Temp Files (*.tmp)|*.tmp"
    CommonDialog1.FilterIndex = 2
```

```
      CommonDialog1.ShowSave
      filename = CommonDialog1.filename
      MsgBox ("Save Filename: " & filename)
    Case 2
      Text1.Text = Text1.Text & " Close "
    End Select
End Sub
```

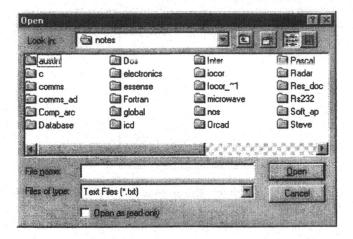

Figure 21.11   Example run

Figure 21.12   Example run

21.3.1 File Open/Save Dialog Box Flags

Various flags can be set before the file open/save dialog box. These are defined in Table 21.1 are set with:

CommonDialog1.Flags= *flag1* + *flag2* + ...

The flags can either be defined with their constant name (such as cdlOFNReadOnly) or by the value (0x01). These values or names can be added together to achieve the required functionality. For example, modifying the previous example to checks the read-only check box for Open dialog box. A sample dialog box is shown in Figure 21.13.

```
      CommonDialog1.Filter = "All Files (*.*)|*.*|
         Text Files (*.txt)|*.txt|Temp Files (*.tmp)|*.tmp"
      CommonDialog1.FilterIndex = 2
      CommonDialog1.Flags = cdlOFNHideReadOnly
      CommonDialog1.ShowOpen
      filename = CommonDialog1.filename
      MsgBox ("Open Filename: " & filename)
```

For example, modifying the previous example to hide the read-only check box. A sample dialog box is shown in Figure 21.14.

```
CommonDialog1.Filter = "All Files (*.*)|*.*|
    Text Files (*.txt)|*.txt|Temp Files (*.tmp)|*.tmp"
CommonDialog1.FilterIndex = 2
CommonDialog1.Flags = cdlOFNHideReadOnly
CommonDialog1.ShowOpen
filename = CommonDialog1.filename
MsgBox ("Open Filename: " & filename)
```

For example, modifying the previous example to Allows the File Name list box to have multiple selections. A sample dialog box is shown in Figure 21.15.

```
CommonDialog1.Filter = "All Files (*.*)|*.*|
    Text Files (*.txt)|*.txt|Temp Files (*.tmp)|*.tmp"
CommonDialog1.FilterIndex = 2
CommonDialog1.Flags = cdlOFNAllowMultiselect
CommonDialog1.ShowOpen
filename = CommonDialog1.filename
MsgBox ("Open Filename: " & filename)
```

The flag settings for a dialog box with the read-only box checking and that the user is not allowed to change the directory can either be set with:

```
CommonDialog1.Flags = cdlOFNReadOnly + cdlOFNNoChangeDir
```

or

```
CommonDialog1.Flags = 9
```

which is 1 (cdlOFNReadOnly) added to 8 (cdlOFNNoChangeDir). The method of using lable constants is preferable as it helps to document the program.

Table 21.1   CommonDialog control constants

Constant	Value	Description
cdlOFNReadOnly	&H1 (1)	Checks Read-Only check box for Open and Save As dialog boxes, see Figure 21.13.
cdlOFNOverwritePrompt	&H2 (2)	Causes the Save As dialog box to generate a message box if the selected file already exists.
cdlOFNHideReadOnly	&H4 (4)	Hides the Read-Only check box, see Figure 21.14.
cdlOFNNoChangeDir	&H8 (8)	Sets the current directory to what it was when the dialog box was invoked.
cdlOFNHelpButton	&H10 (16)	Causes the dialog box to display the Help button.

`cdlOFNNoValidate`	&H100 (256)	Allows invalid characters in the returned filename.
`cdlOFNAllowMultiselect`	&H200 (512)	Allows the File Name list box to have multiple selections, see Figure 21.15.
`cdlOFNExtensionDifferent`	&H400 (1024)	The extension of the returned filename is different from the extension set by the DefaultExt property.
`cdlOFNPathMustExist`	&H800 (2096)	User can enter only valid path names.
`cdlOFNFileMustExist`	&H1000 (4096)	User can enter only names of existing files.

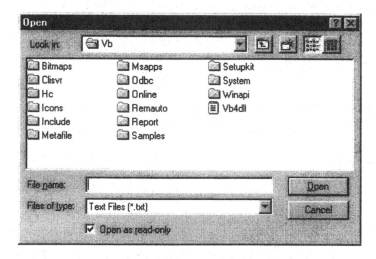

Figure 21.13   Example run

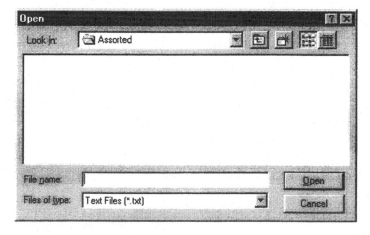

Figure 21.14   Example run

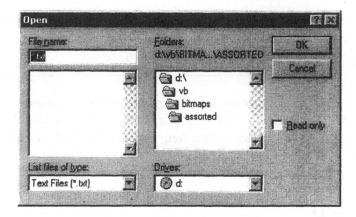

Figure 21.15 Example run

21.3.2 Showing help manuals

The common dialog box can also be used to run the help program
WINHELP.EXE. This is achieved by:

```
CommonDialog1.ShowHelp
```

and the name of the help manual is specified by:

```
CommonDialog1.HelpFile = filename
```

Various options flags can also be set with the help option, these are specified in Table 21.2.

Table 21.2   CommonDialog control constants

Constant	Value	Description
cdlHelpCommand	&H102	Executes a help macro.
cdlHelpContents	&H3	Displays the help contents topic.
cdlHelpContext	&H1	Displays help for a particular context.
cdlHelpContextPopup	&H8	Displays in a pop-up window a particular Help topic.
cdlHelpForceFile	&H9	Ensures that WinHelp displays the correct Help file. If the correct Help file is currently displayed, no action occurs. If the incorrect Help file is displayed, WinHelp opens the correct file.
cdlHelpHelpOnHelp	&H4	Displays Help for using the help application itself.

`cdlHelpIndex`	&H3	Displays the index of the specified Help file.
`cdlHelpKey`	&H101	Displays Help for a particular keyword.

Typically a help menu option is added to a program. An example help event is given in Program 21.4. The help command used is `cdlHelpConents` which shows the index page of the help manual. Figure 21.16 shows an example of a run and Figure 21.17 shows an example for the An example of Vbcbt help manual.

📄 **Program 21.4**

```
Private Sub Help_Click()
Dim filename As String

    CommonDialog1.Filter = "All Files (*.*)|*.*|Help Files (*.hlp)|*.hlp"
    CommonDialog1.FilterIndex = 2
    CommonDialog1.ShowOpen

    filename = CommonDialog1.filename

    CommonDialog1.HelpFile = filename
    CommonDialog1.HelpCommand = cdlHelpContents
    CommonDialog1.ShowHelp
End Sub
```

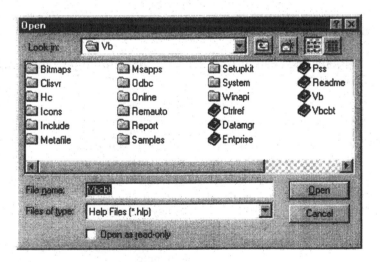

Figure 21.16 Selecting help file

21.4 Running an application program

Visual Basic allows for the execution of applications with the `Shell` command. Its format is:

`Shell` (*pathname[, windowstyle]*)

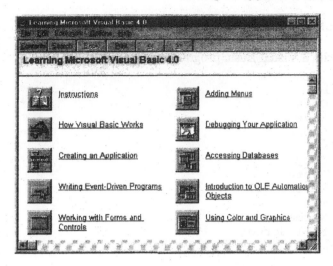

Figure 21.17 Example help manual

where the *windowstyle* is as defined in Table 21.3. If this argument is missing then the *windowstyle* is that the program is started with a minimized focus.

Table 21.3 Windowstyle named argument

Constant	Val	Description
vbHide	0	Window is hidden and focus is passed to the hidden window.
vbNormalFocus	1	Window has focus and is restored to its original size and position.
vbMinimizedFocus	2	Window is displayed as an icon with focus.
vbMaximizedFocus	3	Window is maximized with focus.
vbNormalNoFocus	4	Window is restored to its most recent size and position. The currently active window remains active.
vbMinimizedNoFocus	6	Window is displayed as an icon. The currently active window remains active.

The following shows an example of executing the `Calc.exe` and `Notepad.exe` programs:

📄 **Program 21.5**

```
Private Sub Notepad_Click()
Dim rtn As Integer
  rtn = Shell
     ("c:\windows\notepad.exe", 1)
End Sub

Private Sub Calc_Click()
Dim rtn As Integer
  rtn = Shell
     ("c:\windows\calc.exe", 1)
End Sub
```

21.5 Exercises

21.4.1   Write a Visual Basic program with the following menu system:

```
File
   Open
   Save
   Close
Edit
   Copy
   Paste
   Select All
View
   Normal
   Full Screen
Help
```

21.4.2   Modify the menu system in Exercise 21.4.1 so that the program displays the function of the menu option.

21.4.3   Expand Program 21.5 and its menu system so that it runs other Windows programs. An example could be:

```
Utils
   Calculator
   Notepad
   Paint
WordProcessing
   Word
   AmiPro
Spreadsheets
   Lotus123
   Excel
Exit
```

21.4.4   Integrate some of the programs from previous chapters into a single program with menu options. For example, the menu system could be:

```
Programs
   Temperature Conversion
   Quadratic Equation
   Straight Line
Exit
```

One possible method of implementing this program is to compile the temperature conversion, quadratic equation and straight line programs to an EXE (File→Make EXE file...). Then run the Shell function from the object call.

```
rtn = Shell("tempcon.exe", 1)
```

22 Events

22.1 Introduction

Visual Basic differs from many other programming languages in that it is event-driven where the execution of a program is defined by the events that happen. This is a different approach to many programming languages which follow a defined sequence of execution and the programmer must develop routines which react to events. This chapter discusses the events that happen in Visual Basic.

22.2 Program events

Each object in Visual Basic has various events associated with it. For example, the single click on an object may cause one event but a double click causes another. The events are displayed at the right-hand side of the code window. An example is shown in Figure 22.1 which in this case shows the events: KeyPress, KeyUp, LinkClose, LinkError, LinkExecute, LinkOpen and Load. The name of the routine which contains the code for the event and object is in the form:

```
Private Sub ObjectName_Event()

End
```

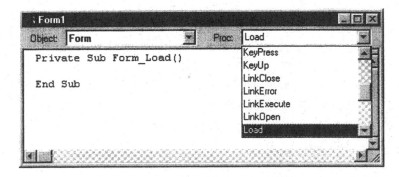

Figure 22.1   Form events

The main events that occur in Visual Basic include:

- **Click**. Which occurs when a user performs a single click of the mouse button on the object.
- **DblClick**. Which occurs when a user performs a double click of the mouse button on the object.
- **MouseUp**. Which occurs when the user releases the mouse button.
- **MouseDown**. Which occurs when the user presses the mouse button down.
- **MouseMove**. Which occurs when the user moves the mouse.
- **KeyUp**. Which occurs when the user releases a key.
- **KeyDown**. Which occurs when the user presses a key.
- **KeyPress**. Which occurs when the user presses and releases a key (the KeyDown and KeyUp events).
- **Load**. Which occurs when a form is loaded.
- **DragDrop**. Which occurs at the begin and end of a drag operation of any control.
- **LostFocus**. Which occurs when an object loses the focus, either by user action, such as tabbing to or clicking another object.
- **Resize**. Which occurs when a form is first displayed or if the object size is changed.
- **Unload**. Which occurs when a form is about to be removed from the screen.

Not all controls and forms (objects) have all the events associated with them. For example a form may have the following:

```
Private Sub Form_Activate()
Private Sub Form_Click()
Private Sub Form_Deactivate()
Private Sub Form_DblClick()
Private Sub Form_DragDrop(Source As Control, X As Single, Y As Single)
Private Sub Form_DragOver(Source As Control, X As Single, Y As Single, State
As Integer)
Private Sub Form_KeyDown(KeyCode As Integer, Shift As Integer)
Private Sub Form_KeyPress(KeyAscii As Integer)
Private Sub Form_KeyUp(KeyCode As Integer, Shift As Integer)
Private Sub Form_Load()
Private Sub Form_LostFocus()
Private Sub Form_MouseDown(Button As Integer, Shift As Integer, X As Single,
Y As Single)
Private Sub Form_MouseMove(Button As Integer, Shift As Integer, X As Single,
Y As Single)
Private Sub Form_MouseUp(Button As Integer, Shift As Integer, X As Single, Y
As Single)
Private Sub Form_Resize()
Private Sub Form_Terminate()
Private Sub Form_Unload(Cancel As Integer)
```

whereas a command button has a reduced set of routines:

```
Private Sub Command1_Click()
Private Sub Command1_DragDrop(Source As Control, X As Single, Y As Single)
Private Sub Command1_DragOver(Source As Control, X As Single, Y As Single,
State As Integer)
Private Sub Command1_GotFocus()
Private Sub Command1_KeyDown(KeyCode As Integer, Shift As Integer)
Private Sub Command1_KeyPress(KeyAscii As Integer)
Private Sub Command1_KeyUp(KeyCode As Integer, Shift As Integer)
Private Sub Command1_LostFocus()
Private Sub Command1_MouseDown(Button As Integer, Shift As Integer, X As
Single, Y As Single)
Private Sub Command1_MouseMove(Button As Integer, Shift As Integer, X As
Single, Y As Single)
Private Sub Command1_MouseUp(Button As Integer, Shift As Integer, X As Sin-
gle, Y As Single)
```

The parameters passed into the routine depend on the actions associated with the events. For example, the Keypress event on a command button causes the routine:

```
Private Sub Command1_KeyPress(KeyAscii As Integer)

End
```

to be called. The value of KeyAscii will contain the ASCII value of the character press.

The mouse down event for a command button has the following routine associated with it:

```
Private Sub Command1_MouseDown(Button As Integer, Shift As Integer, X As
Single, Y As Single)
```

where the value of Button is the value of the button press (0 for none, 1 for the left button and 2 for the right button). Shift specifies if the shift key has been pressed, and X, Y specify the x,y co-ordinates of the mouse point. The following sections discuss some events which occur.

22.2.1 Click

The Click event occurs when the user presses and then releases a mouse button over an object. When the object is a form the event occurs when the user clicks on a blank area or a disabled control. If the object is a control, then the event occurs when the user:

- Clicks on a control with any of the mouse buttons. When the control is a CheckBox, CommandButton, or OptionButton control then the Click event occurs only when the user clicks the left mouse button.
- Presses the ALT-*hotkey* for a control, such as pressing Alt-X for the E&xit control property name.
- Presses the Enter key when the form has a command button.
- Presses the Space key when command button, option button or check box control has the focus.

22.2.2 Dbl Click

The Dbl Click event occurs when the user presses and releases the mouse button twice over an object. In a form it occurs when the user double clicks either on a disabled control or a blank area of a form. On a control the event happens when the user:

- Double-clicks on a control with the left mouse button.
- Double-clicks an item within a ComboBox control whose Style property is set to 1 (Simple).

An example form is shown in Figure 22.2. In this case a text box is added to the form. Then the following code is added to the form:

```
Private Sub Form_DblClick()
   Text1.Text = "MISS"
End Sub
```

which will display the text 'MISS' when the user click on any blank area on the form. The following code is added to the text box:

```
Private Sub Text1_DblClick()
   Text1.Text = "HIT"
End Sub
```

which displays the text 'HIT' when the user double clicks on the text box.

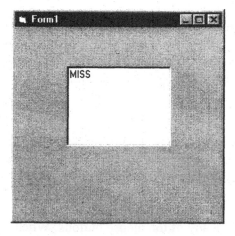

Figure 22.2   Form events

22.2.3 Mouse Up, Mouse Down

The MouseUp event occurs when the user releases the mouse button and the MouseDown event when the user presses a mouse button. The standard format for the associated routines are:

```
Private Sub Form_MouseDown(button As Integer, shift As Integer,
                          x As Single, y As Single)

Private Sub Form_MouseUp(button As Integer, shift As Integer,
                         x As Single, y As Single)
```

where

- `button` identifies which button was pressed (MouseDown) or released (MouseUp). The first bit of the value identifies the left button, the second identifies the right button and the third bit identifies the middle button. Thus a value of 1 identifies the left button, a value of 2 identifies the right and a value of 4 identifies the middle button (if the mouse has one). A combination of these values can be used to identify button press combinations, such as a value of 6 identifies that the middle and right buttons have been pressed. Table 22.1 shows the constant names for the buttons.
- `shift` identifies the state of the Shift, Cntrl, and Alt keys when the button was pressed (or released). The first bit of the value identifies the Shift key, the second bit identifies the Cntrl key and the third bit identifies the Alt key. Thus a value of 1 identifies the Shift key, a value of 2 identifies the Cntrl key and a value of 4 identifies the Alt key. A combination of these values can be used to identify key combinations, such as a value of 6 identifies that the Cntrl and Alt keys are pressed. Table 22.1 shows the constant names for the keys.
- `x, y` identifies the current location of the mouse pointer. The x and y values are relatively to the ScaleHeight, ScaleWidth, ScaleLeft, and ScaleTop properties of the object.

An example of code for the mouse down event is given next and a sample run is shown in Figure 22.3.

```
Private Sub Text1_MouseDown(Button As Integer, Shift As Integer, X As
Single, Y As Single)
    If (Button = vbLeftButton) Then
        Text1.Text = "LEFT BUTTON " & X & " " & Y
    ElseIf (Button = vbRightButton) Then:
        Text1.Text = "RIGHT BUTTON " & X & " " & Y
    ElseIf (Button = vbMiddleButton) Then:
        Text1.Text = "MIDDLE BUTTONS " & X & " " & Y
    End If
End Sub
```

Table 22.1 Button constants

Constant (Button)	Value	Description
vbLeftButton	1	Left button is pressed
vbRightButton	2	Right button is pressed
vbMiddleButton	4	Middle button is pressed

Table 22.2 Button constants

Constant	Value	Description
vbShiftMask	1	SHIFT key is pressed
vbCtrlMask	2	CTRL key is pressed
vbAltMask	4	ALT key is pressed

Figure 22.3 Form events

22.2.4 Mouse move

The MouseMove event occurs when the user moves the mouse. Thus it is continually being called as it moves across an object. The standard format for the associated routines are:

```
Private Sub Form_MouseMove(button As Integer, shift As Integer,
        x As Single , y As Single)

Private Sub object_MouseMove(button As Integer, shift As Integer,
        x As Single, y As Single)
```

where button, shift, x and y have the same settings as the ones defined in the previous section.

An example of code for the MouseMove event is given next and a sample code is shown in Figure 22.4. Figure 22.5 shows a sample run. In this case, the x and y co-ordinate is the mouse pointer and the button pressed are displayed to the text box.

```
Private Sub Form_MouseMove(Button As Integer, Shift As Integer,
        X As Single, Y As Single)

    Text1.Text = "X,Y:" & X & " " & Y & " Button:" & Button

End Sub
```

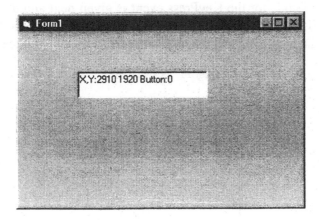

```
Private Sub Form_MouseMove(Button As Integer, Shift As Integer,
    Text1.Text = "X,Y:" & X & " " & Y & " Button:" & Button
End Sub
```

Figure 22.4   Sample code

Figure 22.5   Sample run

22.2.5 Drag and drop

The DragDrop event occurs at the end of a drag and drop operation of any control. The standard format for the associated routines are:

```
Private Sub Form_DragDrop(source As Control, x As Single, y As Single)

Private Sub object_DragDrop([index As Integer,]source As Control,
            x As Single, y As Single)
```

where:

- source is the control being dragged.
- x, y specifies the current x and y co-ordinate of the mouse pointer within the target form or control. The x and y values are relative to the ScaleHeight, ScaleWidth, ScaleLeft, and ScaleTop properties of the object.

22.2.6 Keypress

The KeyPress event occurs when the user presses and releases a key (the KeyDown and KeyUp events). The standard format for the associated routines are:

```
Private Sub Form_KeyPress(Keyascii As Integer)
Private Sub object_KeyPress(Keyascii As Integer)
```

where Keyascii specifies a standard ANSI keycode, this can be converted into an ASCII character with the conversion function:

```
Chr(KeyAscii)
```

An example of code for the KeyPress event is given next and a sample run is shown in Figure 22.6. For this program the user can type in the blank area on the form (the grey area) and the text will be displayed in the text box.

```
Private Sub Form_KeyPress(KeyAscii As Integer)
    Text1.Text = Chr(KeyAscii)
End Sub
```

Figure 22.6   Sample run

22.2.7 Lost focus

The Lost focus occurs when an object loses the focus, either by user action, such as tabbing to or clicking another object. The standard format for the associated routines are:

```
Private Sub Form_LostFocus()
Private Sub object_LostFocus()
```

22.2.8 Resize

The Resize event occurs when a form is first displayed or if the object size is changed. The standard format for the associated routines are:

```
Private Sub Form_Resize()
```

```
Private Sub object_Resize(height As Single, width As Single)
```

where:

- height is a value specifying the new height of the control.
- width is a value specifying the new width of the control.

22.2.9 Unload

The Unload event occurs when a form is about to be removed from the screen. Then if the form is reloaded the contents of all its controls are reinitialized. The standard format for the associated routines are:

```
Private Sub object_Unload(cancel As Integer)
```

where cancel determines whether the form is removed from the screen. If it is 0 then the form is removed else it will not be removed.

22.3 Exercises

22.3.1   Write a Visual Basic program which has a single form and a text box. The text box should show all of the events that occur with the form, that is:

Activate, *Click*, *DblClick*, *Deactivate*, *DragDrop*, *DragOver*, and so on.

Investigate when these events occur. Notice that when an event has code attached then the procedure name in the View Code pulldown menu becomes highlighted.

22.3.2   Write a Visual Basic program which has a single command button and a text box. The text box should show all of the events that occur with the command button, that is:

Click, *KeyDown*, *KeyPress*, *KeyUp*, *MouseDown*, *MouseUp*, and so on.

Investigate when these events occur.

22.3.3   Write a Visual Basic program in which the program displays the message 'IN TEXT AREA' when the mouse is within the textbox area and 'OUT OF TEXT AREA' when it is out of the textbox area.

22.3.4 Write a Visual Basic program which has a command button. If the user presses any lowercase letter then the program ends, else it should continue. A sample event which quits when the letter 'x' is pressed is given next:

```
Private Sub Command1_KeyPress(KeyAscii As Integer)
   If (Chr(KeyAscii) = "x") Then
        End
   End If
End Sub
```

22.3.5 Modify the program in Exercise 22.3.4 so that if the user clicks on the form (and not on the command button) the program automatically prompts the user for his/her name. This name should then appear in the command button caption property (Command1.caption). An outline of the event is given next:

```
Private Sub Form_Click()
Dim username As String

    username = InputBox("Enter your name")
     ::::::

End Sub
```

22.3.6 Write a Visual Basic program which displays the message 'So long and thanks for the fish' when the main form is Unloaded. The form should be removed from the screen.

22.3.7 Write a Visual Basic program with a textbox which displays the current keypress, including Alt-, Ctrl- and Shift- keystrokes.

22.3.8 Modify the program in Exercise 22.3.7 so that it also displays the mouse button press and the coordinates of the mouse.

23 Graphics

23.1 Introduction

This chapter discusses how graphics files are loaded into the program and how graphic objects can be drawn.

23.2 Loading graphics files

Visual Basic allows a graphic file to be loaded into a form, a picturebox or an image control. The standard function is:

```
LoadPicture(graphfile)
```

Where *graphfile* specifies the name of the graphics file, if no name is given then the graphic in the form, picture box or image control is cleared. The standard graphics files supported by Visual Basic are:

- **BMP**. Windows bitmap file.
- **ICO**. Icon file (maximum size of 32×32 pixels).
- **RLE**. Run-length encoded files.
- **WMF**. Windows metafile files.

Normally graphics files are displayed in a PictureBox ▦. Thus to display the graphic file "CLOUD.BMP" to Picture1 then:

```
Picture1.Picture = LoadPicture("CLOUD.BMP")
```

A picture can be loaded put into the clipboard using:

```
Clipboard.SetData LoadPicture(filename)
```

The following example loads a graphics file into a PictureBox. Figure 23.1 shows a sample form which contains a PictureBox, a CommandButton and a DialogBox.

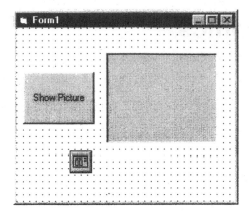

Figure 23.1   Form

The code added to the command button is as follows:

```
Private Sub Command1_Click()
Dim filename As String

    CommonDialog1.Filter = _
    "All Files (*.*)|*.*|Text Files (*.bmp)|*.bmp|Icon Files (*.ico)|*.ico"
        CommonDialog1.FilterIndex = 2
        CommonDialog1.ShowOpen
        filename = CommonDialog1.filename
        Picture1.Picture = LoadPicture(filename)

End Sub
```

This will display a dialog box with the default file setting of *.BMP. After the user has selected a graphic then the LoadPicture function is used to display the graphic file to the picture box (Picture1). Figure 23.2 shows a sample dialog box and Figure 23.3 shows a sample graphic.

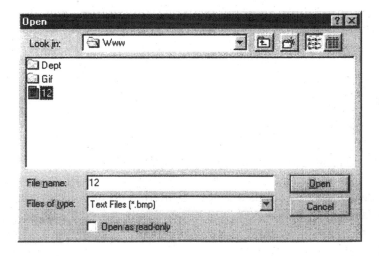

Figure 23.2   Dialog box

Figure 23.3 Sample graphic

One problem with the previous example code is that a non-graphic file could be loaded and the program would not give an error message. The following code overcomes this by testing for an error. This is achieved by testing the Err parameter after the picture has been loaded. If it is set to TRUE then an error message is displayed and, after the user has accepted the error then the picture will be cleared from the picture box.

```
Private Sub Command1_Click()
Dim filename,Msg As String

    CommonDialog1.Filter = _
      "All Files (*.*)|*.*|Text Files (*.bmp)|*.bmp|Icon Files (*.ico)|*.ico"
    CommonDialog1.FilterIndex = 2
    CommonDialog1.ShowOpen
    filename = CommonDialog1.filename
    Picture1.Picture = LoadPicture(filename)

    If Err Then
        Msg = "Error loading graphics file."
        MsgBox Msg
        Picture = LoadPicture()
    End If
End Sub
```

23.3 Colours

The background and foreground colour of an object can be modified with the BackColor or ForeColor property, respectively. The standard form is:

object.BackColor= *colour*
object.ForeColor= *colour*

A colour is defined as its RGB (red/green/blue) strength, with a hexadecimal strength from &H00 to &H77 for each colour. The valid ranges of colour are thus from 0 to &HFFFFFF.

23.3.1 RGB function

The RGBColor function returns the hexadecimal colour value given the three strengths of red, green and blue. Its standard form is:

colour=RGB (*red, green, blue*)

where red, green and blue are a value from 0 to 255. Table 23.1 gives some example colours with their RGB colour. For example:

```
Form.BkColor=QBColor(0,0,255)       ' set background colour to Blue
Form.ForeColor=QBColor(255,255,0)   ' set foreground colour to Yellow
```

Table 23.1   Colour strengths

Colour	Red value	Green value	Blue value
Black	0	0	0
Blue	0	0	255
Green	0	255	0
Cyan	0	255	255
Red	255	0	0
Magenta	255	0	255
Yellow	255	255	0
White	255	255	255

23.3.2 QBColor function

The QBColor function has a limited range of 15, typically colours as specified in Table 23.1. For example, the following sets the

```
Form.BkColor=QBColor(1)     ' set background colour to Blue
Form.ForeColor=QBColor(6)   ' set foreground colour to Yellow
```

Table 23.2   Colour values

Number	Colour	Number	Colour
0	Black	8	Grey
1	Blue	9	Light Blue
2	Green	10	Light Green
3	Cyan	11	Light Cyan
4	Red	12	Light Red
5	Magenta	13	Light Magenta
6	Yellow	14	Light Yellow
7	White	15	Bright White

23.4 Drawing

Visual Basic has a wide range of drawing functions (graphics methods); these include:

- **Line**. Draws lines and rectangles on an object.
- **Circle**. Draws a circle, ellipse, or arc on an object.
- **Cls**. Clears graphics and text from a Form or PictureBox.
- **Fill colour**. Returns or sets the colour used to fill in shapes (drawn by Line and Circle graphics methods).
- **Fill style**. Returns or sets the pattern used to fill Shapes (as created by Line or Circle graphics methods).

23.4.1 Line

The Line graphic method draws lines and rectangles on an object. Its standard form is:

object.Line Step $(x1, y1)$ - Step $(x2, y2)$, *colour*, BF

where:

- Step. Keyword specifying the starting point co-ordinates. (Optional)
- $(x1, y1)$ Define the starting point co-ordinates of a line or rectangle. If they are omitted then the line starts at the current x,y position (CurrentX and CurrentY). Optional
- Step. Keyword specifying the end point co-ordinates. (Optional)
- $(x2, y2)$ Define the end point co-ordinates of a line or rectangle. (Required)
- *color*. Defines the RGB colour used to draw the line. If it is omitted then the ForeColor property setting is used. (Optional)
- B. If the B option is added then a box is drawn using the co-ordinates for opposite corners. (Optional)
- F. If the F option is added then the box is filled with the same colour as the line colour. (Optional)

The CurrentX and CurrentY values are set to the end of the line after a line has been drawn. The DrawWidth property sets the width of the line and DrawMode and DrawStyle properties define the way that the line or box is drawn.

The following code is added to a form and with the Click event. It displays 15 solid rectangles of a random size and random colour. The QBColor function is used to display one of the 16 predefined colours. Note that the Rnd function returns a random number from 0 to 1.

```
Private Sub Form_Click()
Dim x1, x2, y1, y2, i As Integer

    For i = 1 To 15
        ForeColor = QBColor(i)
        x1 = ScaleWidth * Rnd
        y1 = ScaleHeight * Rnd
```

```
        x2 = ScaleWidth * Rnd
        y2 = ScaleHeight * Rnd
        Line (x1, y1)-(x2, y2), , BF
    Next i
End Sub
```

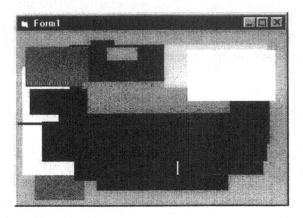

Figure 23.4   Sample graphic

23.4.2 Circle

The Circle graphic method draws a circle, ellipse or an arc on an object. Its standard form is:

object.Circle Step *(x, y)*, *radius*, *colour*, *start*, *end*, *aspect*

where:

- Step. Keyword specifying that the centre of the circle, ellipse, or arc is relative to the current co-ordinates given by the CurrentX and CurrentY properties of object. (Optional)
- *(x, y)*. Value which gives the co-ordinates of the centre of the circle, ellipse or arc. (Required)
- *radius*. Value which specifies the radius of the circle, ellipse or arc. (Required)
- *color*. Value which specifies the colour of the circle's outline. If omitted, the value of the ForeColor property is used. (Optional)
- *start*, *end*. Values specifying the start and end angle (in radians) for an arc or a partial circle or ellipse is drawn, start and end specify (in radians) the beginning and end positions of the arc. (Optional)
- *aspect*. Value specifying the aspect ratio of the circle. The default value is 1.0, which yields a perfect circle. (Optional)

The QBColor or RGB function are typically used to set the colour. A circle is fill with a defined colour with the FillColor property and the fill style is set by the FillStyle properties. The DrawWidth property defines the width of the line used to draw the circle, ellipse, or arc.

The following example code draws 10 circles of increasing size of a random colour. Figure 23.5 shows a sample run.

```
Private Sub Form_Click()
Dim x, y, Radius, RadiusInc As Double
Dim i As Integer

    x = ScaleWidth / 2 ' Set X position.
    y = ScaleHeight / 2    ' Set Y position.
    If (ScaleWidth > ScaleHeight) Then
        RadiusInc = ScaleHeight / 20
    Else
        RadiusInc = ScaleWidth / 20
    End If

    For i = 1 To 10 ' Set radius.
        Radius = Radius + RadiusInc
        Circle (x, y), Radius, RGB(Rnd * 255, Rnd * 255, Rnd * 255)
    Next i

End Sub
```

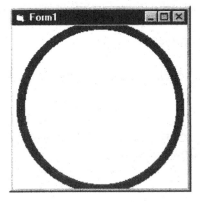

Figure 23.5   Sample graphic

23.4.3 DrawWidth

The DrawWidth property returns or sets the line width of a graphic method. To set the width of line the following is used:

object.DrawWidth = *size*

where the *size* is a value from 1 (the default) to 32,767 and is measured in pixels.

23.4.4 DrawStyle

The DrawWidth property returns or sets the line style of a graphic method.

To set the line style the following is used:

object.DrawWidth = *value*

where *value* is a number from 0 (the default) to 6 which corresponds to the following line styles:

0   Solid.   1   Dash.
2   Dot.   3   Dash Dot.
4   Dash-Dot-Dot.   5   Transparent.
6   Inside Solid.

For example to set a line style on a Form to dashed:

```
DrawWidth=1
```

23.4.5 FillColor

The FillColor property returns or sets the colour used to fill in shapes, circles and boxes. To set the fill colour the following is used:

object.FillColor = *value*

where the *size* is a hexadecimal colour and by default it is Black.

23.4.6 FillStyle

The FillStyle property returns or sets the fill style of a graphic method. To set the line style the following is used:

object.FillStyle = *value*

where *value* is a number from 0 (the default) to 6 which corresponds to the following line styles:

0   Solid.   1   (Default) Transparent.
2   Horizontal Line.   3   Vertical Line.
4   Upward Diagonal.   5   Downward Diagonal.
6   Cross.   7   Diagonal Cross.

The following example displays 15 randomly filled circles with a radius of 200 pixels. It is initiated with the user clicks on the form. Figure 23.6 shows a sample run.

```
Private Sub Form_Click()
Dim x1, x2, y1, y2, i As Integer
```

```
    For i = 1 To 15
        ForeColor = QBColor(i)
        FillStyle = Int(7 * Rnd)
        x1 = ScaleWidth * Rnd
        y1 = ScaleHeight * Rnd
        x2 = ScaleWidth * Rnd
        y2 = ScaleHeight * Rnd
        Circle (x1, y1), 200
    Next i
End Sub
```

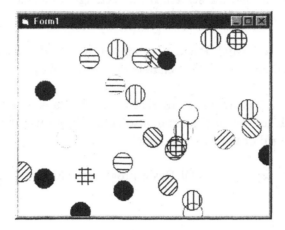

Figure 23.6   Sample graphic

23.4.7 CurrentX, CurrentY

The CurrentX and CurrentY property return or sets the horizontal and vertical (CurrentY) co-ordinates, respectively. The standard format is:

object.CurrentX = *x*
object.CurrentY = *y*

The (0,0) co-ordinate is at the upper-left hand corner of an object.

23.4.8 Cls

The Cls method clears graphics and text generated at run time from a Form or PictureBox. Its standard form is:

object.Cls

23.5 Exercises

23.5.1   Write a program which displays a coloured rectangle in the middle of a form. Each time the user clicks on the form the rectangle should change colour.

23.5.2   Write a Visual Basic program which automatically moves a rectangle from the top left-hand side of the screen to the bottom right-hand side. The program should quit once it reaches the bottom corner.

23.5.3   Write a Visual Basic program which randomly moves a small rectangle around the screen. If the rectangle touches any of the edges it should rebound off the edge.

23.5.4   Write a Visual Basic program in which the user controls the movement of a small rectangle by the arrowkeys.

23.5.5   Write a Visual Basic program which displays a circle on the form which follows the users mouse cursor (note use the `cls` function to get rid of the existing circle).

23.5.6   Write separate Visual Basic programs which draw the following objects:

(i)   A car.
(ii)   A ship.
(iii)   A house.

23.5.7   Write a Visual Basic program which displays the following images. Use the `sleep` and `cls` functions to create a timing delay of 1 second between each image display (the animation should look as if it is winking).

PART E

C/Pascal
C++
Assembly Language
Visual Basic
HTML/Java
DOS
Windows 3.x
Windows 95
UNIX

24 HTML (Introduction)

24.1 Introduction

HTML is a standard hypertext language for the WWW and has several different versions. Most WWW browsers support HTML 2 and most of the new versions of the browsers support HTML 3. WWW pages are created and edited with a text editor, a word processor or, as is becoming more common, within the WWW browser.

HTML tags contain special formatting commands and are contained within a less than (<) and a greater than (>) symbol (which are also known as angled brackets). Most tags have an opening and closing version; for example, to highlight bold text the bold opening tag is <B> and the closing tag is </B>. Table 24.1 outlines a few examples.

Table 24.1 Example HTML tags

Open tag	Closing tag	Description
<HTML>	</HTML>	Start and end of HTML
<HEAD>	</HEAD>	Defines the HTML header
<BODY>	</BODY>	Defines the main body of the HTML
<TITLE>	</TITLE>	Defines the title of the WWW page
<I>	</I>	Italic text
<B>	</B>	Bold text
<U>	</U>	Underlined text
<BLINK>	</BLINK>	Make text blink
<EM>	</EM>	Emphasize text
<FONT SIZE=+1>	</FONT>	Increase font size by one increment
<FONT SIZE=-1>	</FONT>	Reduce font size by one increment
<CENTER>	</CENTER>	Center text
<H1>	</H1>	Section header, level 1
<H2>	</H2>	Section header, level 2
<H3>	</H3>	Section header, level 3
<P>		Create a new paragraph
 		Create a line break
<!-->	-->	Comments
<SUPER>	</SUPER>	Superscript
_		Subscript

HTML script 1 gives an example script and Figure 24.1 shows the output from the WWW browser. The first line is always <HTML> and the last line is </HTML>. After this line the HTML header is defined between <HEAD> and </HEAD>. The title of the window in this case is My first HTML page. The main HTML text is then defined between <BODY> and </BODY>.

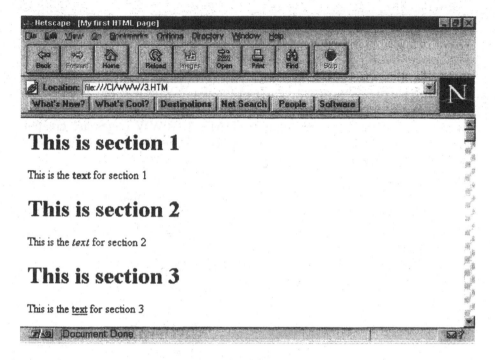

Figure 24.1   Example window from example HTML script

📖 HTML script 24.1

```
<HTML>
<HEAD>
<TITLE>My first HTML page</TITLE>
</HEAD>
<BODY>
<H1> This is section 1</H1>
This is the <b>text</b> for section 1
<H1> This is section 2</H1>
This is the <i>text</i> for section 2
<H1> This is section 3</H1>
This is the <u>text</u> for section 3
<p>
This is the end of the text
</BODY>
</HTML>
```

The WWW browser fits text into the window size and does not interpret line breaks in the HTML source. To force a new line the
 (line break) or a new paragraph (<P>) is used. The example also shows bold, italic and underlined text.

24.2 Links

The topology of the WWW is set-up using links where pages link to other related pages. A reference takes the form:

<A HREF="*url*"> *Reference Name* </A>

where *url* defines the URL for the file, *Reference Name* is the name of the reference and </A> defines the end of the reference name. HTML script 24.2 shows an example of the uses of references and Figure 24.2 shows a sample browser page. The background colour is set using the <BODY BGCOLOR="#FFFFFF"> which set the background colour to white. In this case the default text colour is black and the link is coloured blue.

📖 HTML script 24.2

```
<HTML>

<HEAD>
<TITLE>Fred's page</TITLE>
</HEAD>

<BODY BGCOLOR="#FFFFFF">

<H1>Fred's Home Page</H1>

If you want to access information on
this book <A HREF="softbook.html">click here</A>.

<P>

A reference to the <A REF="http:www.iee.com/">IEE</A>

</BODY>
</HTML>
```

24.2.1 Other links

Links can be set-up to send e-mail and newsgroups. For example:

```
<A HREF="news:sport.tennis"> Newsgroups for tennis</A>
```

to link to a tennis newsgroup and

```
<A HREF="mailto:f.bloggs@fredco.co.uk">Send a mes-
sage to me</A>
```

to send a mail message to the e-mail address: f.bloggs@ fredco.co.uk.

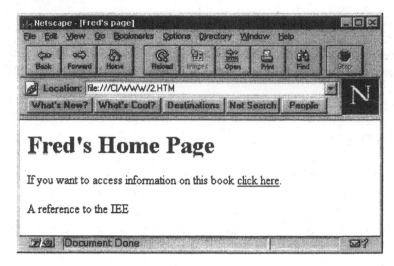

Figure 24.2 Example window from example HTML script 24.2

24.3 Lists

HTML allows ordered and unordered lists. Lists can be declared anywhere in the body of the HTML.

24.3.1 Ordered lists

The start of an ordered list is defined with <OL> and the end of the list by </OL>. Each part of the list is defined after the <LI> tag. Unordered lists are defined between the <UL> and </UL> tags. HTML script 24.3 gives examples of an ordered and an unordered list. Figure 24.3 shows the output from the browser.

📖 HTML script 24.3

```
<HTML>
<HEAD>
<TITLE>Fred's page</TITLE>
</HEAD>
<BODY BGCOLOR="#FFFFFF">
<H1>List 1</H1>
<OL>
<LI>Part 1
<LI>Part 2
<LI>Part 3
</OL>
<H1>List 2</H1>
<UL>
<LI>Section 1
<LI>Section 2
<LI>Section 3
</UL>
</BODY>
</HTML>
```

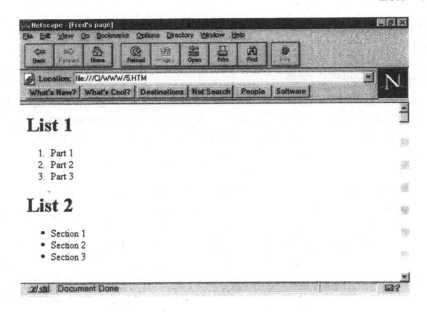

Figure 24.3 WWW browser with an ordered and unordered lists

Some browsers allow the type of numbered list to be defined with the <OL TYPE=x>, where x can either be:

- A for capital letters (such as a, b, c, and so on).
- a for small letters (such as a, b, c, and so on).
- I for capital roman letters (such as I, II, III, and so on).
- i for small roman letters (such as i, ii, iii, and so on).
- I for numbers (which is the default).

```
<OL Type=I>
<LI> List 1
<LI> List 2
<LI> List 3
</OL>

<OL Type=A>
<LI> List 1
<LI> List 2
<LI> List 3
</OL>
```

would be displayed as:

I.   List 1
II.   List 2
III. List 3
A.   List 1
B.   List 2
C.   List 3

The starting number of the list can be defined using the <LI VALUE=*n*>
where *n* defines the initial value of the defined item list.

24.3.2 Unordered lists

Unordered lists are used to list a series of items in no particular order. They
are defined between the <UL> and </UL> tags. Some browsers allow the
type of bullet point to be defined with the <LI TYPE=*shape*>, where
shape can either be:

- *disc* for round solid bullets (which is the default for first level lists).
- *round* for round hollow bullets (which is the default for second level
  lists).
- *square* for square bullets (which is the default for third).

HTML script 24.4 gives an example of an unnumbered list and Figure 24.4
shows the WWW page output for this script. It can be seen from this that the
default bullets for level 1 lists are discs, for level 2 they are round and for
level 3 they are square.

📖 HTML script 24.4

```
<HTML>
<HEAD>
<TITLE>Example list</TITLE>
</HEAD>
<H1> Introduction </H1>
<UL>
<LI> OSI Model
<LI> Networks
   <UL>
   <LI> Ethernet
      <UL>
      <LI> MAC addresses
      </UL>
   <LI> Token Ring
   <LI> FDDI
   </UL>
<LI> Conclusion
</UL>
<H1> Wide Area Networks </H1>
<UL>
<LI> Standards
<LI> Examples
   <UL>
   <LI> EastMan
   </UL>
<LI> Conclusion
</UL>
</BODY>
</HTML>
```

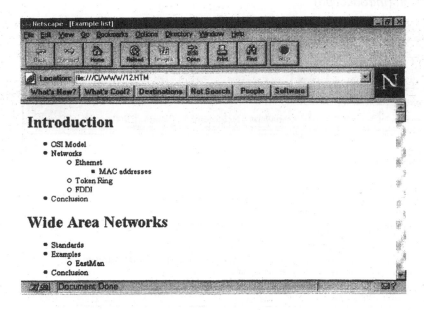

Figure 24.4   WWW page with an unnumbered list

24.3.3 Definition lists

HTML uses the <DL> and </DL> tags for definition lists. These are normally used when building glossaries. Each entry in the definition is defined
by the <DT> tag and the text associated with the item is defined after the
<DD> tag.   The end of the list is defined by </DL>. HTML script 24.5
shows an example with a definition list and Figure 24.5 gives a sample output. Note that it uses the <EM> tag to emphasize the definition subject.

📖 **HTML script 24.5**

```
<HTML>
<HEAD>
<TITLE>Example list</TITLE>
</HEAD>
<H1> Glossary </H1>
<DL>
<DT> <EM> Address Resolution Protocol (ARP) </EM>
<DD> A TCP/IP process which maps an IP address to an Ethernet address.
<DT> <EM> American National Standards Institute (ANSI) </EM>
<DD> ANSI is a non-profit organization which is made up of expert committees
that publish standards for national industries.
<DT> <EM> American Standard Code for Information Interchange (ASCII) </EM>
<DD> An ANSI-defined character alphabet which has since been adopted as a
standard international alphabet for the interchange of characters.
</DL>
</BODY>
</HTML>
```

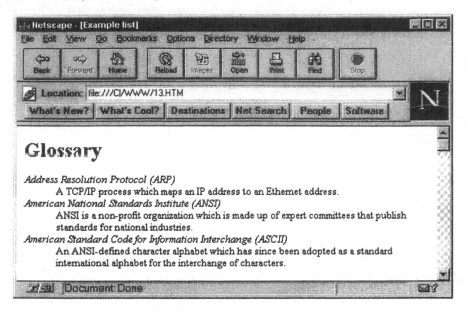

Figure 24.5 WWW page with definition list

24.4 Colours

Colours in HTML are defined in the RGB (red/green/blue) strength. The format is #rrggbb, where rr is the hexadecimal equivalent for the red component, gg the hexadecimal equivalent for the green component and bb the hexadecimal equivalent for the blue component. Table 24.2 lists some of the codes for certain colours.

Individual hexadecimal numbers use base 16 and range from 0 to F (in decimal this ranges from 0 to 15). A two-digit hexadecimal number ranges from 00 to FF (in decimal this ranges from 0 to 255). Table 24.3 outlines hexadecimal equivalents.

Table 24.2 Hexadecimal colours

Colour	Code	Colour	Code
White	#FFFFFF	Dark red	#C91F16
Light red	#DC640D	Orange	#F1A60A
Yellow	#FCE503	Light green	#BED20F
Dark green	#088343	Light blue	#009DBE
Dark blue	#0D3981	Purple	#3A0B59
Pink	#F3D7E3	Nearly black	#434343
Dark grey	#777777	Grey	#A7A7A7
Light grey	#D4D4D4	Black	#000000

Table 24.3 Hexadecimal to decimal conversions

Hex.	Dec.	Hex.	Dec.	Hex.	Dec.	Hex.	Dec.
0	0	1	1	2	2	3	3
4	4	5	5	6	6	7	7
8	8	9	9	A	10	B	11
C	12	D	13	E	14	F	15

HTML uses percentage strengths for the colours. For example, FF represents full strength (100%) and 00 represent no strength (0%). Thus, white is made from FF (red), FF (green) and FF (blue) and black is made from 00 (red), 00 (green) and 00 (blue). Grey is made from equal weighting of each of the colours, such as 43, 43, 43 for dark grey (#434343) and D4, D4 and D4 for light grey (#D4D4D4). Thus, pure red with be #FF0000, pure green will be #00FF00 and pure blue with be #0000FF.

Each colour is represented by 8 bits, thus the colour is defined by 24 bits. This gives a total of 16 777 216 colours (2^{24} different colours). Note that some video displays will not have enough memory to display 16.777 million colours in the certain mode so that colours may differ depending on the WWW browser and the graphics adapter.

The colours of the background, text and the link can be defined with the BODY tag. An example with a background colour of white, a text colour of orange and a link colour of dark red is:

```
<BODY BGCOLOR="#FFFFFF" TEXT="#F1A60A" LINK="#C91F16">
```

and for a background colour of red, a text colour of green and a link colour of blue:

```
<BODY BGCOLOR="#FF0000" TEXT="#00FF00" LINK="#0000FF">
```

When a link has been visited its colour changes. This colour itself can be changed with the VLINK. For example, to set-up a visited link colour of yellow:

```
<BODY VLINK="#FCE503" "TEXT=#00FF00" "LINK=#0000FF">
```

Note that the default link colours are:

    Link: #0000FF   (Blue)
    Visited link: #FF00FF   (Purple)

24.5 Background images

Image (such as GIF and JPEG) can be used as a background to a WWW page. For this purpose the option BACKGROUND='*src.gif*' is added to the

<BODY> tag. An HTML script with a background of CLOUDS.GIF is given in HTML script 24.6. A sample output from a browser is shown in Figure 24.6.

📖 HTML script 24.6

```
<HTML>
<HEAD>
<TITLE>Fred's page</TITLE>
</HEAD>
<BODY BACKGROUND="clouds.gif">
<H1>Fred's Home Page</H1>
If you want to access information on
this book <A HREF="gbook.html">click here</A>.<P>
A reference to the <A HREF="http://www.iee.com/">IEE</A>
</BODY>
</HTML>
```

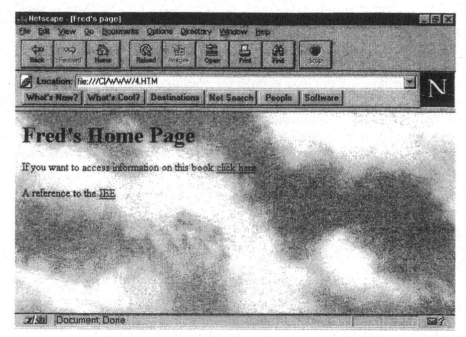

Figure 24.6 WWW page with CLOUD.GIF as a background

24.6 Displaying images

WWW pages can support graphics images within a page. The most common sources of images are either JPEG or GIF files, as these types of images normally have a high degree of compression. GIF images, as was previously mentioned, support only 256 colours from a pallet of 16.7 million colours, whereas JPEG supports more than 256 colours.

24.6.1 Inserting an image

Images can be displayed within a page with the <IMG SRC=*'src.gif'*>

which inserts the graphic *src.gif*. HTML script 24.7 contains three images: `myson.gif`, `me.gif` and `myson2.gif`. These are aligned either to the left or the right using the `ALIGN` option within the `<IMG SRC>` tag. The first image (`myson.gif`) is aligned to the right, while the second image (`me.gif`) is aligned to the left. Figure 24.7 shows a sample output from this script. Note that images are left aligned by default.

📖 HTML script 24.7

```
<HTML>
<HEAD>
<TITLE>My first home page</TITLE>
</HEAD>
<BODY BGCOLOR="#ffffff">
<IMG SRC ="myson.gif" ALIGN=RIGHT>
<H1> Picture gallery </H1>
<P>
<P>
Here are a few pictures of me and my family. To the right
is a picture of my youngest son showing his best smile.
Below to the left is a picture of me at Christmas and to
the right is a picture of me and my son also taken at
Christmas.
<P>
<P>
<IMG SRC ="me.gif" ALIGN=LEFT>
<IMG SRC ="myson2.gif" ALIGN=RIGHT>

</BODY>
</HTML>
```

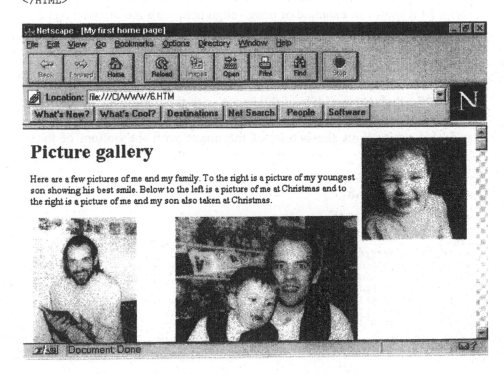

Figure 24.7 WWW page with three images

24.6.2 Alternative text

Often users choose not to view images in a page and select an option on the viewer which stops the viewer from displaying any graphic images. If this is the case then the HTML page can contain substitute text which is shown instead of the image. For example:

```
<IMG SRC ="myson.gif" ALT="Picture of my son" ALIGN=RIGHT>
<IMG SRC ="me.gif" ALT="Picture of me ALIGN=LEFT>
<IMG SRC ="myson2.gif" ALT="Another picture of my son" ALIGN=RIGHT>
```

24.6.3 Other options

Other image options can be added, such as:

- HSPACE=x VSPACE=y defines the amount of space that should be left around images. The x value defines the number of pixels in the x-direction and the y value defines the number of pixels in the y-direction.
- WIDTH= x HEIGHT=y defines the scaling in the x- and y-direction, where x and y are the desired pixel width and height, respectively, of the image.
- ALIGN=*direction* defines the alignment of the image. This can be used to align an image with text. Valid options for aligning with text are *texttop*, *top*, *middle*, *absmiddle*, *bottom*, *baseline* or *absbottom*. HTML script 24.8 shows an example of image alignment with the image a.gif (which is just the letter 'A' as a graphic) and Figure 24.8 shows a sample output. It can be seen that *texttop* aligns the image with highest part of the text on the line, *top* alights the image with the highest element in the line, *middle* aligns with the middle of the image with the baseline, *absmiddle* alights the middle of the image with the middle of the largest item, *bottom* aligns the bottom of the image with the bottom of the text and *absbottom* aligns the bottom of the image with the bottom of the largest item.

📖 HTML script 24.8

```
<HTML>
<HEAD>
<TITLE>My first home page</TITLE>
</HEAD>
<BODY BGCOLOR="#ffffff">
<IMG SRC ="a.gif" ALIGN=texttop>pple<P>
<IMG SRC ="a.gif" ALIGN=top>pple<P>
<IMG SRC ="a.gif" ALIGN=middle>pple<P>
<IMG SRC ="a.gif" ALIGN=bottom>pple<P>
<IMG SRC ="a.gif" ALIGN=baseline>pple<P>
<IMG SRC ="a.gif" ALIGN=absbottom>pple
</BODY>
</HTML>
```

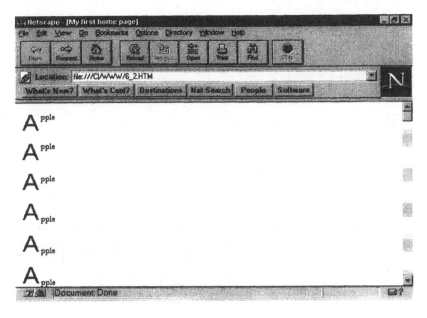

Figure 24.8 WWW page showing image alignment

24.7 Horizontal lines

A horizontal line can be added with the <HR> tag. Most browsers allow extra parameters, such as:

SIZE= *n* – which defines that the height of the rule is *n* pixels.
WIDTH=*w* – which defines that the width of the rule is *w* pixels or as a percentage.
ALIGN=*direction* – where direction refers to the alignment of the rule Valid options for *direction* are *left, right* or *center*.
NOSHADE – which defines that the line should be solid with no shading.

HTML script 24.9 gives some example horizontal lines and Figure 24.9 shows an example output.

HTML script 24.9

```
<HTML><HEAD>
<TITLE>My first home page</TITLE>
</HEAD>
<BODY BGCOLOR="#ffffff">
<IMG SRC ="a.gif">pple<P>
<HR>
<IMG SRC ="a.gif">pple<P>
<HR WIDTH=50% ALIGN=CENTER>
<IMG SRC ="a.gif">pple<P>
<HR SIZE=10 NOSHADE>

</BODY></HTML>
```

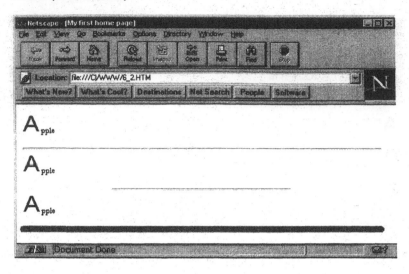

Figure 24.9   WWW page showing horizontal lines

24.8 Exercises

24.8.1   The home page for this book can be found at the URL:

```
http://www.eece.napier.ac.uk/~bill_b/soft.hmtl
```

Access this page and follow any links it contains.

24.8.2   If possible, create a WWW page with the following blinking text:

```
This is some blinking text
```

24.8.3   The last part of the server name normally gives an indication of the country where the server is located (for example www.fredco.co.uk is located in the UK). Determine which countries use the following country names:

(a) de   (b) nl   (c) it   (d) se   (e) dk   (f) sg
(g) ca   (h) ch   (i) tr   (j) jp   (k) au

Determine some other country identifier names.

24.8.4   Determine the HTML colour represent for the following:

(a)   red
(b)   green
(c)   blue

    (d)   white

    (e)   black

24.8.5   Determine the HTML for the background, text and link colour:

    (a) `<BODY BKCOLOR="#00FF00" "TEXT=#FF0000"`
                          `"LINK=#0000FF">`
    (b) `<BODY BKCOLOR="#DC640D" "TEXT=#777777"`
                          `"LINK=#009DBE">`

24.8.6   Determine the error in the following HTML script:

📖 **HTML script 24.10**

```
<HTML>
<HEAD>
<TITLE>Fred's page</TITLE>
</HEAD>
<BODY BGCOLOR="#FFFFFF">
<H1>List 1</H1>
<OL>
<LI>Part 1
<LI>Part 2
<LI>Part 3
<H1>List 2</H1>
<UL>
<LI>Section 1
<LI>Section 2
<LI>Section 3
</UL>
</BODY>
</HTML>
```

25 Further HTML

25.1 Introduction

Chapter 24 introduced HTML; this chapter discusses some of HTML's more advanced features. HTML differs from compiled languages, such as C and Pascal, in that the HTML text file is interpreted by an interpreter (the browser) while languages such as C and Pascal must be precompiled before they can be run. HTML thus has the advantage that it does not matter if it is the operating system, the browser type or the computer type that reads the HTML file, as the file does not contain any computer specific code. The main disadvantage of interpreted files is that the interpreter does less error checking as it must produce fast results.

The basic pages on the WWW are likely to evolve around HTML and while HTML can be produced manually with a text editor, it is likely that, in the coming years, there will be an increase in the amount of graphically-based tools that will automatically produce HTML files. Although these tools are graphics-based they still produce standard HTML text files. Thus a knowledge of HTML is important as it defines the basic specification for the presentation of WWW pages.

25.2 Anchors

An anchor allows users to jump from a reference in a WWW page to another anchor point within the page. The standard format is:

<A NAME="*anchor name*">

where *anchor name* is the name of the section which is referenced. The </A> tag defines the end of an anchor name. A link is specified by:

<A HREF="#*anchor name*">

followed by the </A> tag. HTML script 25.1 shows a sample script with four anchors and Figure 25.1 shows a sample output. When the user selects one of the references, the browser automatically jumps to that anchor. Figure 25.2 shows the output screen when the user selects the #Token reference. Anchors are typically used when an HTML page is long or when a backwards or forwards reference occurs (such as a reference within a published paper).

📖 HTML script 25.1

```
<HTML><HEAD><TITLE>Sample page</TITLE></HEAD><BODY BGCOLOR="#FFFFFF">
<H2>Select which network technology you wish information:</H2>
<P><A HREF="#Ethernet">Ethernet</A></P>
<P><A HREF="#Token">Token Ring</A></P>
<P><A HREF="#FDDI">FDDI</A></P>
<P><A HREF="#ATM">ATM</A></P>
<H2><A NAME="Ethernet">Ethernet</A></H2>
Ethernet is a popular LAN which works at 10Mbps.
<H2><A NAME="Token">Token Ring</A></H2>
Token ring is a ring based network which operates
at 4 or 16Mbps.
<H2><A NAME="FDDI">FDDI</A></H2>
FDDI is a popular LAN technology which uses a ring of
fibre optic cable and operates at 100Mbps.
<H2><A NAME="ATM">ATM</A></H2>
ATM is a ring based network which operates at 155Mbps.</BODY></HTML>
```

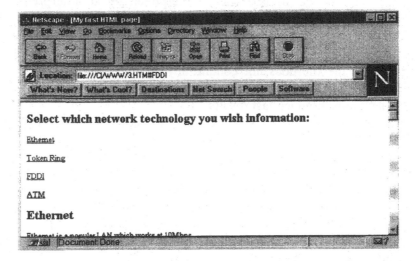

Figure 25.1   Example window with references

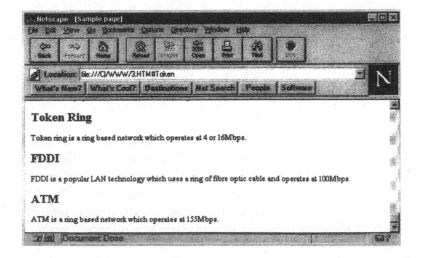

Figure 25.2   Example window with references

25.3 Tables

Tables are one of the best methods to display complex information in a simple way. Unfortunately, in HTML they are relatively complicated to set up. The start of a table is defined with the <TABLE> tag and the end of a table by </TABLE>. A row is defined between the <TR> and </TR>, while a table header is defined between <TH> and </TH>. A regular table entry is defined between <TD> and </TD>. HTML script 25.2 shows an example of a table with links to other HTML pages. The BORDER=*n* option has been added to the <TABLE> tag to define the thickness of the table border (in pixels). In this case the border size has a thickness of 10 pixels.

📖 HTML script 25.2

```
<HTML><HEAD><TITLE> Fred Bloggs</TITLE></HEAD>
<BODY TEXT="#000000" BGCOLOR="#FFFFFF">
<H1>Fred Bloggs Home Page</H1>
I'm Fred Bloggs. Below is a tables of links.<HR><P>
<TABLE BORDER=10>
<TR>
   <TD><B>General</B></TD>
   <TD><A HREF="res.html">Research</TD>
   <TD><A HREF="cv.html">CV</TD>
   <TD><A HREF="paper.html">Papers Published</TD>
</TR>
<TR>
   <TD><B>HTML Tutorials</B></TD>
   <TD><A HREF="intro.html">Tutorial 1</TD>
   <TD><A HREF="inter.html">Tutorial 2</TD>
   <TD><A HREF="adv.html">Tutorial 3</TD>
</TR>
<TR>
   <TD><B>Java Tutorials</B></TD>
   <TD><A HREF="java1.html">Tutorial 1</TD>
   <TD><A HREF="java2.html">Tutorial 2</TD>
   <TD><A HREF="java3.html">Tutorial 3</TD>
</TR>
</TABLE></BODY></HTML>
```

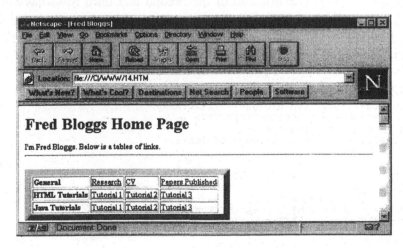

Figure 25.3 Example window from example Java script

Other options in the <TABLE> tag are:

- WIDTH=*x*, HEIGHT=*y* – defines the size of the table with respect to the full window size. The parameters *x* and *y* are either absolute values in pixels for the height and width of the table or are percentages of the full window size.
- CELLSPACING=*n* – defines the number of pixels desired between each cell where *n* is the number of pixels (note that the default cell spacing is 2 pixels).

An individual cell can be modified by adding options to the <TH> or <TD> tag. These include:

- WIDTH=*x*, HEIGHT=*y* – defines the size of the table with respect to the table size. The parameters *x* and *y* are either absolute values in pixels for the height and width of the table or are percentages of the table size.
- COLSPAN=*n* – defines the number of columns the cell should span.
- ROWSPAN=*n* – defines the number of rows the cell should span.
- ALIGN=*direction* – defines how the cell's contents are aligned horizontally. Valid options are *left*, *center* or *right*.
- VALIGN=*direction* – defines how the cell's contents are aligned vertically. Valid options are *top*, *middle* or *baseline*.
- NOWRAP – informs the browser to keep the text on a single line (that is, with no line breaks).

HTML script 25.3 shows an example use of some of the options in the <TABLE> and <TD> options. In this case the text within each row is centre aligned. On the second row the second and third cells are merged using the COLSPAN=2 option. The first cell of the second and third rows have also been merged using the ROWSPAN=2 option. Figure 25.4 shows an example output. The table width has been increased to 90% of the full window, with a width of 50%.

📖 HTML script 25.3

```
<HTML><HEAD><TITLE> Fred Bloggs</TITLE></HEAD>
<BODY TEXT="#000000" BGCOLOR="#FFFFFF">
<H1>Fred Bloggs Home Page</H1>
I'm Fred Bloggs. Below is a table of links.
<HR>
<P>
<TABLE BORDER=10 WIDTH=90% LENGTH=50%>
<TR>
   <TD><B>General</B></TD>
   <TD><A HREF="res.html">Research</TD>
   <TD><A HREF="cv.html">CV</TD>
   <TD><A HREF="paper.html">Papers Published</TD>
   <TD></TD>
</TR>
```

```
<TR>
   <TD ROWSPAN=2><B>HTML/Java Tutorials</B></TD>
   <TD><A HREF="intro.html">Tutorial 1</TD>
   <TD COLSPAN=2><A HREF="inter.html">Tutorial 2</TD>
</TR>
<TR>
   <TD><A HREF="java1.html">Tutorial 1</TD>
   <TD><A HREF="java2.html">Tutorial 2</TD>
   <TD><A HREF="java3.html">Tutorial 3</TD>
</TR>
</TABLE></BODY></HTML>
```

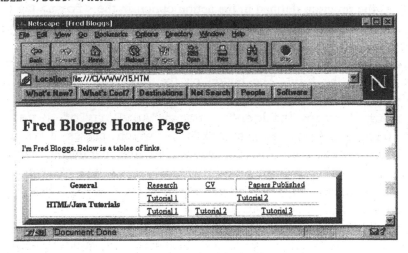

Figure 25.4   Example window from example script

25.4 CGI scripts

CGI (Common Gateway Interface) scripts are normally written in either C or Perl and are compiled to produce an executable program. They can also come precompiled or in the form of a batch file. Perl has the advantage in that it is a script that can be easily run on any computer, while a precompiled C program requires to be precompiled for the server computer.

CGI scripts allow the user to interact with the server and store and request data. They are often used in conjunction with forms and allow an HTML document to analyze, parse and store information received from a form. On most UNIX-type systems the default directory for CGI scripts is `cgi-bin`.

25.5 Forms

Forms are excellent methods of gathering data and can be used in conjunction with CGI scripts to collect data for future use.

A form is identified between the `<FORM>` and `</FORM>` tags. The method used to get the data from the form is defined with the `METHOD="POST"`. The `ACTION` option defines the URL script to be run when the form is submitted. Data input is specified by the <INPUT TYPE> tag. HTML script 25.4 form has the following parts:

- <form action="/cgi-bin/AnyForm2" method="POST"> – which defines the start of a form and when the "submit" option is selected the cgi script /cgi-bin/AnyForm2 will be automatically run.
- <input type="submit" value="Send Feedback"> – which causes the program defined in the action option in the <form> tag to be run. The button on the form will contain the text "Send Feedback", see Figure 25.5 for a sample output screen.
- <input type="reset" value="Reset Form"> – which resets the data in the form. The button on the form will contain the text "Reset Form", see Figure 25.5 for a sample output screen.
- <input type="hidden" name="AnyFormTo" value= "f.bloggs @toytown.ac.uk"> – which passes a value of f.bloggs@toytown.ac.uk which has the parameter name of "AnyFormTo". The program AnyForm2 takes this parameter and automatically sends it to the email address defined in the value (that is, f.bloggs @toytown.ac.uk).
- <input type="hidden" name="AnyFormSubject" value="Feedback form"> – which passes a value of Feedback form which has the parameter name of "AnyFormSubject". The program AnyForm2 takes this parameter and adds the text "Feedback form" in the text sent to the email recipient (in this case, f.bloggs @toytown.ac.uk).
- Surname <input name="Surname"> – which defines a text input and assigns this input to the parameter name Surname.
- <textarea name="Address" rows=2 cols=40> </textarea> – which defines a text input area which has two rows and has a width of 40 characters. The thumb bars appear at the right-hand side of the form if the text area exceeds more than 2 rows, see Figure 25.5.

📖 HTML script 25.4

```
<HTML>
<HEAD>
<TITLE>Example form</TITLE>
</HEAD>
<H1><CENTER>Example form</CENTER></H1><P>
<form action="/cgi-bin/AnyForm2" method="POST">

<input type="hidden" name="AnyFormTo" value="f.bloggs@toytown.ac.uk">
<input type="hidden" name="AnyFormSubject"
value="Feedback form">

Surname <input name="Surname"> First Name/Names <input name="First Name"><P>
Address (including country)<P>
<textarea name="Address" rows=2 cols=40></textarea><P>
```

```
Business Phone <input name="Business Phone">Place of study (or company)
<input name="Study"><P>
E-mail    <input name="E-mail">  Fax Number <input name="Fax Number"><P>

<input type="submit" value="Send Feedback"> <input type="reset"
             value="Reset Form">
</Form>
</HTML>
```

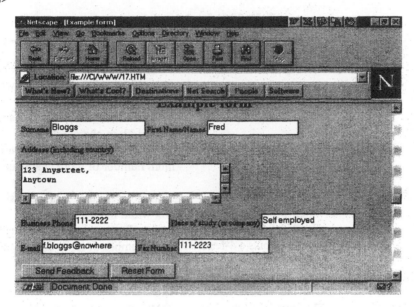

Figure 25.5 Example window showing an example form

In this case the recipient (f.bloggs@toytown.ac.uk) will receive an email with the contents:

```
Anyform Subject=Example form
Surname=Bloggs
First name=Fred
Address=123 Anystreet, Anytown
Business Phone=111-222
Place of study (or company)=Self employed
Email= f.bloggs@nowhere
Fax Number=111-2223
```

The extra options to the <input> tag are size="*n*", where *n* is the width of the input box in characters, and maxlength="*m*", where *m* is the maximum number of characters that can be entered, in characters. For example:

```
<input type="text"  size="15" maxlength="10">
```

defines that the input type is text, the width of the box is 15 characters and the maximum length of input is 10 characters.

25.5.1 Input types

The type options to the <input> tag are defined in Table 25.1. HTML
script 25.5 gives a few examples of input types and Figure 25.6 shows a
sample output.

Table 25.1 Input type options

TYPE=	Description	Options
"text"	The input is normal text.	NAME="*nm*" where *nm* is the name that will be sent to the server when the text is entered. SIZE="*n*" where *n* is the desired box width in characters. SIZE="*m*" where *m* is the maximum number of input characters.
"password"	The input is a password which will be displayed with *'s. For example if the user inputs a 4-letter password then only **** will be displayed.	SIZE="*n*" where *n* is the desired box width in characters. SIZE="*m*" where *m* is the maximum number of input characters.
"radio"	The input takes the form of a radio button (such as ⊙ or ◯). They are used to allow the user to select a single option from a list of options.	NAME="*radname*" where *radname* defines the name of the button. VALUE="*val*" where *val* is the data that will be sent to the server when the button is selected. CHECKED is used to specify that the button is initially set.
"checkbox"	The input takes the form of a checkbox (such as ☒ or ☐). They are used to allow the user to select several options from a list of options.	NAME="*chkname*" where *chkname* defines the common name for all the checkbox options. VALUE="*defval*" where *defval* defines the name of the option. CHECKED is used to specify that the button is initially set.

📖 HTML script 25.5

```
<HTML><HEAD><TITLE>Example form</TITLE> </HEAD>
<FORM METHOD="Post" >
<H2>Enter type of network:</H2><P>
<INPUT TYPE="radio" NAME="network" VALUE="ethernet"
CHECKED>Ethernet <INPUT TYPE="radio" NAME="network" VALUE="token">
```

```
Token Ring <INPUT TYPE="radio" NAME="network" VALUE="fddi" >
FDDI <INPUT TYPE="radio" NAME="network" VALUE="atm" >
ATM
<H2>Enter usage:</H2><P>
<INPUT TYPE="checkbox" NAME="usage" VALUE="multi"
>Multimedia
<INPUT TYPE="checkbox" NAME="usage" VALUE="word"
>Word Processing
<INPUT TYPE="checkbox" NAME="usage" VALUE="spread"
>Spread Sheets
<P>Enter Password<INPUT TYPE="password" NAME="passwd" SIZE="10">
</FORM></HTML>
```

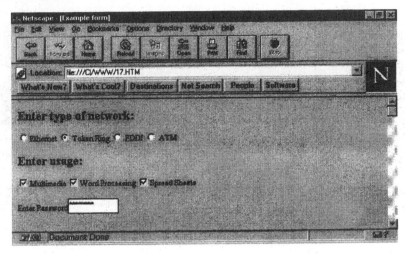

Figure 25.6   Example window with different input options

25.5.2 Menus

Menus are a convenient method of selecting from multiple options. The `<SELECT>` tag is used to define start of a list of menu options and the `</SELECT>` tag defines the end. Menu elements are then defined with the `<OPTION>` tag. The options defined within the `<SELECT>` are:

- `NAME="name"` – which defines that *name* is the variable name of the menu. This is used when the data is collected by the server.
- `SIZE="n"` – which defines the number of options which are displayed in the menu.

HTML script 25.6 shows an example of a menu. The additional options to the `<OPTION>` tag are:

- `SELECTED` – which defines the default selected option.
- `VALUE="val"` – where *val* defines the name of the data when it is collected by the server.

📖 HTML script 25.6

```
<HTML><HEAD><TITLE>Example form</TITLE> </HEAD>
<FORM METHOD="Post" >
Enter type of network:
<select Name="network" size="1">
<option>Ethernet
<option SELECTED>Token Ring
<option>FDDI
<option>ATM
</select></FORM></HTML>
```

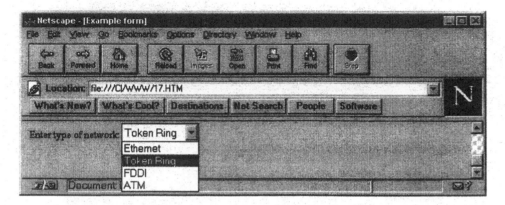

Figure 25.7 Example window showing an example form

25.6 Multimedia

If the browser cannot handle all the file types it may call on other application helpers to process the file. This allows other 'third-party' programs to integrate into the browser. Figure 25.8 shows an example of the configuration of the helper programs. The options in this case are:

- View in browser.
- Save to disk.
- Unknown: prompt user.
- Launch an application (such as an audio playback program or MPEG viewer).

For certain applications the user can select as to whether the browser processes the file or another application program processes it. Helper programs make upgrades in helper applications relatively simple and also allow new file types to be added with an application helper. Typically when a program is installed which can be used with a browser it will prompt the user to automatically update the helper application list so that it can handle the given file type(s).

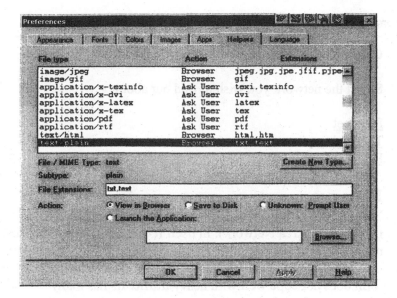

Figure 25.8   Example window showing an example form

Each file type is defined by the file extension, such as .ps for postscript files, .exe for a binary executable file, and so on. These file extensions have been standardized in MIME (Multipurpose Internet Mail Extensions) specfi-cation. Table 25.2 shows some typical file extensions.

Table 25.2   Input type options

Mime type	Extension	Typical action
application/octet-stream	exe, bin	Save
application/postscript	ps, ai, eps	Ask user
application/x-compress	Z	Compress program
application/x-gzip	gz	GZIP compress program
application/x-javascript	js, mocha	Ask user
application/x-msvideo	avi	Audio player
application/x-perl	pl	Save
application/x-tar	tar	Save
application/x-zip-compressed	zip	ZIP program
audio/basic	au, snd	Audio player
image/gif	gif	Browser
image/jpeg	jpeg, jpg, jpe	Browser
image/tiff	tif, tiff	Graphics viewer
image/x-MS-bmp	bmp	Graphics viewer
text/html	htm, html	Browser
text/plain	text, txt	Browser
video/mpeg	mpeg, mpg, mpe, mpv, vbs, mpegv	Video player
video/quicktime	qt, mov, moov	Video player

25.7 Exercises

25.7.1   Construct a WWW page with anchor points for the following:

Select the network you wish to find out about:

Ethernet
Token ring
FDDI

Ethernet

Ethernet is the most widely used networking technology used in LAN (Local Area Network). In itself it cannot make a network and needs some other protocol such as TCP/IP or SPX/IPX to allow nodes to communicate. Unfortunately, Ethernet in its standard form does not cope well with heavy traffic. Its has many advantages, though, including:

- Networks are easy to plan and cheap to install.
- Network components are cheap and well supported.
- It is well-proven technology which is fairly robust and reliable.
- Simple to add and delete computers on the network.
- Supported by most software and hardware systems.

Token ring

Token ring networks were developed by several manufacturers, the most prevalent being the IBM Token Ring. Token ring networks cope well with high network traffic loadings. They were at one time extremely popular but their popularity has since been overtaken by Ethernet. Token ring networks have, in the past, suffered from network management problems and poor network fault tolerance.

Token ring networks are well suited to situations which have large amounts of traffic and also work well with most traffic loadings. They are not suited to large networks or networks with physically remote stations. Their main advantage is that they cope better with high traffic rates than Ethernet, but require a great deal of maintenance especially when faults occur or when new equipment is added to or removed from the network. Many of these problems have now been overcome by MAUs (multi-station access units), which are similar to the hubs using in Ethernet.

FDDI

A token-passing mechanism allows orderly access to a network. Apart from token ring the most commonly used token-passing network is the Fiber Distributed Data Interchange (FDDI) standard. This operates at 100 Mbps and, to overcome the problems of line breaks, has two concentric token rings. Fibre optic cables have a much high specification over copper cables and allow extremely long interconnection lengths. The maximum circumference of the ring is 100 km (62 miles), with a maximum 2 km between stations (in FDDI stations are also known as stations). It is thus an excellent mechanism for connecting interconnecting networks over a city or a campus. Up to 500 stations can connect to each ring with a maximum of 1,000 stations for the complete network. Each station connected to the FDDI highway can be a normal station or a bridge to a conventional local area network, such as Ethernet or token ring.

25.7.2   Construct a WWW glossary page with the following terms:

Address	A unique label for the location of data or the identity of a communications device.
Address Resolution Protocol (ARP)	A TCP/IP process which maps an IP address to an Ethernet address.
American National Standards Institute (ANSI)	ANSI is a non-profit organization which is made up of expert committees that publish standards for national industries.
American Standard Code for Information Interchange (ASCII)	An ANSI-defined character alphabet which has since been adopted as a standard international alphabet for the interchange of characters.
Amplitude modulation (AM)	Information is contained in the amplitude of a carrier.
Amplitude-Shift Keying (ASK)	Uses two, or more, amplitudes to represent binary digits. Typically used to transmit binary data over speech-limited channels.
Application layer	The highest layer of the OSI model.
Asynchronous transmission	Transmission where individual characters are sent one-by-one. Normally each character is delimited by a start and stop bit. With asynchronous communication the transmitter and receiver only have to be roughly synchronized.

25.7.3   Construct a WWW page which can be used to enter a person's CV (note, use a form). The basic fields should be:

Name:
Address:
Email address:
Telephone number:
Experience:
Interests:
Any other information:

25.7.4   Write an HTML script which displays the following timetable.

	9–11	11–1	1–3	3–5
Monday	Data Comms		Networking	
Tuesday	Software Systems		Networking	Data Comms
Wednesday	Networking	FREE	Java	FREE
Thursday	Software Systems	C++	Networking	FREE
Friday	FREE		Networking	

25.7.5   Design your own home page with a basic user home page (index.html) which contains links to a basic CV page (for example, it could be named cv.html) and a page which lists your main interests (myinter.html). Design one of the home pages with a list of links and another with a table of links. If possible incorporate graphics files into the pages.

26 JavaScript

26.1 Introduction

Computer systems contain a microprocessor which controls the operation of the computer. The microprocessor only understands binary information and operates on a series of binary commands known as machine code. It is extremely difficult to write large programs in machine code, so that high-level languages are used instead. A low-level language is one which is similar to machine code and normally involves the usage of keyword macros to replace machine code instructions. High-level languages have a syntax that is almost like written English and thus make programs easy to read and to modify. In most programs the actual operation of the hardware is invisible to the programmer. A compiler changes the high-level language into machine code. Typical high-level languages include C/C++, BASIC, COBOL, FORTRAN and Pascal; an example of a low-level language is 80486 Assembly Language.

Java is a high-level language that has been developed specifically for the WWW and is well suited to networked applications. It was originally developed by Sun Microsystems and is based on C++ (but with less of the difficulties of C++). Most new versions of Web browsers now support its usage. Java's main attributes are:

- It runs either as a stand-alone program or it can run within the Web browser. When run within the browser the Java program is known as an applet.
- Java is a portable language and applets can run on any type of microprocessor type (such as a PC based on Intel 80486 or Pentium, or a Motorola-based computer).
- Java applets are hardware and operating system independent. For example, the program itself does not have to interface directly to the hardware such as a video adapter or mouse. Typical high-level languages, such as C/C++ and Pascal, produce machine-dependent machine code, and can thus only be executed on a specific computer or operating systems.
- Java allows for a client/server approach where the applet can run on the remote computer which thus reduces the loading on the local computer

(typically the remote computer will be a powerful multitasking computer with enhanced computer architecture).

- A Java compiler creates stand-alone programs or applets. Many new versions of browsers have an integrated Java compiler.

Figure 26.1 shows the main functional differences between a high-level language, a Java applet and JavaScript. JavaScript is interpret by the browser, whereas a Java applet is compiled to a virtual machine code which can be run on any computer system. The high-level language produces machine-specific code.

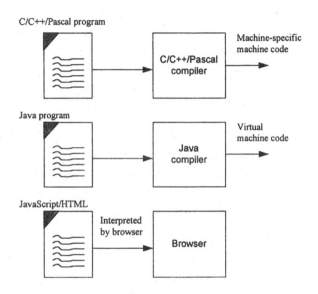

Figure 26.1   Differences between C++/Java and JavaScript

A normal C++ program allows access to hard-disk drives. This would be a problem on the Web as unsolicited users ('hackers') or novice users could cause damage on the Web server. To overcome this Java does not have any mechanism for file input/output (I/O). It can read standard file types (such as GIF and JPG) but cannot store changes to the Web server. A Java developers kit is available, free of charge, from `http://java.sun.com`.

The following is an example HTML script and highlighted JavaScript. Figure 26.2 gives the browser output.

📖 JavaScript 26.1

```
<HTML> <HEAD><TITLE>My Java</TITLE></HEAD>
<BODY>
<SCRIPT language="javascript">
document.writeln("This is my first JavaScript");
for (i=0;i<10;i++)
 document.write("<center><font size=+1><b>Loop</b> ",i);
</SCRIPT></BODY></HTML>
```

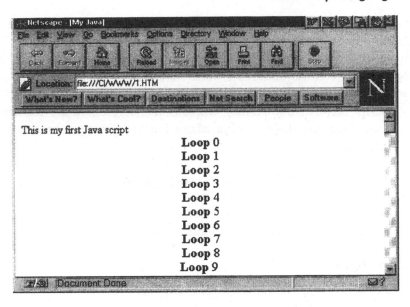

Figure 26.2 Example window from example JavaScript

26.2 JavaScript language .

Programming languages can either be compiled to produce an executable program or they can be interpreted while the user runs the program. Java is a program language which needs to be compiled before it is used. It thus cannot be used unless the user has the required Java compiler. JavaScript, on the other hand, is a language which is interpreted by the browser. It is similar in many ways to Java but allows the user to embed Java-like code into an HTML page. JavaScript supports a small number of data types representing numeric, Boolean, and string values and is supported by most modern WWW browsers, such as Microsoft Internet Explorer and Netscape.

HTML is useful when pages are short and do not contain expressions, loops or decisions. JavaScript allows most of the functionality of a high-level language for developing client and server Internet applications. It can be used to respond to user events such as mouse clicks, form input, and page navigation.

A major advantage that JavaScript has over HTML is that it supports the use of functions without any special declarative requirements. It is also simpler to use than Java because it has easier syntax, specialized built-in functionality, and minimal requirements for object creation.

Important concepts in Java and JavaScript are objects. Objects are basically containers for values. The main differences between JavaScript and Java are:

- JavaScript is interpreted by the client, while Java is compiled on the server before it is executed.

- JavaScript is embedded into HTML pages, while Java applets are distinct from HTML and accessed from HTML pages.
- JavaScript has loose typing for variables (that is, a variables data type does not have to be declared), while Java has strong typing (that is, a variables data type must always be declared before it is used).
- JavaScript has dynamic binding where object references are checked at run-time. Java has static binding where object references must exist at compile-time.

26.3 JavaScript values, variables and literals

JavaScript values, variables and literals are similar to the C programming language. Their syntax is discussed in this section.

26.3.1 Values

The four different types of values in JavaScript are:

- Numeric value, such as 12 or 91.5432.
- Boolean values which are either TRUE or FALSE.
- String types, such as 'Fred Bloggs'.
- A special keyword for a NULL value

Numeric values differ from those in most programming languages in that there is no explicit distinction between a real value (such as 91.5432) and an integer (such as 12).

26.3.2 Data type conversion

JavaScript differs from Java in that variables do not need to have their data types defined when they are declared (loosely typed). Data types are then automatically converted during the execution of the program. Thus a variable could be declared with a numeric value as:

```
var value
    value = 19
```

and then in the same script it could be assigned a string value, such as:

```
    value = "Enter your name >>"
```

The conversion between numeric values and strings in JavaScript is easy, as numeric values are automatically converted to an equivalent string. For example:

```
<HTML><HEAD><TITLE>My Java</TITLE></HEAD>
```

```
<BODY BGCOLOR="#ffffff">
<SCRIPT language="javascript">
var x,y,str
        x=13
        y=10
        str= x + " added to " + y + " is " + x+y
        document.writeln(str)

        z=x+y
        str= x + " added to " + y + " is " + z
        document.writeln(str)
</SCRIPT>
</BODY></HTML>
```

Sample run 26.1 gives the output from this script. It can be seen that x and y have been converted to a string value (in this case, "13" and "10") and that x+y in the string conversion statement has been converted to "1310". If a mathematical operation is carried out (z=x+y) then z will contain 23 after the statement is executed.

🖳 **Sample run 26.1**

```
13 added to 10 is 1310   13 added to 10 is 23
```

JavaScript provides several special functions for manipulating string and numeric values:

- The eval (*string*) function which converts a string to a numerical value.
- The parseInt (*string [,radix]*) function which converts a string into an integer of the specified radix (number base). The default radix is base-10.
- The parseFloat (*string*) function which converts a string into a floating-point value.

26.3.3 Variables

Variables are symbolic names for values within the script. A JavaScript identifier must either start with a letter or an underscore ('_'), followed by letters, an underscore or any digit (0–9). Like C, JavaScript is case sensitive so that variables with the same character sequence but with different cases for one or more characters are different. The following are different variable names:

```
i=5
I=10
valueA=3.543
VALUEA=10.543
```

26.3.4 Variable scope

A variable can be declared by either simply assigning it a value or by using the var keyword. For example, the following declares to variables Value1 and Value2:

```
var    Value1;
       Value2=23;
```

A variable declared within a function is taken as a local variable and can only be used within that function. A variable declared outside a function is a global variable and can be used anywhere in the script. A variable which is declared locally which is already declared as global variable needs to be declared with the var keyword, otherwise the use of the keyword is optional.

26.3.5 Literals

Literal values have fixed values within the script. Various reserved forms can be used to identify special types, such as hexadecimal values, exponent format, and so on. With an integer the following are used:

- If the value is preceded by a 0x then the value is a hexadecimal value (that is, base 16). Examples of hexadecimal values are 0x1FFF, 0xCB.
- If the value is preceded by a 0 then the value is an octal value (that is, base 8). Examples of octal values are 0777, 010.
- If it is not preceded by either a 0x or a 0 then it is a decimal integer.

Floating-point values

Floating-point values are typically represented as a real value (such as 1.342) or in exponent format. Some exponent format values are:

Value	*Exponent format*
0.000001	1e-6
1342000000	1.342e9

Boolean

The true and false literals are used with Boolean operations.

Strings

In C a string is represented with double quotes ("*str*") whereas JavaScript accepts a string within double (") or single (') quotation marks. Examples of strings are:

```
"A string"
'Another string'
```

26.4 Expressions and operators

The expressions and operators used in Java and JavaScript are based on C and C++. This section outlines the main expressions and operators used in JavaScript.

26.4.1 Expressions

As with C, expressions are any valid set of literals, variables, operators and expressions that evaluate to a single value. There are basically two types of expression, one which assigns a value to a variable and the other which simply gives a single value. A simple assignment is:

```
value = 21
```

which assigns the value of 21 to `value` (note that the result of the expression is 21).

The result from a JavaScript expression can either be:

- A numeric value.
- A string.
- A logical value (true or false).

26.5 JavaScript operators

Both Java and JavaScript have a rich set of operators, of which there are four main types:

- Arithmetic.
- Logical.
- Bitwise.
- Relational.

26.5.1 Arithmetic

Arithmetic operators operate on numerical values. The basic arithmetic operations are add (+), subtract (−), multiply (*), divide (/) and modulus division (%). Modulus division gives the remainder of an integer division. The following gives the basic syntax of two operands with an arithmetic operator.

```
operand operator operand
```

The assignment operator (=) is used when a variable 'takes on the value' of an operation. Other short-handed operators are used with it, including add equals (+=), minus equals (−=), multiplied equals (*=), divide equals (/=) and modulus equals (%=). The following examples illustrate their uses.

Statement	Equivalent
x+=3.0	x=x+3.0
voltage/=sqrt(2)	voltage=voltage/sqrt(2)
bit_mask *=2	bit_mask=bit_mask*2

In many applications it is necessary to increment or decrement a variable by 1. For this purpose Java has two special operators; ++ for increment and -- for decrement. These can either precede or follow the variable. If they precede, then a pre-increment/decrement is conducted, whereas if they follow it, a post-increment/decrement is conducted. The following examples show their usage.

Statement	Equivalent
no_values++	no_values=no_values+1
i--	i=i-1

Table 26.1 summarizes the arithmetic operators.

Table 26.1 Arithmetic operators

Operator	Operation	Example
-	subtraction or minus	5-4→1
+	addition	4+2→6
*	multiplication	4*3→12
/	division	4/2→2
%	modulus	13%3→1
+=	add equals	x += 2 is equivalent to x=x+2
-=	minus equals	x -= 2 is equivalent to x=x-2
/=	divide equals	x /= y is equivalent to x=x/y
*=	multiplied equals	x *= 32 is equivalent to x=x*32
=	assignment	x = 1
++	increment	Count++ is equivalent to Count=Count+1
--	decrement	Sec-- is equivalent to Sec=Sec-1

26.5.2 Relationship

The relationship operators determine whether the result of a comparison is TRUE or FALSE. These operators are greater than (>), greater than or equal to (>=), less than (<), less than or equal to (<=), equal to (==) and not equal to (!=). Table 26.2 lists the relationship operators.

26.5.3 Logical (TRUE or FALSE)

A logical operation is one in which a decision is made as to whether the operation performed is TRUE or FALSE. If required, several relationship operations can be grouped together to give the required functionality. C assumes that a numerical value of 0 (zero) is FALSE and that any other value is TRUE. Table 26.3 lists the logical operators.

The logical AND operation will yield a TRUE only if all the operands are TRUE. Table 26.4 gives the result of the AND (&&) operator for the operation A&& B. The logical OR operation yields a TRUE if any one of the operands is TRUE. Table 26.4 gives the logical results of the OR (||) operator for the statement A||B and also gives the logical result of the NOT (!) operator for the statement !A.

Table 26.2 Relationship operators

Operator	Function	Example	TRUE condition
>	greater than	(b>a)	when b is greater than a
>=	greater than or equal	(a>=4)	when a is greater than or equal to 4
<	less than	(c<f)	when c is less than f
<=	less than or equal	(x<=4)	when x is less than or equal to 4
==	equal to	(x==2)	when x is equal to 2
!=	not equal to	(y!=x)	when y is not equal to x

Table 26.3 Logical operators

Operator	Function	Example	TRUE condition
&&	AND	((x==1) && (y<2))	when x is equal to 1 *and* y is less than 2
||	OR	((a!=b) || (a>0))	when a is not equal to b *or* a is greater than 0
!	NOT	(!(a>0))	when a is *not* greater than 0

Table 26.4 Logical operations

A	B	AND (&&)	OR (||)	NOT (!A)
FALSE	FALSE	FALSE	FALSE	TRUE
FALSE	TRUE	FALSE	TRUE	TRUE
TRUE	FALSE	FALSE	TRUE	FALSE
TRUE	TRUE	TRUE	TRUE	FALSE

26.5.4 Bitwise

The bitwise logical operators work conceptually as follows:

- The operands are converted to 32-bit integers, and expressed a series of bits (zeros and ones).
- Each bit in the first operand is paired with the corresponding bit in the second operand: first bit to first bit, second bit to second bit, and so on.
- The operator is applied to each pair of bits, and the result is constructed bitwise.

The bitwise operators are similar to the logical operators but they should not be confused as their operation differs. Bitwise operators operate directly on the individual bits of an operand(s), whereas logical operators determine whether a condition is TRUE or FALSE.

Numerical values are stored as bit patterns in either an unsigned integer format, signed integer (2's complement) or floating-point notation (an exponent and mantissa). Characters are normally stored as ASCII characters.

The basic bitwise operations are AND (&), OR (|), 1s complement or bitwise inversion (~), XOR (^), shift left (<<) and shift right (>>). Table 26.5 gives the results of the AND bitwise operation on two bits $Bit1$ and $Bit2$.

The Boolean bitwise instructions operate logically on individual bits. The XOR function yields a 1 when the bits in a given bit position differ, the AND function yields a 1 only when the given bit positions are both 1's. The OR operation gives a 1 when any one of the given bit positions are a 1. For example:

```
          00110011              10101111              00011001
AND       11101110      OR      10111111      XOR     11011111
          00100010              10111111              11000110
```

Table 26.5   Bitwise operations

A	B	AND	OR	EX-OR
0	0	0	0	0
0	1	0	1	1
1	0	0	1	1
1	1	1	1	0

To perform bit shifts, the <<, >> and >>> operators are used. These operators shift the bits in the operand by a given number defined by a value given on the right-hand side of the operation. The left shift operator (<<) shifts the bits of the operand to the left and zeros fill the result on the right. The sign-propagating right shift operator (>>) shifts the bits of the operand to the right and zeros fill the result if the integer is positive; otherwise it will fill with 1s. The zero-filled right shift operator (>>>) shifts the bits of the operand to the right and fills the result with zeros. The standard format is:

```
operand >>  no_of_bit_shift_positions
operand >>> no_of_bit_shift_positions
operand <<  no_of_bit_shift_positions
```

26.5.5 Precedence

There are several rules for dealing with operators:

- Two operators, apart from the assignment, should never be placed side by side. For example, x * % 3 is invalid.
- Groupings are formed with parentheses; anything within parentheses will be evaluated first. Nested parentheses can also be used to set priorities.
- A priority level or precedence exists for operators. Operators with a higher precedence are evaluated first; if two operators have the same precedence, then the operator on the left-hand side is evaluated first. The priority levels for operators are as follows:

HIGHEST PRIORITY

() [] .	primary
! ~ ++ -- -	unary
* / %	multiply
+ -	additive
<< >> >>>	shift
< > <= >=	relation
== !=	equality
&	
^	bitwise
|	
&&	logical
||	
= += -=	assignment

LOWEST PRIORITY

The assignment operator has the lowest precedence.

26.5.6 Conditional expressions

Conditional expressions can result in one of two values based depending on a condition. The syntax is:

```
(expression) ? value1 : value2
```

If the expression is true then `value1` is executed else `value2` is executed. For example:

```
(val >= 0) ? sign="positive" : sign="negative"
```

This will assign the string 'positive' to `sign` if the value of `val` is greater than or equal to 0, else it will assign 'negative'.

26.5.7 String operators

The normal comparison operators, such as <, >, >=, ==, and so on, can be used with strings. In addition, the concatenation operator (+) can be used to concatenate two string values together. For example:

```
str="This is " + "an example"
```

will result in the string

```
"This is an example"
```

26.6 JavaScript statements

JavaScript statements are similar to C and allow a great deal of control of the execution of a script. The basic categories are:

- Conditional statements, such as `if...else`.
- Repetitive statements, such as `for`, `while`, `break` and `continue`.
- Comments, using either the C++ style for single-line comments (`//`) or standard C multi-line comments (`/*...*/`).
- Object manipulation statements and operators, such as `for...in`, `new`, `this`, and `with`.

26.7 Conditional statements

Conditional statements allow a program to make decisions on the route through a program.

26.7.1 `if...else`

A decision is made with the `if` statement. It logically determines whether a conditional expression is TRUE or FALSE. For a TRUE, the program executes one block of code; a FALSE causes the execution of another (if any). The keyword `else` identifies the FALSE block. Braces are used to define the start and end of the block.

Relationship operators (>,<,>=,<=,==,!=) yield a TRUE or FALSE from their operation. Logical statements (&&, ||, !) can then group these together to give the required functionality. If the operation is not a relationship, such as bitwise or an arithmetic operation, then any non-zero value is TRUE and a zero is FALSE.

The following is an example syntax of the `if` statement. If the statement block has only one statement the braces (`{  }`) can be excluded.

```
if (expression)
{
    statement block
}
```

The following is an example format with an `else` extension.

```
if (expression)
{
    statement block1
}
else
{
    statement block2
}
```

It is possible to nest `if...else` statements to give a required functionality. In the next example, *statement block1* is executed if `expression1` is TRUE. If it is FALSE then the program checks the next expression. If this is TRUE the program executes *statement block2*, else it checks the next expression, and so on. If all expressions are FALSE then the program executes the final `else` statement block, in this case, *statement block 4*:

```
if (expression1)
{
    statement block1
}
else if (expression2)
{
    statement block2
}
else
{
    statement block4
}
```

26.8 Loops

26.8.1 `for()`

Many tasks within a program are repetitive, such as prompting for data, counting values, and so on. The `for` loop allows the execution of a block of code for a given control function. The following is an example format; if there is only one statement in the block then the braces can be omitted.

```
for (starting condition; test condition; operation)
{
    statement block
}
```

where:

starting condition — the starting value for the loop;
test condition   — if test condition is TRUE the loop will
                       continue execution;
operation   — the operation conducted at the end of the loop.

26.8.2 while()

The while() statement allows a block of code to be executed while a specified condition is TRUE. It checks the condition at the start of the block; if this is TRUE the block is executed, else it will exit the loop. The syntax is:

```
while (condition)
{
    :        :   statement block
    :        :
}
```

If the statement block contains a single statement then the braces may be omitted (although it does no harm to keep them).

26.9 Comments

Comments are author notations that explain what a script does. Comments are ignored by the interpreter. JavaScript supports Java-style comments:

- Comments on a single line are preceded by a double-slash (//).
- Multiline comments can be preceded by /* and followed by */.

The following example shows two comments:

```
// This is a single-line comment.
/* This is a multiple-line comment. It can be of any
length, and you can put whatever you want here. */
```

26.10 Functions

JavaScript supports modular design using functions. A function is defined with a JavaScript with the function reserved word and the code within the function is defined within curly brackets. The standard format is:

```
function myfunct(param1, param2 ...)
{
   statements
   return(val)
}
```

where the parameters (param1, param2, and so on) are the values passed into the function. Note that the return value (`val`) from the function is only required when a value is returned from the function.

JavaScript 26.2 gives an example with two functions (`add()` and `mult()`). In this case the values `value1` and `value2` are passed into the variables `a` and `b` within the `add()` function, the result is then sent back from the function into `value3`.

Table 26.6   Example JavaScript

JavaScript	Output
📖 JavaScript 26.2	

```
<HTML><TITLE>Example</TITLE>
<BODY BGCOLOR="#FFFFFF">

<SCRIPT>
var value1,value2,value3,value4;

value1=15;
value2=10;
value3=add(value1,value2)
value4=mult(value1,value2)
document.write("Added is ",value3)
document.write("<P>Multiplied is ",value4)

function add(a,b)
{
var   c
      c=a+b
      return(c)
}
function mult(a,b)
{
var   c
      c=a*b
      return(c)
}
</SCRIPT></FORM></HTML>
```

Output:
Added is 25
Multiplied is 150

26.11 Objects and properties

JavaScript is based on a simple object-oriented paradigm, where objects are a construct with properties that are JavaScript variables. Each object has properties associated with it and can be accessed with the dot notation, such as:

objectName . propertyName

26.12 Document objects

The document object contains information on the currently opened document. HTML expressions can be displayed with the `document.write()` or `document.writeln()` functions. The standard format is:

```
document.write(exprA, [,exprB], ... [,exprN])
```

which displays one or more expressions to the specified window. To display to the current window the `document.write()` is used. If a display to a specified window then the window reference is defined, for example:

```
mywin=window.open("fred.html")
mywin.document.write("Hello")
```

is used to output to the `mywin` window.

The document object can also be used to display HTML properties. The standard HTML format is:

```
<BODY BACKGROUD="bgndimage" BGCOLOR="bcolor" TEXT="fcolor" LINK="ufcolor"
ALINK="actcolor" VLINK="fcolor" </BODY>
```

These and other properties can be accessed within a JavaScript with:

```
document.alinkColor          document.anchors
document.bgColor             document.fgColor
document.lastModified        document.linkColor
document.title               document.URL
document.vlinkColor
```

Table 26.7 shows an example.

Table 26.7   Example JavaScript

JavaScript	Output

📖 JavaScript 26.3
 	
`<HTML><TITLE>Example</TITLE>`
`<BODY BGCOLOR="#FFFFFF">`
 	
`<SCRIPT>`
`document.write('ALINK color is ',`
`    document.alinkColor)`
`document.write('<P>BGCOLOR is ',`
`    document.bgColor)`
`document.write('<P>URL is ',document.URL)`
`document.write('<P>Title is',`
`    document.title)`
`</SCRIPT></HTML>`
	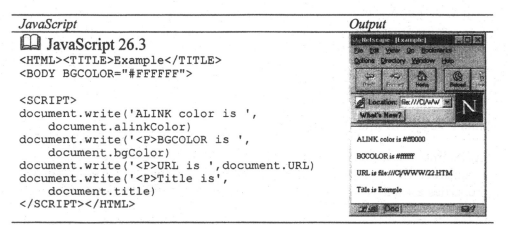

26.13 Event handling

JavaScript has event handlers which, on a certain event, causes other events to occur. For example, if a user clicks the mouse button on a certain menu option then the event handler can be made to carry-out a particular action, such as adding to numbers together. Table 26.8 outlines some event handlers.

Table 26.8 Example event handers

Event	Description	Example	Caused by
onBlur	Blur events occur when the select or text field on a form loses focus.	`<INPUT TYPE="text" VALUE="" NAME="userName" on-Blur="check(this.value)">` When the onBlur event occurs the JavaScript code required is executed (in this case the function `check()` is called.	select, text, textarea
onChange	The change event occurs when the select or text field loses focus and its value has been modified.	`<INPUT TYPE="button" VALUE="Compute" on-Click="Calc(this.form)">` When the onChange event occurs the required JavaScript code is executed (in this case the function `Calc()` is called).	button, checkbox, radio, link, reset, submit
onClick	The click event occurs when an object on a form is clicked.	`<INPUT TYPE="button" VALUE="Calculate" on-Click="go(this.form)">`	button, checkbox, radio, link, reset, submit
onFocus	The focus event occurs when a field receives input focus by tabbing with the keyboard or clicking with the mouse. Selecting within a field results in a select event and not a focus event.	`<INPUT TYPE="textarea" VALUE="" NAME="valueField" onFo-cus="valueCheck()">`	select, text, textarea
onLoad	The load event occurs when the browser finishes loading a window.	`<BODY on-Load="window.alert("Hello to my excellent page")>`	
onMouse Over	A mouseover event occurs each time the mouse pointer moves over an object. Note that a true value must be returned within the event handler.	`<A HREF="http://home.html/" onMouseOver="window.status= 'Select this to go home'; return true"> Go Home</A>`	
onSelect	A select event occurs when a user selects some of the text within a text or textarea field. The onSelect event handler executes JavaScript code when a select event occurs.	`<INPUT TYPE="text" VALUE="" NAME="valueField" onSe-lect="selectState()">`	
onUnload	An Unload event occurs when the browser quits from a window.	`<BODY onUnload="goodbye()">`	

26.14 Window objects

The window object is the top-level object for documents and can be used to open and close windows.

26.14.1 Window alert

The alert window shows an alert message to the users. Its format is:

```
window.alert("Message")
```

or

```
alert("Message");
```

Table 26.9 Example JavaScript

JavaScript	Output
📖 JavaScript 26.4	

```
<HTML><HEAD><TITLE>Example form</TITLE></HEAD>

<BODY BGCOLOR="#ffffff">

<FORM>
Enter name<INPUT TYPE="text" NAME="myname"
   onBlur="testname(myname.value)">
</FORM>

<SCRIPT>
function testname(name)
{
  if (name!="fred") alert("You are not fred");
}
</SCRIPT></HTML>
```

26.14.2 Opening and closing windows

Windows are opened with `window.open()` and closed with `window.close()`. Examples are:

```
fredwin=window.open("fred.html");
```

```
fredwin.close(fredwin);
```

or to open a window it is possible to simply use `open()` and to close the current window the `close()` function is used. The standard format is:

[*winVar* =][window].open("*url*", "*winName*", ["*features*"])

where

winVar is the name of a new window which can be used to refer to a given window.

winName is the window name given to the window;

features is a comma-separated list with any of the following:

toolbar[=yes|no]	location[=yes|no]	directories[=yes|no]
status[=yes|no]	menubar[=yes|no]	scrollbars[=yes|no]
resizable[=yes|no]	width=pixels	height=pixels

26.14.3 Window confirm

The window confirm is used to display a confirm dialogue box with a specified message and the OK and Cancel buttons. If the user selects the OK button then the function returns a TRUE, else it returns a FALSE. Table 26.10 gives an example of the confirm window. In this case when the Exit button is selected then the function ConfirmExit() is called. In this function the user is asked to confirm the exit with the confirm window. If the user selects OK then the window is closed (if it is the only window open then the browser quits).

Table 26.10   Example JavaScript with confirm window

JavaScript	*Output*

```<HTML><TITLE>Example</TITLE>```
```<BODY BGCOLOR="#FFFFFF">```
 	
```<form name="ExitForm">```
```<INPUT TYPE="button" VALUE="Exit" on-```
```Click="ConfirmExit()">```
```</FORM>```
 	
```<SCRIPT>```
```function ConfirmExit()```
```{```
```   if (confirm("Do you want to exit"))```
```   {```
```   prompt("Enter your password")```
```   }```
```}```
```</SCRIPT></HTML>```
	Netscape
	
	(?) JavaScript Confirm:
	Do you want to exit
	
	OK Cancel

26.14.4 Window prompt

The window prompt displays a prompt dialog box which contains a message and an input field. Its standard format is:

```
prompt("Message");
```

26.15 Object manipulation statements and operators

JavaScript has several methods in which objects can be manipulated. These include: the new operator, the this keyword, the for...in statement, and the with statement.

26.15.1 this keyword

The this keyword is used to refer to the current object. The general format is:

> this[.*propertyName*]

JavaScript 26.5 gives an example of the this keyword. In this case this is used to pass the property values of the input form. This is then passed to the function checkval() when the onBlur event occurs.

📖 JavaScript 26.5

```
<HTML><TITLE>Example</TITLE>
<BODY BGCOLOR="#FFFFFF">

<FORM>
Enter a value<INPUT TYPE = "text" NAME = "inputvalue"
onBlur="checkval(this, 0,10)">

<SCRIPT>
function checkval(val, minval, maxval)
{
   if ((val.value < minval) || (val.value > maxval))
      alert("Invalid value (0-10)")
}

</SCRIPT></FORM>
</HTML>
```

26.15.2 new operator

The new operator is used to define a new user-defined object type or of one of the pre-defined object types, such as array, Boolean, date, function and math. JavaScript 26.6 gives an example which creates an array object with 6 elements and then assigns strings to each of the array. Note that in Java the first element of the array is indexed as 0.

📖 JavaScript 26.6

```
<HTML><HEAD><TITLE>Java Example</TITLE></HEAD>
<BODY BGCOLOR="#ffffff">
<SCRIPT language="javascript">

no_of_networks=6;
Networks = new Array(no_of_networks);

   Networks[0]="Ethernet"; Networks[1]="Token Ring"
   Networks[2]="FDDI";  Networks[3]="ISDN"
```

```
Networks[4]="RS-232";    Networks[5]="ATM"

for (i=0;i<no_of_networks;i++)
{
   document.writeln("Network type "+Networks[i]);
   document.writeln("<P>");
}
```

`</SCRIPT></BODY></HTML>`

```
Network type Ethernet
Network type Token Ring
Network type FDDI
Network type ISDN
Network type RS-232
Network type ATM
```

Typically the new operator is used to create new data objects. For example:

```
today = new Date()
Xmasday = new Date("December 25, 1997 00:00:00")
Xmasday = new Date(97,12,25)
```

26.15.3 for...in

The for...in statement is used to iterate a variable through all its properties. In general its format is:

```
for (variable in object)
{
    statements
}
```

26.15.4 with

The with statement defines a specified object for a set of statements. A with statement looks as follows:

```
with (object)
{
    statements
}
```

For example JavaScript 26.7 contains calls to the Math object for the PI property and cos and sin methods. JavaScript 26.8 then uses the with statement to define the Match object is the default object.

📖 JavaScript 26.7

```
<HTML><TITLE>Example</TITLE>
<BODY BGCOLOR="#FFFFFF">
<SCRIPT>
var area,x,y,radius
```

```
        radius=20
        area=Math.PI*radius*radius
        x=radius*Math.cos(Math.PI/4)
        y=radius*Math.sin(Math.PI/4)
        document.write("Area is ",area)
        document.write("<P>x is ",x, "<P>y is ",y)
</SCRIPT></FORM></HTML>
```

📖 JavaScript 26.8

```
<HTML><TITLE>Example</TITLE>
<BODY BGCOLOR="#FFFFFF">

<SCRIPT>
var area,x,y,radius
    radius=10
    with (Math)
    {
        area=PI*radius*radius
        x=radius*cos(PI/4)
        y=radius*sin(PI/4)
        document.write("Area is ",area)
        document.write("<P>x is ",x, "<P>y is ",y)
    }
</SCRIPT></FORM></HTML>
```

26.16 Exercises

26.16.1 Explain how Java differs from JavaScript.

26.16.2 Explain the main advantages of using Java rather than a high-level language, such as C++ or Pascal.

26.16.3 Implement the JavaScripts in the text and test their operation.

26.16.4 Write a JavaScript in which the user enters a value and the script calculates the square of that value.

26.16.5 Write a JavaScript in which the user initially enters their name. The script should then test the entered name and if it is not 'FRED', 'BERT' or 'FREDDY' then the browser exits.

27 Java (Introduction)

27.1 Introduction

Java has the great advantage over conventional software languages in that it produces code which is computer hardware independent. This is because the compiled code (called bytecodes) is interpreted by the WWW browser. Unfortunately this leads to slower execution, but as much of the time is spent in graphical user interface programs, to update the graphics, then the overhead is, as far as the user is concerned, not a great one.

The other advantages that Java has over conventional software languages include:

- It is a more dynamic language than C/C++ and Pascal, and was designed to adapt to an evolving environment. It is extremely easy to add new methods and extra libraries without affecting existing applets. It is also useful in Internet applications as it supports most of the standard image audio and video formats.
- It has networking facilities built into the language (support for TCP/IP sockets, URLs, IP addresses and datagrams).
- While Java is based on C and C++ it avoids some of the difficult areas of C/C++ code (such as pointers and parameter passing).
- It supports client/server applications where the Java applets run on the server and the client receives the updated graphics information. In the most extreme case the client can simply be a graphics terminal which runs Java applets over a network. The small 'black-box' networked computer is one of the founding principles of Java, and it is hoped in the future that small Java-based computers could replace the complex PC/workstation for general purpose applications, like accessing the Internet or playing network games. This 'black-box' computer concept is illustrated in Figure 27.1.

Most existing Web browsers are enabled for Java applets (such as Internet Explorer 3.0 and Netscape 2.0 and later versions). Figure 27.2 shows how Java applets are created. First the source code is produced with an editor, next a Java compiler compiles the Java source code into bytecode (normally appending the file name with .class). An HTML page is then constructed which has the reference to the applet. After this a Java-enabled browser or applet viewer can then be used to run the applet.

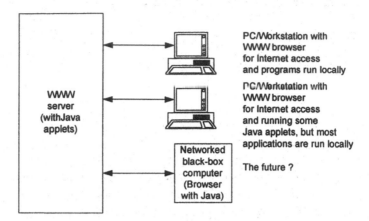

Figure 27.1 Internet accessing

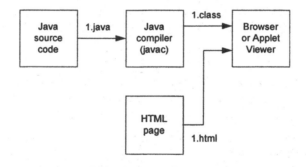

Figure 27.2 Constructing Java applets

The Java Development Kit (JDK) is available, free, from Sun Microsystems from the WWW site http://java.sun.com. This can be used to compile Java applets and stand alone programs. There are versions for Windows NT/95, Mac or UNIX-based systems with many sample applets.

Table 27.1 shows the main files used in the PC version. Figure 27.3 shows the directory structure of the JDK tools. The Java compiler, Java interpreter and applet viewer programs are stored in the bin directory. On the PC, this directory is normally set up in the PATH directory, so that the Java compiler can be called while the user is in another directory. The following is a typical setup (assuming that the home directory is C:\JAVA):

```
PATH=C:\WINDOWS;C:\WINDOWS\COMMAND;C:\JAVA\BIN
CLASSPATH=C:\JAVA\LIB;.;C:\JAVA
```

Table 27.1 JDK programs

File	Description
Javac.exe	Java compiler
Java.exe	Java interpreter
AppletViewer.exe	Applet viewer for testing and running applets
classes.zip	It is needed by the compiler and interpreter
javap.exe	Java class disassembler
javadoc.exe	Java document generator
jbd.exe	Java debugger

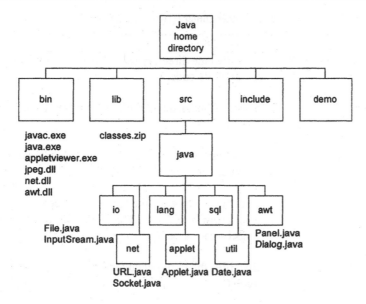

Figure 27.3 Directory structure of JDK

The lib directory contains the classes.zip file which is a zipped-up version of the Java class files. These class files are stored in the directories below the src/java directory. For example, the io classes (such as File.java and InputStream.java) are used for input/output in Java, the awt classes (such as Panel.java and Dialog.java) are used to create and maintain windows. These and other classes will be discussed later.

The include directory contains header files for integrating C/C++ programs with Java applets and the demo directory contains some sample Java applets.

27.1.1 Applet tag

An applet is called from within an HTML script with the APPLET tag, such as:

```
<applet code="Test.class" width=200 height=300></applet>
```

which loads an applet called Test.class and sets the applet size to 200 pixels wide and 300 pixels high. Table 27.2 discusses some optional parameters.

27.1.2 Applet viewer

A useful part of the JDK tools is an applet viewer which is used to test applets before they are run within the browse. The applet viewer on the PC version is AppletViewer.exe and the supplied argument is the HTML file that contains the applet tag(s). It then runs all the associated applets in separate windows.

Table 27.2   Other applet HTML parameters

Applet parameters	Description
CODEBASE=*codebaseURL*	Specifies the directory (*codebaseURL*) that contains the applet's code.
CODE=*appletFile*	Specifies the name of the file (*appletFile*) of the compiled applet.
ALT=*alternateText*	Specifies the alternative text that is displayed if the browser cannot run the Java applet.
NAME=*appletInstanceName*	Specifies a name for the applet instance (*appletInstanceName*). This makes it possible for applets on the same page to find each other.
WIDTH= *pixels* HEIGHT=*pixels*	Specifies the initial width and height (in *pixels*) of the applet.
ALIGN=*alignment*	Specifies the *alignment* of the applet. Possible values are: left, right, top, texttop, middle, absmiddle, baseline, bottom and absbottom.
VSPACE=*pixels* HSPACE=*pixels*	Specifies the number of *pixels* above and below the applet (VSPACE) and on each side of the applet (HSPACE).

27.2 Creating an applet

Java applet 27.1 shows a simple Java applet which displays two lines of text and HTML script 27.1 shows how the applet integrates into an HTML script.

   First the Java applet (j1.java) is created. In this case the edit program is used. The directory listing below shows that the files created are j1.java and j1.html (Note that Windows NT/95 displays the 8.3 filename format on the left hand side of the directory listing and the long filename on the right hand side).

📖 Java applet 27.1 (`j1.java`)

```
import java.awt.*;
import java.applet.*;

public class j1 extends Applet
{
  public void paint(Graphics g)
  {
    g.drawString("This is my first Java",5,25);
    g.drawString("applet.....",5,45);
  }
}
```

📖 HTML script 27.1 (`j1.html`)

```
<HTML>
<TITLE>First Applet</TITLE>
<APPLET CODE=j1.class WIDTH=200 HEIGHT=200>
</APPLET>
</HTML>
```

💻 **Sample run 27.1**

```
C:\DOCS\notes\INTER\java>edit j1.java

C:\DOCS\notes\INTER\java>dir
 Directory of C:\DOCS\notes\INTER\java
J1~1      JAV      263   07/04/97   13:35 j1.java
J1~1      HTM      105   07/04/97   13:31 j1.html
          2 file(s)          368 bytes
          2 dir(s)    136,085,504 bytes free
```

Next the Java applet is compiled using the `javac.exe` program. It can be seen from the listing that, if there are no errors, the compiled file is named `j1.class`. This can then be used, with the HTML file, to run as an applet.

💻 **Sample run 27.2**

```
C:\DOCS\notes\INTER\java>javac j1.java
C:\DOCS\notes\INTER\java>dir
 Directory of C:\DOCS\notes\INTER\java
J1~1      JAV      263   07/04/97   13:35 j1.java
J1~1      HTM      105   07/04/97   13:31 j1.html
J1~1      CLA      440   07/04/97   23:00 j1.class
          3 file(s)          808 bytes
          2 dir(s)    130,826,240 bytes free

C:\DOCS\notes\INTER\java>appletviewer j1.html
```

27.3 Applet basics

Java applet 27.1 recaps the previous Java applet. This section analyzes the main parts of this Java applet.

📖 Java applet 27.1 (`j1.java`)

```
import java.awt.*;
import java.applet.*;
public class j1 extends Applet
{
  public void paint(Graphics g)
  {
    g.drawString("This is my first Java",5,25);
    g.drawString("applet.....",5,45);
  }
}
```

27.3.1 Import statements

The `import` statement allows previously written code to be included in the applet. This code is stored in class libraries (or packages), which are compiled Java code. For the JDK tools, the Java source code for these libraries is stored in the `src/java` directory.

Each Java applet created begins with:

```
import java.awt.*;
import java.applet.*;
```

These include the `awt` and `applet` class libraries. The `awt` class provide code that handles windows and graphics operations. The applet in Java script 27.1 uses `awt` code, for example, displays the 'This is ..' message within the applet window. Likewise the applet uses the applet code to let the browser run the applet.

The default Java class libraries are stored in the `classes.zip` file in the `lib` directory. This file is in a compressed form and should not be unzip before it is used. The following is an outline of the file.

```
Searching ZIP: CLASSES.ZIP
Testing: java/
Testing: java/lang/
Testing: java/lang/Object.class
Testing: java/lang/Exception.class
Testing: java/lang/Integer.class
   ::          ::
Testing: java/lang/Win32Process.class
Testing: java/io/
Testing: java/io/FilterOutputStream.class
Testing: java/io/OutputStream.class
   ::          ::
Testing: java/io/StreamTenizer.class
Testing: java/util/
Testing: java/util/Hashtable.class
Testing: java/util/Enumeration.class
   ::          ::
Testing: java/util/Stack.class
Testing: java/awt/
Testing: java/awt/Toolkit.class
Testing: java/awt/peer/
Testing: java/awt/peer/WindowPeer.class
```

```
Testing: java/awt/peer/DialogPeer.class
Testing: java/awt/Image.class
Testing: java/awt/MenuItem.class
Testing: java/awt/MenuComponent.class
Testing: java/awt/image/
   ::          ::
   ::          ::
Testing: java/awt/ImageMediaEntry.class
Testing: java/awt/AWTException.class
Testing: java/net/
Testing: java/net/URL.class
Testing: java/net/URLStreamHandlerFactory.class
   ::          ::
Testing: java/net/URLEncoder.class
Testing: java/applet/
Testing: java/applet/Applet.class
Testing: java/applet/AppletContext.class
Testing: java/applet/AudioClip.class
Testing: java/applet/AppletStub.class
```

Table 27.3 lists the main class libraries and some sample libraries.

Table 27.3 Class libraries

Class libraries	Description	Example libraries
java.lang.*	Java language	java.lang.Class java.lang.Number java.lang.Process java.lang.String
java.io.*	I/O routines	java.io.InputStream java.io.OutputStream
java.util.*	Utilities	java.util.BitSet java.util.Dictionary
java.awt.*	Windows, menus and graphics	java.awt.Point java.awt.Polygon java.awt.MenuComponent java.awt.MenuBar java.awt.MenuItem
java.net.*	Networking (such as sockets, URL support, ftp, telnet, SMTP and HTTP).	java.net.ServerSocket java.net.Socket java.net.SocketImpl
java.applet.*	Code required to run an applet.	java.applet.AppletContext java.applet.AppletStub java.applet.AudioClip

It can be seen that upgrading the Java compiler is simple, as all that is required is to replace the class libraries with new ones. For example, if the basic language is upgraded then java.lang.* files is simply replaced with a new version. The user can also easily add new class libraries to the standard ones. A complete listing of the classes is given in Appendix A.

27.3.2 Applet class

The start of the applet code is defined in the form:

```
public class j1 extends Applet
```

which informs the Java compiler to create an applet named j1 that extends the existing Applet class. The public keyword at the start of the statement allows the Java browser to run the applet, while if it is omitted the browser cannot access your applet.

The class keyword is used to creating a class object named j1 that extends the applet class. After this the applet is defined between the left and right braces (grouping symbols).

27.3.3 Applet functions

Functions allow Java applets to be split into smaller sub-tasks called functions. These functions have the advantage that:

• They allow code to be reused.
• They allow for top-level design.
• They make applet debugging easier as each function can be tested in isolation to the rest of the applet.

A function has the public keyword, followed by the return value (if any) and the name of the function. After this the parameters passed to the function are defined within rounded brackets. Recapping from the previous example:

```
public void paint(Graphics g)
{
   g.drawString("This is my first Java",5,25);
   g.drawString("applet.....",5,45);
}
```

This function has the public keyword which allows any user to execute the function. The void type defines that there is nothing returned from this function and the name of the function is paint(). The parameter passed into the function is g which has the data type of Graphics. Within the paint() function the drawString() function is called. This function is defined in java.awt.Graphics class library (this library has been included with the import java.awt.* statement. The definition for this function is:

```
public abstract void drawString(String str, int x, int y)
```

which draws a string of characters using the current font and colour. The x,y position is the starting point of the baseline of the string (str).

It should be noted that Java is case sensitive and the names given must be refered to in the case that they are defined as.

27.4 Stand-alone programs

A Java program can also be run as a stand-alone program. This allows the Java program to be run without a browser and is normally used when testing a Java applet. The method of output to the screen is:

```
System.out.println("message");
```

which prints a message (message) to the display. This type of debugging is messy as these statements need to be manually inserted in the program. It is likely that later versions of the JDK toolkit will contain a run-time debugger which will allow developers to view the execution of the program.

To run a stand-alone program the java.exe program is used and the user adds output statements with the System.out.println() function. Note that there is no output from applet with the System.out.println() function.

Java stand-alone program 27.1 gives a simple example of a stand-alone program. The public static void main(Strings[] args) defines the main function. Sample run 27.3 shows how the Java program is created (with edit) and then compiler (with javac.exe), and then finally run (with java.exe).

📖 Java stand-alone program 27.1 (j2.java)

```
public class j2
{
    public static void main(String[] args)
    {
    int I;

       i=10;

       System.out.println("This is an example of the ");
       System.out.println("output from the standalone");
       System.out.println("program");
       System.out.println("The value of i is " + i);
    }
}
```

💻 **Sample run 27.3**
```
C:\DOCS\notes\INTER\java>edit j2.java
C:\DOCS\notes\INTER\java>javac j2.java
C:\DOCS\notes\INTER\java>java j2
This is an example of the
output from the standalone
program
The value of i is 10
```

27.5 Java reserved words

Like any programming language, Java has various reserved words which cannot be used as a variable name. These are given next:

```
abstract     boolean    break       byte        case           cast
catch        char       class       cons        continue       default
do           double     else        extends     final          finally
float        for        future      generic     goto           if
implements   import     inner       instanceof  in             interface
long         native     new         null        operator       outer
package      private    protected   public      rest           return
short        static     super       switch      synchronized   this
throw        throws     transient   try         var            unsigned
virtual      void       volatile    while
```

27.6 Applet variables

Variables are used to stored numeric values and characters. In Java all variables must be declared with their data type before they can be used. The Java data types are similar to C/C++ types, and are stated in Table 27.4.

27.6.1 Converting numeric data types

Java is a strongly typed language and various operations follow standard conversions for data types. If the developer wants to convert from one data type to another (such as from an integer to a double) then the data type conversion is used where the converted data type is defined within rounded brackets. For example:

```
double   y,z;
int      x;
   x=(int) (y+z);
```

converts the addition of y and z to an integer.

27.6.2 The paint() object

The paint() object is the object that is called whenever the applet is redrawn. It will thus be called whenever the applet is run and then it is called whenever the applet is redisplayed.

Table 27.4 Java data types and their range

Type	Storage (bytes)	Range
boolean		True or False
byte	1	−128 to 127
char	2	Alphabetic characters
int	4	−2,147,483,648 to 2,147,483,647
short	2	−32,768 to 32,767
long	4	−2,147,483,648 to 2,147,483,647
float	4	$\pm3.4\times10^{-38}$ to $\pm3.4\times10^{38}$
double	8	$\pm1.7\times10^{-308}$ to $\pm1.7\times10^{308}$

27.7 Java operators

The table below recaps the Java operators from the previous chapter.

HIGHEST PRIORITY

() [] .	primary
! ~ ++ -- -	unary
* / %	multiply
+ -	additive
<< >> >>>	shift
< > <= >=	relation
== !=	equality
&	
^	bitwise
|	
&&	logical
||	
= += -=	assignment

LOWEST PRIORITY

27.8 Mathematical operations

Java has a basic set of mathematics functions which are defined in the `java.lang.Math` class library. Table 27.5 outlines these functions. An example of the functions in this library is `abs()` which can be used to return the absolute value of either a double, an int or a long value. Java automatically picks the required format and the return data type will be of type of the value to be operated on.

As the functions are part of the Math class they are preceded with the `Math.` class method. For example:

```
val2=Math.sqrt(val1);

val3=Math.abs(val2);

z=Math.min(x,y);
```

Java stand-alone program 27.2 shows a few examples of mathematical operations and Sample run 27.4 shows a sample compilation and run session.

Java has also two pre-defined mathematical constants. These are:

- `Pi` is equivalent to 3.14159265358979323846.
- `E` is equivalent to 2.7182818284590452354.

Table 27.5 Functions defined in java.lang.Math

Function	Description
`double abs(double a)`	Returns the absolute double value of a.
`float abs(float a)`	Returns the absolute float value of a.
`int abs(int a)`	Returns the absolute integer value of a.
`long abs(long a)`	Returns the absolute long value of a.
`double acos(double a)`	Returns the arc cosine of a, in the range of 0.0 through Pi.
`double asin(double a)`	Returns the arc sine of a, in the range of Pi/2 through Pi/2.
`double atan(double a)`	Returns the arc tangent of a, in the range of –Pi/2 through Pi/2.
`double atan2(double a, double b)`	Converts rectangular coordinates (a, b) to polar (r, theta).
`double ceil(double a)`	Returns the 'ceiling' or smallest whole number greater than or equal to a.
`double cos(double a)`	Returns the trigonometric cosine of an angle.
`double exp(double a)`	Returns the exponential number e(2.718…) raised to the power of a.
`double floor(double a)`	Returns the 'floor' or largest whole number less than or equal to a.
`double IEEEremainder(double f1, double f2)`	Returns the remainder of f1 divided by f2 as defined by IEEE 754.
`double log(double a)`	Returns the natural logarithm (base e) of a.
`double max(double a, double b)`	Takes two double values, a and b, and returns the greater number of the two.
`double max(float a, float b)`	Takes two float values, a and b, and returns the greater number of the two.
`int max(int a, int b)`	Takes two int values, a and b, and returns the greater number of the two.
`max(long a, long b)`	Takes two long values, a and b, and returns the greater number of the two.
`double min(double a, double b)`	Takes two double values, a and b, and returns the smallest number of the two.
`float min(float a, float b)`	Takes two float values, a and b, and returns the smallest number of the two.
`int min(int a, int b)`	Takes two integer values, a and b, and returns the smallest number of the two.
`long min(long a, long b)`	Takes two long values, a and b, and returns the smallest number of the two.
`double pow(double a, double b)`	Returns the number a raised to the power of b.
`double random()`	Generates a random number between 0.0 and 1.0.
`double rint(double b)`	Converts a double value into an integral value in double format.
`long round(double a)`	Rounds off a double value by first adding 0.5 to it and then returning the largest integer that is less than or equal to this new value.
`int round(float a)`	Rounds off a float value by first adding 0.5 to it and then returning the largest integer that is less than or equal to this new value.
`double sin(double a)`	Returns the trigonometric sine of an angle.
`double sqrt(double a)`	Returns the square root of a.
`double tan(double a)`	Returns the trigonometric tangent of an angle.

📖 Java stand-alone program 27.2 (j3.java)

```
import java.lang.Math;
public class j2
{
    public static void main(String[] args)
    {
    double x,y,z;
    int i;
        i=10;
        y=Math.log(10.0);
        x=Math.pow(3.0,4.0);
        z=Math.random(); // random number from 0 to 1
        System.out.println("Value of i is " + i);
        System.out.println("Value of log(10) is " + y);
        System.out.println("Value of 3^4 is " + x);
        System.out.println("A random number is " + z);
        System.out.println("Square root of 2 is " +
            Math.sqrt(2));
    }
}
```

💻 **Sample run 27.4**

```
C:\DOCS\notes\INTER\java>javac j2.java
C:\DOCS\notes\INTER\java>java j2
Value of i is 10
Value of log(10) is 2.30259
Value of 3^4 is 81
A random number is 0.0810851
Square root of 2 is 1.41421
```

27.9 Loops

27.9.1 `for()`

As with C/C++ and JavaScript the standard format for a `for()` loop is:

```
for (starting condition;test condition;operation)
{
      statement block
}
```

where:

`starting condition`	–	the starting value for the loop;
`test condition`	–	if `test condition` is TRUE the loop will continue execution;
`operation`	–	the operation conducted at the end of the loop.

Java applet 27.2 shows how a `for()` loop can be used to display the square and cube of the values from 0 to 9. Notice that the final value of `i` within the `for()` loop is 9 because the end condition is `i<10` (while `i` is less than 10).

📖 Java applet 27.2 (j4.java)

```
import java.awt.*;
import java.applet.*;

public class j4 extends Applet
{
  public void paint(Graphics g)
  {
  int    i;

    g.drawString("Value Square Cube",5,10);
    for (i=0;i<10;i++)
    {
      g.drawString(""+ i,5,20+10*i);
      g.drawString(""+ i*i ,45,20+10*i);
      g.drawString(""+ i*i*i,85,20+10*i);
    }
  }
}
```

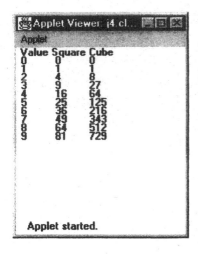

📖 HTML script 27.2 (j4.html)

```
<HTML>
<TITLE>First Applet</TITLE>
<APPLET CODE=j4.class WIDTH=200 HEIGHT=200>
</APPLET>
</HTML>
```

27.9.2 while()

The while() statement allows a block of code to be executed while a specified condition is TRUE. It checks the condition at the start of the block; if this is TRUE the block is executed, else it will exit the loop. The syntax is:

```
while (condition)
{
  :        :    statement block
  :        :
}
```

If the statement block contains a single statement then the braces may be omitted (although it does no harm to keep them).

27.10 Conditional statements

Conditional statements allow a program to make decisions on the route through a program.

27.10.1 if...else

As with C/C++ and JavaScript the standard format for a if() descision is:

```
if (expression)
{
  statement block
}
```

The following is an example format with an `else` extension.

```
if (expression)
{
   statement block1
}
else
{
   statement block2
}
```

Java applet 27.3 uses a `for()` loop and the `if()` statement to test if a value is less than, equal to or greater than 5. The loop is used to repeat the test 10 times.

The `random()` function is used to generate a value between 0 and 1, the returned value is then multiplied by 10 so as to convert to into a value between 0 and 10. Then it is converted to an integer using the data type modifier `(int)`. The `if()` statement is then used to test the value.

📖 Java applet 27.3 (`j5.java`)

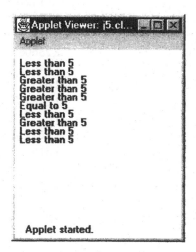

```
import java.awt.*;
import java.applet.*;
import java.lang.Math;

public class j5 extends Applet
{
  public void paint(Graphics g)
  {
    int      i,x;
    double   val;
    for (i=0;i<10;i++)
    {
      val=Math.random();

      x=(int)(val*10.0);
        // Convert value between 0 and 10

      if (x<5)
       g.drawString("Less than 5",5,20+i*10);
      else if (x==5)
       g.drawString("Equal to 5",5,20+i*10);
      else
       g.drawString("Greater than 5",5,20+i*10);
    }
  }
}
```

📖 HTML script 27.3 (`j5.html`)

```
<HTML>
<TITLE>First Applet</TITLE>

<APPLET CODE=j5.class WIDTH=200 HEIGHT=200>

</APPLET>
</HTML>
```

27.11 Exercises

27.11.1 Write a Java applet which displays the following table of powers.

Value	Square	Cube	Fourth power
1	1	1	1
2	4	8	16
3	9	27	81
4	16	64	256
5	25	125	625
6	36	216	1296

27.11.2 Write a Java applet which displays the following table of square root values from 1 to 15.

Value	Square root
1	1
2	1.414214
3	1.732051
4	2
5	2.236068
6	2.44949
7	2.645751
8	2.828427
9	3
10	3.162278
11	3.316625
12	3.464102
13	3.605551
14	3.741657
15	3.872983

27.11.3 Write a Java applet which display 20 random numbers from between 0 and 20.

27.11.4 Write a Java applet that simulates the rolling of two dice. A sample output is:

```
Dice 1:   3
Dice 2:   5
Total:    8
```

Java (Extended functions)

28.1 Introduction

Chapter 27 discussed the Java programming language. This chapter investigates event-driven programs. The main events are:

- Initialization and exit functions (init(), start(), stop() and destroy()).
- Repainting and resizing (paint()).
- Mouse events (mouseUp(), mouseDown() and mouseDrag()).
- Keyboard events (keyUp() and keyDown()).

28.2 Initialization and exit functions

Java has various reserved functions which are called when various event occur. Table 28.1 shows typical initialization functions and their events, and Figure 28.1 illustrates how they are called.

Table 28.1 Java initialization and exit functions

Function	Description
public void init()	This function is called each time the applet is started. It is typically used to add user-interface components.
public void start()	This function is called after the init() function is called. It is also called whenever the user returns to the page containing the applet and thus can be called many times as opposed to the init() function which will only be called when the applet is first started. Thus, code which is to be executed only once is normally put in the init() function, and code which must be executed every time the applet is accessed should be inserted into the start() function.

`public void stop()`	This function is called when the user moves away from the page on which the applet resides. It is thus typically used to stop processing while the user is not accessing the applet. Typically it is used to stop animation or audio files, or mathematical processing. The `start()` function normally restarts the processing.
`public void paint(Graphics g)`	This function is called when the applet is first called and whenever the user resizes or moves the windows.
`public void destroy()`	This function is called when the applet is stopped and is normally used to release associated resources, such as freeing memory, closing files, and so on.

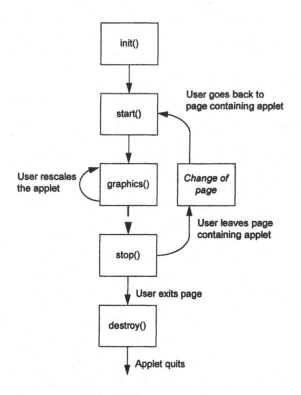

Figure 28.1   Java initialization and exit functions

Java applet 28.1 gives an example using the `init()` and `start()` functions. The variable `i` is declared within the applet and it is set to a value of 5 in the `init()` function. The `start()` function then adds 6 onto this value. After this the `paint()` function is called so that it displays the value of `i` (which should equal 11).

📖 Java applet 28.1 (j6.java)

```
import java.awt.*;
import java.applet.*;

public class j6 extends Applet
{
int      i;

  public void init()
  {
    i=5;
  }
  public void start()
  {
    i=i+6;
  }
  public void paint(Graphics g)
  {
   g.drawString("The value of i is " + i,5,25);
  }
}
```

Applet Viewer: j6.cl...
Applet

The value of i is 11

Applet started.

📖 HTML script 28.1 (j6.html)

```
<HTML>
<TITLE>Applet</TITLE>
<APPLET CODE=j6.class WIDTH=200 HEIGHT=200>
</APPLET>
</HTML>
```

28.3 Mouse events

Most Java applets require some user interaction, normally with the mouse or from the keyboard. A mouse operation causes mouse events. The three basic events which are supported in Java are:

- mouseUp().
- mouseDown().
- mouseDrag().

Java applet 28.2 uses the three mouse events to display the current mouse cursor. Each of the functions must return a true value to identify that the event has been handled successfully (the return type is of data type Boolean thus the return could only be a true or a false). In the example applet, on moving the mouse cursor with the left mouse key pressed down the mouse-Drag() function is automatically called. The x and y co-ordinate of the cursor is stored in the x and y variable when the event occurs. This is used in the functions to build a message string (in the case of the drag event the string name is MouseDragMsg).

📖 Java applet 28.2 (j7.java)

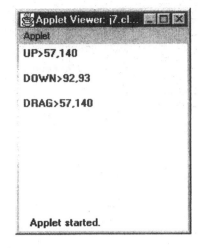

```
import java.awt.*;
import java.applet.*;
public class j7 extends Applet
{
String  MouseDownMsg=null;
String  MouseUpMsg=null;
String  MouseDragMsg=null;

  public boolean mouseUp(Event event, int x,
    int y)
  {
    MouseUpMsg = "UP>" +x + "," + y;
    repaint();    // call paint()
    return(true);
  }
  public boolean mouseDown(Event event, int x,
      int y)
  {
    MouseDownMsg = "DOWN>" +x + "," + y;
    repaint();    // call paint()
    return(true);
  }

  public boolean mouseDrag(Event event, int x,
      int y)
  {
    MouseDragMsg = "DRAG>" +x + "," + y;
    repaint();    // call paint()
    return(true);
  }

  public void paint(Graphics g)
  {
    if (MouseUpMsg !=null)
      g.drawString(MouseUpMsg,5,20);
    if (MouseDownMsg !=null)
      g.drawString(MouseDownMsg,5,40);
    if (MouseDragMsg !=null)
      g.drawString(MouseDragMsg,5,60);
  }
}
```

📖 HTML script 28.2 (j7.html)

```
<HTML>
<TITLE>Applet</TITLE>
<APPLET CODE=j7.class WIDTH=200 HEIGHT=200>
</APPLET></HTML>
```

28.4 Mouse selection

In many applets the user is prompted to select an object using the mouse. To achieve this the x and y position of the event is tested to determine if the cursor is within the defined area. Java applet 28.3 is a program which allows the user to press the mouse button on the applet screen. The applet then uses the mouse events to determine if the cursor is within a given area of the screen (in this case between 10,10 and 100,100). If the user is within this defined area then the message displayed is HIT, else it is MISS. The graphics function g.drawRect(x1,y1,x2,y2) draws a rectangle from (x1,y1) to (x2,y2).

📖 Java applet 28.3 (j8.java)

```
import java.awt.*;
import java.applet.*;
public class j8 extends Applet
{
String  Msg=null;
int     x_start,y_start,x_end,y_end;
  public void init()
  {
    x_start=10;   y_start=10;
    x_end=100;    y_end=50;
  }

  public boolean mouseUp(Event event, int x,
    int y)
  {
    if ((x>x_start) && (x<x_end) &&
          (y>y_start) && (y<y_end))
              Msg = "HIT";
    else Msg="MISS";
    repaint();   // call paint()
    return(true);
  }
  public boolean mouseDown(Event event, int x,
    int y)
  {
    if ((x>x_start) && (x<x_end) &&
          (y>y_start) && (y<y_end))
              Msg = "HIT";
    else Msg="MISS";
    repaint();   // call paint()
    return(true);
  }

  public void paint(Graphics g)
  {
    g.drawRect(x_start,y_start,x_end,y_end);
    g.drawString("Hit",30,30);
    if (Msg !=null)
      g.drawString("HIT OR MISS: " + Msg,5,80);
  }
}
```

📖 HTML script 28.3 (j8.html)

```
<HTML>
<TITLE>Applet</TITLE>
<APPLET CODE=j8.class WIDTH=200 HEIGHT=200>
</APPLET></HTML>
```

28.5 Keyboard input

Java provides for two keyboard events, these are:

- keyUp().Is called when a key has been released.
- keyDown(). Is called when a key has been pressed.

The parameters passed into these functions are event (which defines the keyboard state) and an integer Keypressed which describes the key pressed.

📖 Java applet 28.4 (j9.java)

```
import java.awt.*;
import java.applet.*;

public class j9 extends Applet
{
String  Msg=null;

 public boolean keyUp(Event event, int KeyPress)
  {
   Msg="Key pressed="+(char)KeyPress;
   repaint();    // call paint()
   return(true);
  }
 public void paint(Graphics g)
  {
   if (Msg !=null)
        g.drawString(Msg,5,80);
  }
}
```

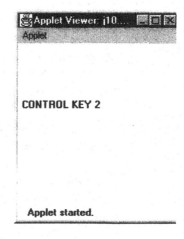

📖 HTML script 28.4 (j9.html)

```
<HTML>
<TITLE>Applet</TITLE>
<APPLET CODE=j9.class WIDTH=200 HEIGHT=200>
</APPLET></HTML>
```

The event contains an identification as to the type of event it is. When one of the function keys is pressed then the variable event.id is set to the macro Event.KEY_ACTION (as shown in Java applet 28.5). Other keys, such as the Ctrl, Alt and Shift keys, set bits in the event.modifier variable. The test for the Ctrl key is:

```
if ((event.modifiers & Event.CTRL_MASK)!=0)
    Msg="CONTROL KEY "+KeyPress;
```

This tests the CTRL_MASK bit; if it is a 1 then the CTRL key has been pressed. Java applet 28.5 shows its uses.

📖 Java applet 28.5 (j10.java)

```
import java.awt.*;
import java.applet.*;

public class j10 extends Applet
{
String  Msg=null;

 public boolean keyDown(Event event,
                        int KeyPress)
 {
   if (event.id == Event.KEY_ACTION)
     Msg="FUNCTION KEY "+KeyPress;
   else if ((event.modifiers & Event.SHIFT_MASK)!=0)
     Msg="SHIFT KEY "+KeyPress;
   else if ((event.modifiers & Event.CTRL_MASK)!=0)
     Msg="CONTROL KEY "+KeyPress;
   else if ((event.modifiers & Event.ALT_MASK)!=0)
     Msg="ALT KEY "+KeyPress;
   else Msg=""+(char)KeyPress;
```

```
      repaint();   // call paint()
      return(true);
   }
   public void paint(Graphics g)
   {
      if (Msg!=null)
         g.drawString(Msg,5,80);
   }
}
```

📖 HTML script 28.5 (`j10.html`)

```
<HTML>
<TITLE>Applet</TITLE>
<APPLET CODE=j10.class WIDTH=200 HEIGHT=200>
</APPLET></HTML>
```

For function keys the `KeyPress` variable has the following values:

Key	Value	Key	Value	Key	Value	Key	Value	Key	Value
F1	1008	F2	1009	F3	1010	F4	1011	F5	1012
F7	1014	F8	1015	F9	1016	F10	1017	F11	1018

Thus, to test for the function keys the following routine can be used:

```
if (event.id == Event.KEY_ACTION)
   if (KeyPress==1008) Msg="F1";
   else if (KeyPress==1009) Msg="F2";
   else if (KeyPress==1010) Msg="F3";
   else if (KeyPress==1011) Msg="F4";
   else if (KeyPress==1012) Msg="F5";
   else if (KeyPress==1013) Msg="F6";
   else if (KeyPress==1014) Msg="F7";
   else if (KeyPress==1015) Msg="F8";
   else if (KeyPress==1016) Msg="F9";
   else if (KeyPress==1017) Msg="F10";
```

For control keys the `KeyPress` variable has the following values:

Key	Value	Key	Value	Key	Value	Key	Value
Cntrl-A	1	Cntrl-B	2	Cntrl-C	3	Cntrl-D	4
Cntrl-E	5	Cntrl-F	6	Cntrl-G	7	Cntrl-H	8

Thus, to test for the control keys the following routine can be used:

```
if ((event.modifiers & Event.CTRL_MASK)!=0)
   if (KeyPress==1) Msg="Cntrl-A";
   else if (KeyPress==2) Msg="Cntrl-B";
   else if (KeyPress==3) Msg="Cntrl-C";
   else if (KeyPress==4) Msg="Cntrl-D";
```

28.6 Graphics images

Java has excellent support images and sound. For graphics files it has support for GIF and JPEG files, each of which is in a compressed form. The image object is declared with:

```
Image mypic;
```

Next the graphics image is associated with the image object with the `getImage()` function:

```
mypic=getImage(getCodeBase(),"myson.gif");
```

where the `getCodeBase()` function returns the applets URL (such as `www.eece.napier.ac.uk`) and the second argument is the name of the graphics file (in this case, `myson.gif`). After this the image can be displayed with:

```
g.drawImage(mypic,x,y,this);
```

where mypic is the name of the image object, and the x and y values are the co-ordinates of the upper-left hand corner of the image. The `this` keyword associates the current object (in this case it is the graphics image) and the current applet. Java applet 28.6 gives an applet which displays an image.

📖 Java applet 28.6 (`j11.java`)

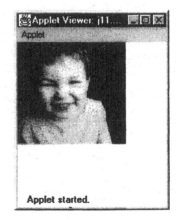

```
import java.awt.*;
import java.applet.*;

public class j11 extends Applet
{
Image   mypic;
  public void init()
  {
   mypic = getImage(getCodeBase(),"myson.gif");
  }

  public void paint(Graphics g)
  {
    g.drawImage(mypic,0,0,this);
  }
}
```

📖 HTML script 28.6 (`j11.html`)

```
<HTML>
<TITLE>Applet</TITLE>
<APPLET CODE=j11.class WIDTH=200 HEIGHT=200>
</APPLET>
</HTML>
```

28.7 Graphics

The `java.awt.Graphics` class contains a great deal of graphics-based functions; these are stated in Table 28.2.

Table 28.2　Java graphics functions

Graphics function	Description
```public abstract void translate(int x,int y)```	Translates the specified parameters into the origin of the graphics context. All subsequent operations on this graphics context will be relative to this origin. Parameters: x - the x co-ordinate y - the y co-ordinate
```public abstract Color get-Color()```	Gets the current colour.
```public abstract void setColor(Color c)```	Set current drawing colour.
```public abstract Font getFont()```	Gets the current font.
```public abstract void setFont(Font font)```	Set the current font.
```public FontMetrics getFontMetrics()```	Gets the current font metrics.
```public abstract FontMetrics getFontMetrics(Font f)```	Gets the current font metrics for the specified font.
```public abstract void copyArea(int x, int y, int width, int height, int dx,int dy)```	Copies an area of the screen where (x,y) is the co-ordinate of the top left-hand corner, width and height and the width and height, and dx is the horizontal distance and dy the vertical distance.
```public abstract void drawLine(int x1,int y1, int x2, int y2)```	Draws a line between the (x1,y1) and (x2,y2).
```public abstract void fillRect(int x, int y, int width,int height)```	Fills the specified rectangle with the current colour.
```public void drawRect(int x,int y, int width, int height)```	Draws the outline of the specified rectangle using the current colour.
```public abstract void clearRect(int x, int y, int width, int height)```	Clears the specified rectangle by filling it with the current background colour of the current drawing surface.
```public void draw3DRect(int x, int y, int width, int height,boolean raised)```	Draws a highlighted 3-D rectangle where raised is a boolean value that defines whether the rectangle is raised or not.
```public void fill3DRect(int x, int y,int width, int height,boolean raised)```	Paints a highlighted 3-D rectangle using the current colour.
```public abstract void drawOval(int x,int y, int width, int height)```	Draws an oval inside the specified rectangle using the current colour.
```public abstract void fillOval(int x,int y, int width, int height)```	Fills an oval inside the specified rectangle using the current colour.

`public abstract void`
`  drawArc(int x, int y,`
`    int width, int height,`
`    int startAngle,int arcAngle)`
	Draws an arc bounded by the specified rectangle starting. Zero degrees for startAngle is at the 3-o'clock position and arcAngle specifies the extent of the arc. A positive value for arcAngle indicates a counter-clockwise rotation while a negative value indicates a clockwise rotation. The parameter (x,y) specifies the centre point, and width and height specifies the width and height of a rectangle

`public abstract void`
`  fillArc(int n, int y,`
`    int width, int height,`
`    int startAngle,`
`    int arcAngle)`
	Fills a pie-shaped arc using the current colour.

`public abstract void`
`  drawPolygon(int xPoints[],`
`    int yPoints[],int nPoints)`
	Draws a polygon using an array of x and y points (xPoints[ ] and yPoints[ ]). The number of points within the array is specified by nPoints.

`public abstract void`
`  fillPolygon(int xPoints[],`
`    int yPoints[],int nPoints)`
	Fills a polygon with the current colour.

`public abstract void`
`  drawString(String str,`
`    int x, int y)`
	Draws the specified String using the current font and colour.

`public abstract boolean`
`  drawImage(Image img,int x,`
`    int y)`
	Draws the specified image at the specified co-ordinate (x, y).

`public abstract void dispose()`	Disposes of this graphics context.

28.7.1 Setting the colour

The current drawing colour is set using the `setColor()` function. It is used as follows:

```
g.setColor(Color.yellow);
```

Colours are defined in the `java.awt.Color` class and valid colours are:

`Color.black`	`Color.blue`	`Color.cyan`	`Color.darkGray`
`Color.gray`	`Color.green`	`Color.lightGray`	`Color.magenta`
`Color.orange`	`Color.pink`	`Color.red`	`Color.white`
`Color.yellow`			

Any other 24-bit colour can be generated with the function Color which has the format:

```
public Color(int r, int g, int b)
```

where `r`, `g` and `b` are values of strength from 0 to 255. For example:

`Color(255,0,0)` gives red;   `Color(255,255,255)` gives white;
`Color(0,128,128)` gives blue/green; `Color(0,0,0)` gives black.

28.7.2 Drawing lines and circles

Normally to draw a graphics object the user must plan its layout for the dimension within the object. Figure 28.2 shows an example graphic with the required dimensions. The `drawOval()` function uses the top level hand point for the x and y parameters in the function and the width and height define the width and height of the oval shape. Thus the `drawOval()` function can be used to draw circles (if the width is equal to the height) or ovals (if the width is not equal to the height). Java applet 28.7 shows the Java code to draw the object. This applet uses the `setColor()` function to make the circle yellow and the other shapes blue.

📖 Java applet 28.7 (`j11.java`)

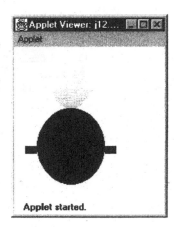

```java
import java.awt.*;
import java.applet.*;
public class j12 extends Applet
{
  public void paint(Graphics g)
  {
    g.setColor(Color.yellow);
    g.fillOval(50,30,50,50);
    g.setColor(Color.blue);
    g.fillOval(30,80,90,100);
    g.fillRect(15,130,15,10);
    g.fillRect(120,130,15,10);
  }
}
```

📖 HTML script 28.7 (`j11.html`)

```html
<HTML><TITLE>Applet</TITLE>
<APPLET CODE=j12.class WIDTH=200 HEIGHT=200>
</APPLET></HTML>
```

28.7.3 Drawing polygons

The `drawPolygon()` function can be used to draw complex objects where the object is defined as a group of (x,y) co-ordinates. Java applet 28.8 draws a basic picture of a car and the xpoints array holds the x co-ordinates and ypoints hold the y co-ordinates. Figure 28.3 illustrates the object.

📖 Java applet 28.8 (`j13.java`)

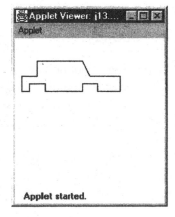

```java
import java.awt.*;
import java.applet.*;
public class j13 extends Applet
{
int
xpoints[]={10,30,30,90,100,140,140,110,110,90,
           90,40,40,20,20,10,10},
 ypoints[]={50,50,30,30,50,50,70,70,60,60,70,
            70,60,60,70,70,50};
public void paint(Graphics g)
{
  g.drawPolygon(xpoints,ypoints,17);
}
}
```

📖 HTML script 28.8 (j13.html)
```
<HTML><TITLE>Applet</TITLE>
<APPLET CODE=j13.class WIDTH=200 HEIGHT=200>
</APPLET></HTML>
```

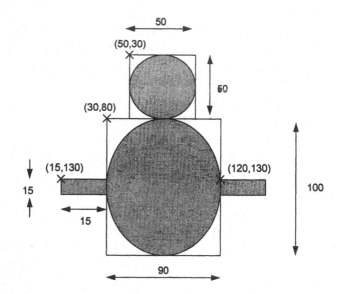

Figure 28.2 Dimensions for graphic

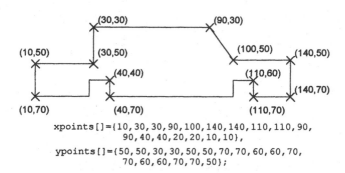

```
xpoints[]={10,30,30,90,100,140,140,110,110,90,
           90,40,40,20,20,10,10},
ypoints[]={50,50,30,30,50,50,70,70,60,60,70,
           70,60,60,70,70,50};
```

Figure 28.3 Co-ordinates of graphic

28.8 Sound

The playing of sound files is similar to displaying graphics files. Java applet 28.9 shows a sample applet which plays an audio file (in this case, test.au). Unfortunately the current version of the Java compiler only supports the AU format, thus WAV files need to be converted into AU format.

The initialization process uses the getAudioClip() function and the audio file is played with the loop() function. This function is contained in the java.applet.AudioClip class, these functions are:

```
public abstract void play()     Plays the audio file and finishes at the end.
public abstract void loop()     Starts playing the clip in a loop.
public abstract void stop()     Stops playing the clip.
```

📖 Java applet 28.9 (j14.java)
```
import java.awt.*;
import java.applet.*;

public class j14 extends Applet
{
AudioClip     audClip;
  public void paint(Graphics g)
  {
    audClip=getAudioClip(getCodeBase(),"hello.au");
    audClip.loop();
  }
}
```

📖 HTML script 28.9 (j14.html)
```
<HTML>
<TITLE>Applet</TITLE>
<APPLET CODE=j14.class WIDTH=200 HEIGHT=200>
</APPLET></HTML>
```

28.9 Dialog boxes

One of the features of Java is that it supports dialog boxes and checkboxes.
These are used with event handlers to produce event-driven options.

28.9.1 Buttons and events

Java applet 28.10 creates three Button objects. These are created with the
add() function which displays the button in the applet window.

📖 Java applet 28.10 (j15.java)
```
import java.awt.*;
import java.applet.*;

public class j15 extends Applet
{
  public void init()
  {
    add(new Button("Help"));
    add(new Button("Show"));
    add(new Button("Exit"));
  }
}
```

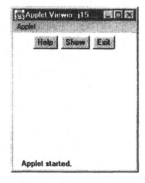

📖 HTML script 28.10 (j15.html)
```
<HTML><TITLE>Applet</TITLE>
<APPLET CODE=j15.class WIDTH=200 HEIGHT=200>
</APPLET></HTML>
```

Applet 28.10 creates three buttons which do not have any action associated with them. Java applet 28.11 uses the action function which is called when an event occurs. Within this function the event variable is tested to see if one of the buttons caused the event. This is achieved with:

```
if (event.target instanceof Button)
```

If this tests to the true then the Msg string takes on the value of the Object, which hold the name of the button that caused the event.

📖 Java applet 28.11 (j15.java)
```
import java.awt.*;
import java.applet.*;

public class j16 extends Applet
{
String  Msg=null;

  public void init()
  {
   add(new Button("Help"));
   add(new Button("Show"));
   add(new Button("Exit"));
  }
  public boolean action(Event event, Object object)
  {
   if (event.target instanceof Button)
   {
     Msg = (String) object;
     repaint();
   }
   return(true);
  }
  public void paint(Graphics g)
  {
   if (Msg!=null)
   g.drawString("Button:" + Msg,30,80);
  }
}
```

📖 HTML script 28.11 (j15.html)
```
<HTML>
<TITLE>Applet</TITLE>
<APPLET CODE=j15.class WIDTH=200 HEIGHT=200>
</APPLET></HTML>
```

28.9.2 Checkboxes

Typically checkboxes are used to select from a number of options. Java applet 28.12 shows how an applet can use checkboxes. As before, the action function is called when a checkbox changes its state and within the function event.target parameter is tested for the checkbox with:

```
if (event.target instanceof Checkbox)
```

If this is true, then the function DetermineCheckState() is called which tests event.target for the checkbox value and its state (true or false).

Java applet 28.12 (j17.java)

```java
import java.awt.*;
import java.applet.*;
public class j17 extends Applet
{
String  Msg=null;

 public void init()
 {
  add(new Checkbox("FAX"));
  add(new Checkbox("Telephone"));
  add(new Checkbox("Email"));
  add(new Checkbox("Post",null,true));
 }
 public void DetermineCheckState(Checkbox Cbox)
 {
  Msg=Cbox.getLabel() + " " + Cbox.getState();
  repaint();
 }
 public boolean action(Event event, Object object)
 {
  if (event.target instanceof Checkbox)
    DetermineCheckState((Checkbox)event.target);
  return(true);
 }
 public void paint(Graphics g)
 {
  if (Msg!=null)
    g.drawString("Check box:" + Msg,30,80);
 }
}
```

HTML script 28.12 (j17.html)

```html
<HTML>
<TITLE>Applet</TITLE>
<APPLET CODE=j17.class WIDTH=200 HEIGHT=200>
</APPLET>
</HTML>
```

28.9.3 Radio buttons

The standard checkboxes allow any number of options to be selected. A radio button allows only one option to be selected at a time. The previous example can be changed as follows:

```java
public void init()
{
    add(new Checkbox("FAX",RadioGroup,false));
    add(new Checkbox("Telephone",RadioGroup,false));
    add(new Checkbox("Email",RadioGroup,false));
    add(new Checkbox("Post",RadioGroup,true));
}
```

This sets the checkbox type to RadioGroup and it can be seen that only one of the checkboxes is initally set (that is, 'Post').

28.9.4 Pop-up menu choices

To create a pop-up menu the Choice object is initally created with:

```
Choice mymenu = new Choice();
```

After this the menu options are defined using the addItem method. Java applet 28.13 shows an example usage of a pop-up menu.

📖 **Java applet 28.13 (j15.java)**

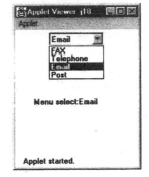

```
import java.awt.*;
import java.applet.*;

public class j18 extends Applet
{
String  Msg=null;
Choice  mymenu= new Choice();

  public void init()
  {
    mymenu.addItem("FAX");
    mymenu.addItem("Telephone");
    mymenu.addItem("Email");
    mymenu.addItem("Post");
    add(mymenu);
  }
  public void DetermineCheckState(Choice mymenu)
  {
    Msg=mymenu.getItem(mymenu.getSelectedIndex());
    repaint();
  }
  public boolean action(Event event, Object object)
  {
   if (event.target instanceof Choice)
     DetermineCheckState((Choice)event.target);
   return(true);
  }
  public void paint(Graphics g)
  {
    if (Msg!=null)
      g.drawString("Menu select:" + Msg,30,120);
  }
}
```

📖 **HTML script 28.13 (j15.html)**

```
<HTML>
<TITLE>Applet</TITLE>
<APPLET CODE=j15.class WIDTH=200 HEIGHT=200>
</APPLET></HTML>
```

28.9.5 Text input

Text can be entered into a Java applet using the TextField action. In Java applet 28.14 the TextField(20) defines a 20-character input field.

📖 **HTML script 28.14 (j15.html)**

```
<HTML>
<TITLE>Applet</TITLE>
<APPLET CODE=j15.class WIDTH=200 HEIGHT=200>
</APPLET></HTML>
```

📖 Java applet 28.14 (j15.java)

```
import java.awt.*;
import java.applet.*;

public class j19 extends Applet
{
String  Msg=null;

 public void init()
 {
   add(new Label("Enter your name"));
   add(new TextField(20));
 }
 public void DetermineText(TextField mytext)
 {
   Msg=mytext.getText();
   repaint();
 }

 public boolean action(Event event, Object object)
 {
   if (event.target instanceof TextField)
      DetermineText((TextField)event.target);
   return(true);
 }
 public void paint(Graphics g)
 {
   if (Msg!=null)
      g.drawString("Your name is:" + Msg,30,120);
 }
}
```

28.10 Fonts

Java is well supported with different fonts. The class library java.awt.Font defines the Font class and the general format for defining the font is:

```
Font font = new Font(font_type,font_attrib,font_size)
```

The main font_types are:

```
"TimesRoman"     "Helvetica"     "Courier"     "Symbol"
```

This book is written in Times Roman. Helvetica looks good as a header, such as **Header** 1. Courier produces a monospace font where all of the characters have the same width. The Java applets in this chapter use the Courier font. Symbol is normally used when special symbols are required. The *font_attrib* can either be BOLD, ITALIC or NORMAL. and the font_size is an integer value which is supported by the compiler. The font size of this text is 11 and most normal text varies between 8 and 12.

Java applet 28.15 shows an example applet using different fonts.

📖 Java applet 28.15 (j20.java)

```
import java.awt.*;
import java.applet.*;

public class j20 extends Applet
{
Font   TimesRoman= new Font("TimesRoman",Font.BOLD,24);
Font   Courier= new Font("Courier",Font.BOLD,24);
Font   Helvetica= new Font("Helvetica",Font.BOLD,24);
Font   Symbol= new Font("Symbol",Font.BOLD,24);

  public void paint(Graphics g)
  {
    g.setFont(TimesRoman);
    g.drawString("Sample text",10,40);
    g.setFont(Courier);
    g.drawString("Sample text",10,60);
    g.setFont(Helvetica);
    g.drawString("Sample text",10,80);
    g.setFont(Symbol);
    g.drawString("Sample text",10,100);
  }
}
```

📖 HTML script 28.15 (j20.html)

```
<HTML><TITLE>Applet</TITLE>
<APPLET CODE=j20.class WIDTH=200 HEIGHT=200>
</APPLET></HTML>
```

28.11 Exercises

28.11.1   Explain how the three mouse events occur.

28.11.2   Write a Java applet that contains a target which has areas with different point values. These point values are 50, 100 and 150. The program should accumulate the score so far. A sample screen is given in Figure 28.4 (refer to Java applet 28.3 for the outline of the program).

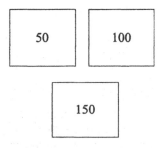

Points: 400

Figure 28.4

28.11.3   Modify the program in 28.11.2 so that a RESET button is displayed. When selected the points value should be reset to zero.

28.11.4 Write a Java applet which displays a square at a starting position (50,50) with a width and length of 50 units. The user should then be able to move the rectangle up, down, left or right using the u key (for up), d key (for down), l key (for left) and r key (for right). The program should continue until the x key is pressed.

28.11.5 Write a Java applet which displays which function key or control key (Cntrl) has been pressed. The program should run continuously until the Cntrl-Z keystroke is pressed.

28.11.6 Write separate Java applets, using simple rectangles and circles, to display the following graphics:

(a) a television (a sample graphic is shown in Figure 28.5)
(b) a face
(c) a house
(d) a robot

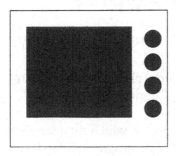

Figure 28.5

28.11.7 Locate three GIF or JPEG files. Then write a Java applet which allows the user to choose which one should be displayed. The function key F1 selects the first image, F2 the second, and F3 the third. The function key F4 should exit the applet.

28.11.8 Using an applet which displays the text TEST TEXT, determine the approximate colour of the following colour settings:

(a)   color(100, 50, 10)
(b)   color(200,200,0)
(c)   color(10,100,100)
(d)   color(200,200,200)
(e)   color(10,10,100)

28.11.9 Write an applet using a polygon for the following shapes:

(a) a ship (a sample is shown in Figure 28.6)
(b) a tank
(c) a plane

Figure 28.6

28.11.10 Locate a sound file and write an application which uses it.

28.11.11 Write a Java applet which displays the following:

Sample Applet

`Please press the x key to exit`

Note that Sample Applet is bold Arial text with a size of 20 and the other text is Courier of size of 14.

28.11.12 Write a Java applet in which the user enters some text. If the user enters EXIT then the program will exit.

PART F

C/Pascal
C++
Assembly Language
Visual Basic
HTML/Java
DOS
Windows 3.x
Windows 95
UNIX

29 Introduction

29.1 Introduction

In 1947 the invention of the transistor caused a great revolution. In fact scientists at the Bell Laboratories kept its invention secret for over seven months so that they could fully understand its operation. On 30 June 1948 the transistor was finally revealed to the world. Unfortunately, as with many other great inventions, it received little public attention and even less press coverage (the *New York Times* gave it 4½ inches on page 46).

Transistors had, initially, been made from germanium, which is not a robust material and cannot withstand high temperatures. The first company to propose a method of using silicon transistors was a geological research company named Texas Instruments (which had diversified into transistors). Soon many companies were producing silicon transistors and by 1955 the electronic valve market had peaked, while the market for transistors was rocketing. The larger electronic valve manufacturers, such as Western Electric, CBS, Raytheon and Westinghouse failed to adapt to the changing market and quickly lost their market share to the new transistor manufacturing companies, such as Texas Instruments, Motorola, Hughes and RCA.

In 1959, IBM built the first commercial transistorized computer named the IBM 7090/7094 series. It was so successful that it dominated the computer market for many years. In 1965, they produced the famous IBM system 360 which was built with integrated circuits. Then in 1970 IBM introduced the 370 system, which included semiconductor memories. Unfortunately, these computers were extremely expensive to purchase and maintain.

Around the same time the electronics industry was producing cheap pocket calculators. The development of affordable computers happened when the Japanese company Busicon commissioned a small, at the time, company named Intel to produce a set of eight to twelve ICs for a calculator. Instead of designing a complete set of ICs, Intel produced a set of ICs which could be programmed to perform different tasks. These were the first ever microprocessors. Soon Intel (short for *Int*egrated *El*ectronics) produced a general-purpose 4-bit microprocessor, named the 4004 and a more powerful 8-bit version, named the 8080. Other companies, such as Motorola, MOS Technologies and Zilog were soon also making microprocessors.

IBM's virtual monopoly on computer systems soon started to slip as many companies developed computers based around the newly available 8-bit microprocessors, namely MOS Technologies 6502 and Zilog's Z-80. IBM's main contenders were Apple and Commodore who introduced a new type of computer – the personal computer (PC). The leading systems were the Apple I and the Commodore PET. These spawned many others, including the Sinclair ZX80/ZX81, the BBC microcomputer, the Sinclair Spectrum, the Commodore Vic-20 and the classic Apple II (all of which were based on or around the 6502 or Z-80).

IBM realized the potential of the microprocessor and used Intel's 16-bit 8086 microprocessor in their version of the PC. It was named the IBM PC and has since become the parent of all the PCs ever produced. IBM's main aim was to make a computer which could run business applications, such as word processors, spread-sheets and databases. To increase the production of this software they made information on the hardware freely available. This resulted in many software packages being developed and helped clone manufacturers to copy the original design. So the term 'IBM-compatible' was born and it quickly became an industry standard by sheer market dominance.

On previous computers IBM had written most of their programs for their systems. For the PC they had a strict time limit, so they went to a small computer company called Microsoft to develop the operating system program. This program was named the Disk Operating System (DOS) because of its original purpose of controlling the disk drives. It accepted commands from the keyboard and displayed them to the monitor. The language of DOS consisted of a set of commands which were entered directly by the user and interpreted to perform file management tasks, program execution and system configuration. The main functions of DOS were to run programs, copy and remove files, create directories, move within a directory structure and to list files.

Microsoft has since gone on develop industry-standard software such as Microsoft Windows Version 3, Microsoft Office and Microsoft Windows 95. Intel has also benefited greatly from the development of the PC and has developed a large market share for their industry-standard microprocessors, such as the 80286, 80386, 80486, Pentium and Pentium Pro processors.

29.2 Introduction to DOS

Most modern PCs either run DOS or allow access to an emulated version of it. All version of Microsoft Windows up to, and including Microsoft Windows 3.1, required DOS to be running before Windows could run. Windows 95 and Windows NT are complete operating systems within themselves and display an emulated version of DOS.

29.2.1 Checking the Version of DOS (*VER*)

The first version of DOS was released in 1981 and each subsequent release has been assigned a new version number. When there is a major change in DOS then the first number is changed, such as, Versions 1.0, 2.0, 3.0, and so on. The second number changes with a relatively minor change, such as version 1.1, 1.2, 1.3, and so on. Most versions of DOS are compatible with previous versions.

To determine the version of DOS that a computer is running then enter the VER command, as shown in Test run 29.1. Many current systems use Version 6 or run an emulated DOS from Windows 95 or Windows NT.

DOS versions later than Version 4 have an on-line help manual. To display the help page on a command then HELP *command_name* is entered, as shown in Test run 29.2.

🖳 **Test run 29.1**
```
C:\>ver
MS-DOS Version 5.00
C:\>
```

🖳 **Test run 29.2**
```
C:\>help ver
Displays the MS-DOS version.
VER
```

29.2.2 Checking the date and time (DATE and TIME)

Most PCs have an on-board clock which is powered by a rechargeable battery. The DATE command displays the current date and allows the user to change the system data, if required. Test run 29.3 shows an example. If the date is displayed correctly then the *<ENTER>* key is pressed, as shown in Test run 29.3.

🖳 **Test run 29.3**
```
C:\>date
Current date is Wed 13/09/1995
Enter new date (dd-mm-yy):
```

If this date is incorrect then the correct date is entered at the Enter new date (dd-mm-yy) prompt. The example in Test run 29.4 shows that the current date has been changed to Sunday 8/9/96. Care must be taken when entering the current date as the system may be set up to display the date in USA format, that is, MM-DD-YY (MONTH, DAY and YEAR).

The TIME command displays the current time. As with the date command the user is prompted as to whether to change the current time, or not. Pressing the *ENTER* key does not change the current time. Test run 29.6 shows an example of changing the system time from 10:55 am to 11:15 pm.

⌨ **Test run 29.4**
```
C:\>date
Current date is Sat 07/09/1996
Enter new date (dd-mm-yy):
C:\>date
Current date is Sat 07/09/1996
Enter new date (dd-mm-yy): 8/9/96
C:\>date
Current date is Sun 08/09/1996
Enter new date (dd-mm-yy):
```

⌨ **Test run 29.5**
```
C:\>time
Current time is  18:40:24.62
Enter new time:
```

⌨ **Test run 29.6**
```
C:\>time
Current time is  10:55:21.76
Enter new time: 11:15
C:\>time
Current time is  11:15:05.40
Enter new time:
```

29.2.3 Clearing the screen (CLS)

The CLS command clears the screen and leaves only a prompt and the cursor.

29.2.4 Starting and stopping DOS

When the PC is restarted, DOS is normally started automatically from the hard-disk (unless the system has no hard-disks). Some PCs try to read the floppy disks drives for the DOS system files. If this happens make sure there are no floppy disks in the drives.

The method of starting DOS is described as the boot procedure. There are two main methods of starting (or re-starting DOS):

- A cold boot – occurs at power-up and causes the system to start a self-test program;
- A warm boot – is executed when the Ctrl, Alt and Del keys are pressed down simultaneously (described as Ctrl-Alt-Del). This is normally required when a program has crashed and the system has hung-up.

The Ctrl-Break keys (or Cntrl-C) are used to interrupt a DOS command. Test run 29.7 shows how the TIME command is interrupted using the Ctrl and 'C' keys.

🖳 **Test run 29.7**

```
C:\>time
Current time is  11:44:11.21
Enter new time: ^C
```

29.3 Disks

Computers store data on floppy disks, hard-disks and optical disks. The disk drives on a PC are identified by the labels from A: to Z:. It is convention that the primary hard-disk drive is drive C:, the primary floppy drive is A:, and the secondary floppy drive is B:., as given in Table 29.1.

Floppy and hard-disks store of information using magnetically fields on separate concentric rings – known as tracks. These are subdivided into blocks of 512 bits – known as sectors. There are various different formats for these tracks and sectors with varying capacities; these are given in Figure 29.2. Differing capacities may lead to compatibility problems.

Table 29.1 Disk drive allocation

Drive Allocation	Description
A:	Primary floppy disk drive
B:	Secondary floppy disk drive
C:	Primary hard-disk drive
D:	Secondary hard-disk drive or CD-ROM drive

Table 29.2 Floppy disk capacity

	Format	Tracks	Sectors	Size
Double-sided	360 KB	40	9	5.25"
Double-sided	720 KB	80	9	3.5"
Double-sided	1.2 MB	80	15	5.25"
Double-sided	1.4 MB	160	9	3.5"

29.4 Formatting floppy disks (*FORMAT*)

The primary hard-disk is named C: and the primary disk drive is named A:. If the PC has another floppy disk drive it is given the name B:. Other drives can exist, such as for network drives, a secondary hard-disk, a CD-ROM drive, and so on.

There are two main types of floppy disks. These are usually referred to by the disks dimensions, as shown in Figure 29.2. Originally PCs used the 5.25" floppy disks. They are extremely sensitive to physical damage, especially when they are bent. Data on the disk can also be corrupted by external particles or finger prints. The disk can be put into a dust cover – although it provides little protection to being bent. A better floppy disk uses a sliding metal shutter to protect the magnetic disk, this is the 3.5" disk. It also allows a greater amount of data storage.

A 3.5" disk can be protected from over-writing stored data by sliding the write protect button, as shown in Figure 29.2. There are two types 3.5" disk types, these differ in the amount of data that can be safely stored on them. They are named the double density (DD) and high density (HD) disk. The DD disk has a maximum capacity of 720 KB and the HD has 1.4 MB. Normally, HD disks are more reliable than DD disks, but they cost more to buy. An HD disk can be identified either by the HD symbol or by the notch at the other side of the write protect slider.

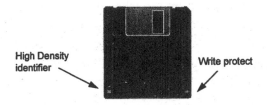

High Density
identifier

Write protect

Figure 29.1   3.5" floppy disks.

29.4.1 Formatting a 3.5" disk

The FORMAT command sets up the necessary format on the disk so that files can be stored. A sample session is shown in Test run 29.8.

```
Test run 29.8
C:\>format a:
Insert new diskette for drive A:
and press ENTER when ready...
Checking existing disk format.
Saving UNFORMAT information.
Verifying 1.44M
Format complete.

Volume label (11 characters, ENTER for none)? fred

   1457664 bytes total disk space
   1457664 bytes available on disk

       512 bytes in each allocation unit.
      2847 allocation units available on disk.

Volume Serial Number is 1720-18EE

Format another (Y/N)?n
C:\>
```

Notice that the message Saving UNFORMAT information is displayed. This allows the user to unformat a disk that has been formatted accidentally (but only if no new files have been written to the disk).

It is also possible to force the FORMAT command to format with a different capacity. The /f switch extension is used for this purpose. Table 29.3 outlines how it is modified for different capacities.

Table 29.3 Floppy disk capacity

Capacity	Command	Notes
720K	format a: /f:720	DD 3.5" floppy disk
1.44M	format a: /f:1.44	HD 3.5" floppy disk
360K	format a: /f:360	DD 5 1/4" floppy disk
1.2M	format a: /f:1.2	HD 5 1/4" floppy disk

29.5 File system structure

Files store data in the form of programs, documents, spread-sheets, and so on. They are organized into a tree structure. The top of the structure is the root directory and each branch is called a subdirectory. This structure makes finding files easier than having all files stored in the one directory.

The route through the tree structure to a given file is known as the path-name and reference to files external to the current directory is possible by using the correct path. The pathname for a given file is traced through the tree structure from the root directory to the location of the file.

Test run 29.9 shows an example list of a top-level directory. The prompt should be set-up to display the current directory. In this case, the DOS prompt is C:\>.

The top level directory in this case contains various sub-directories such as DOCS, WINDOWS, PSP and TURBO. Each of these directories contains files and/or sub-directories. The function of each directory is normally obvious from its name. For example, the DOS directory contains DOS program, help manuals and system files, the WINDOWS directory contains programs for Microsoft Windows.

___

🖳 **Test run 29.9**

```
C:\> dir /w
 Volume in drive C is THINKPAD
 Volume Serial Number is 3A40-09E8
 Directory of C:\
[APPL]        [ASYMPRES]    [BORLANDC]    [CCMOBILE]    [CSW]
[DISKETTS]    [DOCS]        [DOS]         [FACTORY]     [GMOUSE]
[IPFWIN]      [LLW]         [MSIMEV60]    [OLDFILES]    [ORGANIZE]
[PCAUDIO]     [PRIDE]       [PSP]         [RTT]         [SCRNCAM]
[STEVE_M]     [TALKWRKS]    [TANNER]      [THINKPAD]    [TRANXIT]
[TURBO]       [UNIVBE]      [UTILS]       [WINDOWS]     [WINZIP16]
AUTOEXEC.BAT COMMAND.COM  CONFIG.SYS    SYSLEVEL.TP  WIN.BAT
     36 file(s)    65,786 bytes    42,008,576 bytes free
```
___

The top-level directory is the root directory, and is given the name \. A small section of the directory hierarchy is given in Figure 29.2. Notice that there are four sub-directories below the root directory and that below the DOCS directory there are three sub-directories NOTES, CLASSES and ADMIN. There is also one sub-directory below WINDOWS. To identify the directories a pathname must be given. The full pathname each directory is:

```
\DOCS           \WINDOWS    \DOCS\ADMIN   \WINDOWS\SYSTEM
\DOCS\CLASSES   \DOS        \DOCS\NOTES   \DOCS\ADMIN\MEMO
```

All these directories are stored on the hard disk which is labelled as C:, thus the full pathname, with the disk drive, of the directories are:

```
C:\DOCS          C:\WINDOWS C:\DOCS\ADMIN   C:\WINDOWS\SYSTEM
C:\DOCS\CLASSES  C:\DOS     C:\DOCS\NOTES   C:\DOCS\ADMIN|MEMO
```

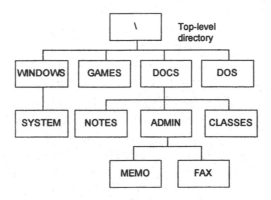

Figure 29.2   Sample file structure

These pathnames are the full pathnames and are absolute pathnames. It is also possible to specify a relative pathname. With relative pathname the name given does not have the preceding top-level directory (\). For example, if we are in the GAMES directory the relative pathname for the DATA directory is CHESS\DATA and if we are in the WINDOWS directory the relative pathname for the SYSTEM directory is simply SYSTEM.

29.6 DOS Filenames

Files are stored on disks with a filename and an extension. The filename can be up to eight characters long and the extension up to three characters. A period ('.') separates the filename from the extension. Figure 29.3 shows the standard format. The extension name gives an indication about what type of file it is.

The valid characters that can be used for the filename are alphabetic characters ('A' – 'Z'), numbers ('0' – '9'), underscores ('_'), braces ('{','}'), parentheses ('(',')'), tilde ('~'), caret ('^'), ampersand ('&'), exclamation point ('!'), pound sign ('£'), hyphen ('-'), and dollar sign ('$'). Other special characters such as ('*'), ('?'), (':'), (';'), (','), ('='), and so on. are reserved for special functions. Some valid and invalid filenames are given in Table 29.4.

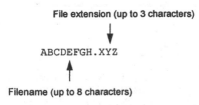

File extension (up to 3 characters)

```
ABCDEFGH.XYZ
```

Filename (up to 8 characters)

Figure 29.3   DOS filenames

29.7 File types

Most files created have a certain purpose; for example documents from a word processor, spread-sheets, text files. The filename extension adds extra information about what type of file it is. Common filename extensions are given in Table 29.5.

Table 29.4   Examples of valid and invalid DOS filenames.

Valid DOS file-names	Invalid DOS file-name	Valid DOS file-names	Invalid DOS file-name
FRED_1.DOC	FRED_?.DOC	SET{1}.BAT	SET{1}.???
ACC9192.WK1	ACC91/92.WK1	INFO^10.SYS	INFO,SYS
INCOME-1.TXT	INCOME:1.TXT	$$$1.$$$	FRED 1.C

Test run 29.10 shows a sample DOS listing. Notice that this directory contains System Files (.SYS), DOS Commands (.COM and .EXE), Text Files (.TXT) and Help Files (.HLP). The other typical files include Basic Language Files (.BAS), Initialization Files (.INI) and Listings (.LST). Programs with the .COM, .EXE or .BAT extension can be executed.

Table 29.5   Example file extensions

File extension	File type	File extension	File type
.ASC	ASCII Text	.PAS	Pascal file
.BAK	Backup File	.PCX	Picture file
.BAT	DOS Batch File	.PRN	Print File
.C	C language File	.SYS	System File
.COM	DOS Program File	.TXT	Text File
.EXE	DOS Executable program	.WK1	123 Ver 1/2 File
.HLP	Help File	.WK3	1-2-3 Ver 3 File
.OVL	Overlay File used by program	.TMP	Temporary File

29.8 Listing files (*DIR*)

The DIR command displays the contents of a directory. A help manual on DIR is given in Test run 29.11.

Various switches modify the way the DIR command displays the directory listing. Refer to the user manual shown in Test run 29.11 for a complete listing. Test run 29.12 shows a sample listing without switches.

🖥 **Test run 29.10**

```
C:\DOS> dir /w
 Volume in drive C is THINKPAD
 Volume Serial Number is 3A40-09E8
 Directory of C:\DOS
[.]            [..]           [DATA]         [SYSTEM]       4201.CPI
4208.CPI       ADDHELP.OVL    ADMIN.PRF      ANSI.SYS       APPEND.EXE
ASSIGN.COM     ATTRIB.EXE     AUTORUN.PRF    DOSSHELL.INI   CHKDSK.COM
  :    :    :    :    :    :    :
WNSCHEDL.EXE   WNTOOLS.GRP    WNUNDEL.HLP    WNUNDEL.EXE    WNVC1.DLL
WNVCVEC1.DLL   WNVE1.DLL      WNVF1.DLL      XCOPY.EXE      CHECKUP.DB
       176 file(s)      6,891,171 bytes
                       41,385,984 bytes free
```

🖥 **Test run 29.11**

```
C:\> help dir
DIR [drive:][path][filename] [/P] [/W] [/A[[:]attributes]]
[/O[[:]sortorder]] [/S] [/B] [/L]
[drive:][path][filename] Specifies drive, directory, or files to list
  /P       Pauses after each screenful of information.
  /W       Uses wide list format.
  /A       Displays files with specified attributes.
  attributes D  Directories              R  Read-only files
             H  Hidden files             A  Files ready for archiving
             S  System files             -  Prefix meaning "not"
  /O          List by files in sorted order.
  sortorder  N  By name (alphabetic)     S  By size (smallest first)
             E  By extension (alphabetic) D  By date & time
             G  Group directories first  -  Prefix to reverse order
  /S    Displays files in specified directory and all subdirectories.
```

Table 29.6 Example switches for the DIR command

Command	Description
DIR /w	display directory in five columns
DIR /p	displays one screen of listing at a time
DIR /s	display all subdirectories and files
DIR /od	display directory in order of date
DIR /l	displays directories in lower case

Test run 29.12 shows an example session using some of the formats. Notice that the basic DIR command displays not only the filename and extension, but also the following:

- The volume label of the disk (in this case, THINKPAD).
- The volume serial number, every disk drive has a unique volume serial number (3A40-09E8).
- The current directory name (C:\).
- The directory name or filename.
- The date file or subdirectory was created, or was last modified (for example, for the DOS directory it is 05/01/95).
- The time the file or directory was created, or was last modified (for example, for the DOS directory it is 12:00).
- The number of files in the directory (in this case, 37).

- The disk space used up by the files in the current directory (in this case, 65,786 bytes).
- The amount of available disk space (in this case, 41,353,216 bytes).

Test run 29.12

```
C:\ dir
 Volume in drive C is THINKPAD
 Volume Serial Number is 3A40-09E8
 Directory of C:\
APPL          <DIR>          05/02/95    15:22
ASYMPRES      <DIR>          05/01/95    12:00
BORLANDC      <DIR>          19/02/95     8:20
CCMOBILE      <DIR>          05/01/95    12:00
CSW           <DIR>          05/01/95    12:00
DISKETTS      <DIR>          05/01/95    12:00
DOCS          <DIR>          08/08/95    10:54
DOS           <DIR>          05/01/95    12:00
FACTORY       <DIR>          05/01/95    12:00
GMOUSE        <DIR>          18/08/96    19:50
  :     :     :
TANNER        <DIR>          27/08/96     9:45
THINKPAD      <DIR>          05/01/95    12:00
TRANXIT       <DIR>          05/01/95    12:00
TURBO         <DIR>          20/08/96    19:12
UNIVBE        <DIR>          18/08/96    20:44
UTILS         <DIR>          09/08/96    13:31
WINDOWS       <DIR>          05/01/95    12:00
WINZIP16      <DIR>          18/08/96    19:56
AUTOEXEC BAT           992   18/08/96    19:55
COMMAND  COM        54,654   25/01/94     1:00
CONFIG   SYS           648   04/01/95    12:39
SYSLEVEL TP             10   05/01/95    13:04
WIN      BAT           133   01/01/95    10:49
WINA20   386         9,349   25/01/94     1:00
        37 file(s)          65,786 bytes
                        41,353,216 bytes free
```

The DIR /p displays one screen of directory information and the user is prompted to *Press and key to continue....* This continues until all the directory information is displayed. The DIR /s command is useful for finding files in subdirectories. Several switches can be used at a single time, for example to display all subdirectories with a pause between each screenful the command used is DIR /p /s.

Worksheet 14

Note, if you are using Windows NT/95 then select either the MSDOS Prompt or shutdown Windows and select Restart the computer in MSDOS mode.

W14.1 Determine the DOS version of the PC you are using.

Version:

W14.2 Display the current date on the PC.

Date:

W14.3 Display the current time of the PC.

Time:

W14.4 Modify the date and time and redisplay.

New date:
New time:

W14.5 Change the date and time back so that they show the correct date and time.

Completed successfully:	YES/NO

W14.6 Use the `Ctrl-C` (or `Ctrl-Break`) keystrokes to quit from the TIME command.

Completed successfully:	YES/NO

W14.7 Clear the screen.

Completed successfully:	YES/NO

W14.8 Locate the following keys (tick, if found):

Keys	✓		✓
Functions keys F1-F10		Page Up/ Page Down	
Alt key		Num Lock	
Cntrl key		Home/End	
Caps Lock		ESC	
Horizontal TAB		Scroll Lock	
Break		Delete	

W14.9 Display the following characters to the screen (tick, if found):

Keys	✓	Keys	✓
£		\	
<		:	
>		(	
?		)	
#		*	
/		|	

W14.10 List all the files in the top level directory on the C: drive.

Note some of contents:

FILES:

DIRECTORIES:

W14.11 Which of the following are valid names for files (tick):

Filename	Valid(✓)	Invalid(✓)
MY_FILE.DAT		
DOC1,DOC		
TEMP.$$$		
WORK~1.BAT		
WORK??.BAT		
TAX:1.DOC		
$1.WP		

W14.12 For the file structure given in Figure W14.1 complete the table given next (the first one has been completed):

Directory	Full pathname
ADMIN	C:\DOCS\ADMIN
MEMO	
BERT	
BATCH	
WINDOWS	
SYSTEM	

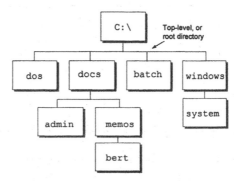

Figure W14.1 File structure

W14.13 For the file structure given in Figure W14.2 determine the full pathnames for the files given next (the first one has been completed):

FILE	Full pathname
T_SHEET.DOC	C:\DOCS\ADMIN\T_SHEET.DOC
P_SHEET.DOC	
MEMO1.DOC	
XCOPY.EXE	
MESSAGE.BAT	
MOUSE.SYS	

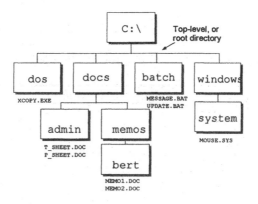

Figure W14.2 File structure

W14.14 Using the `DIR` command list all the files on the hard disk (`C:`). Use the `CNTRL-C` keystroke to exit.

Completed successfully:	YES/NO

W14.15 What is the probable function of the following files:

File	Probable Function
ADMIN.DOC	Document File
TAX91.WK1	
DOSHELP.BAK	
CONFIG.SYS	
XCOPY.HLP	
FIG1.PCX	
GORILLA.BAS	
HELP.TXT	

W14.16 Determine from the `DIR` listing the size of the following files on the hard disk:

FILE	Size of file (Bytes)
COMMAND.COM	
CONFIG.SYS	
AUTOEXEC.BAT	
\DOS\XCOPY.EXE	
\DOS\FC.EXE	

W14.17 Determine from the `DIR` listing the date that the following files were last modified:

FILE	Date last modified
COMMAND.COM	
CONFIG.SYS	
AUTOEXEC.BAT	

\DOS\XCOPY.EXE

\DOS\FC.EXE

W14.18 Determine the amount of disk space that is remaining on the hard disk.

Disk space:

W14.19 Determine the amount of disk space used up by all the files in the \DOS directory:

Disk space used by DOS directory:

W14.20 Determine the number of files in the \DOS directory (or the \WINDOWS directory if there is no \DOS directory):

Number of files:

W14.21 State what the following DIR commands do (try them out):

Command	Function
dir /b	*Lists directories and displays only the name of the file/directory (bare listing)*
dir /l	
dir /on	
dir /os	
dir /od	

W14.22 Reboot the computer using a warm-boot.

Completed successfully:	YES/NO

W14.23 Reboot the computer using a cold-boot.

Completed successfully:	YES/NO

DOS File System/Editor

30.1 Changing directory (CD or CHDIR)

The change directory command, CD (or CHDIR) allows the user to move
around the file system. Table 30.1 shows a few sample commands and
Sample session 30.1 shows a sample session. At the beginning of this session the user is in the top level directory (C:\). The DIR/W command lists
contents of the directory across the screen in five columns. In this case the
main directory contains a subdirectory named \DOCS. To change the current directory to this directory the CD DOCS command line is used. Next,
contents of this directory are listed using DIR/W.

In the \DOCS directory there are nine subdirectories. Each of these sub-
directories have their own function; for example, lecture notes are stored in
the NOTES directory and general administration documents are stored in the
ADMIN directory. Note that the full pathname for the NOTES directory is
C:\DOCS\NOTES.

Next, the user changes the current working directory to the DOCS directory and then to the DOS subdirectory. There are several files in this
directory including the files DOS.DOC and DOS_S1.DOC. The full pathname of the file DOS.DOC is C:\DOS\DOCS\NOTES\DOS\DOS.DOC.

Table 30.1 Examples of the CD command

Example	Description
CD ..	change directory to the one above the current directory
CD \	change directory to the top-level directory
CD \docs\notes\dos	change directory to the directory \DOCS\NOTES\DOS
CD notes	change directory to the sub-directory NOTES

🖳 **Sample session 30.1**

```
C:\>dir /w
 Volume in drive C is HARD-DISK
 Directory of C:\
[DOS]        [WINDOWS]     [VGAUTIL]       [DOCS]       [COMBIOSW]
[GWS]        [RESEARCH]    [TC]            [WINWORD]    [VIRUS]
[MASM]       [RTT]         [T]             [123]        [INTER]
[MSOFFICE]   [CIRRUS]      [DISCWLD.CD]    [CIE]        [ROUTE]
[GAMES]      [PANA]        [SOUND]         [EXECUTOR]   [BILL]
```

```
[FAXWORKS]    [NOTES]           CONFIG.SYS      AUTOEXEC.BAT    LIST
CONFIG.BAK
        31 file(s)          1,497 bytes
                      112,345,088 bytes free
C:\>cd docs
C:\DOCS>dir /w
 Volume in drive C is HARD-DISK
 Directory of C:\DOCS
[.]           [..]              [NOTES]         [CLASSES]       [ADMIN]
[RTT]         [TEMPLATE]        [RESEARCH]      [ABSTRACT]      [OLDDOC2]
[PAPERS]      DOCS.ZIP
        12 file(s)          118,444 bytes
                      112,345,088 bytes free
C:\DOCS>cd notes
C:\DOCS\NOTES>dir /w
 Volume in drive C is HARD-DISK
 Directory of C:\DOCS\NOTES
[.]           [..]              [COMMS]         [C]             [PASCAL]
[DOS]         [MICROP]          [IOCOR]         [ORCAD]         [UNIX]
[FORTRAN]     [ICD]             [TRANSP]        [VAX]           [RS232]
[SOFT_AP]     [COMP_ARC]        [ETHER]         [X]             [TESTS]
[RADAR]       [WINDOWS]         [RES_DOC]       NET.DOC         BILL.REC
ELECT.DOC
        26 file(s)        1,534,592 bytes
                      112,345,088 bytes free
C:\DOCS\NOTES>cd dos
C:\DOCS\NOTES\DOS>dir
 Volume in drive C is HARD-DISK
 Directory of C:\DOCS\NOTES\DOS
.              <DIR>           01/01/80   23:53
..             <DIR>           01/01/80   23:53
DOC_S2    DOC     152,028 17/10/93   19:38
DOS       LZH      73,728 29/07/94   15:50
DOC_S1    DOC     219,067 15/07/94   11:33
DOS       DOC     372,624 17/10/93   10:00
DOSNEW    DOC     438,272 14/09/95    1:48
         7 file(s)        1,255,719 bytes
                      112,345,088 bytes free

C:\DOCS\NOTES\DOS>cd ..
C:\DOCS\NOTES>dir /w
 Volume in drive C is HARD-DISK
 Directory of C:\DOCS\NOTES
[.]           [..]              [COMMS]         [C]             [PASCAL]
[DOS]         [MICROP]          [IOCOR]         [ORCAD]         [UNIX]
[FORTRAN]     [ICD]             [TRANSP]        [VAX]           [RS232]
[SOFT_AP]     [COMP_ARC]        [ETHER]         [X]             [TESTS]
[RADAR]       [WINDOWS]         [RES_DOC]       NET.DOC         BILL.REC
ELECT.DOC
        26 file(s)        1,534,592 bytes
                      111,845,376 bytes free

C:\DOCS\NOTES>cd \
C:\>dir/w

 Volume in drive C is HARD-DISK
 Directory of C:\

[DOS]         [WINDOWS]         [VGAUTIL]       [DOCS]          [COMBIOSW]
[GWS]         [RESEARCH]        [TC]            [WINWORD]       [VIRUS]
[MASM]        [RTT]             [T]             [123]           [INTER]
[MSOFFICE]    [CIRRUS]          [DISCWLD.CD]    [CIE]           [ROUTE]
[GAMES]       [PANA]            [SOUND]         [EXECUTOR]      [BILL]
[FAXWORKS]    [NOTES]           CONFIG.SYS      AUTOEXEC.BAT    CONFIG.BAK
        31 file(s)          1,497 bytes
                      111,845,376 bytes free
```

30.2 Making a directory (*MKDIR* or *MD*)

The MKDIR (or MD) command creates subdirectories. Sample session 30.2 shows a sample session where the subdirectory TEMP is created in the top-level directory.

```
      Sample session 30.2
C:\>mkdir temp
C:\>dir /w
 Volume in drive C is HARD-DISK
 Directory of C:\
[DOS]        [WINDOWS]     [VGAUTIL]       [DOCS]         [COMBIOSW]
[GWS]        [RESEARCH]    [TC]            [WINWORD]      [VIRUS]
[MASM]       [RTT]         [T]             [123]          [INTER]
[MSOFFICE]   [CIRRUS]      [DISCWLD.CD]    [CIE]          [ROUTE]
[GAMES]      [PANA]        [SOUND]         [EXECUTOR]     [BILL]
[FAXWORKS]   [NOTES]       CONFIG.SYS      AUTOEXEC.BAT   LIST
CONFIG.BAK   [TEMP]
        32 file(s)          1,497 bytes
                    112,082,944 bytes free
C:\>cd temp
C:\TEMP>dir
 Volume in drive C is HARD-DISK
 Directory of C:\TEMP
.            <DIR>         14/09/95    1:55
..           <DIR>         14/09/95    1:55
         2 file(s)             0 bytes
                    112,082,944 bytes free
C:\TEMP>
```

30.3 Viewing a file (TYPE)

The TYPE command views the contents of a file. A text file uses a standard alphabet known as ASCII. Non-ASCII files contain data which cannot be viewed by the user. Sample session 30.3 shows the listing of the file CONFIG.SYS which is in the top-level directory. Note that this file contains information on the initial startup environment of the PC.

```
      Sample session 30.3
C:\>type config.sys
DEVICEHIGH=C:\DOS\HIMEM.SYS /TESTMEM:OFF
DEVICEHIGH=C:\DOS\EMM386.EXE RAM
BUFFERS=20,0
FILES=30
DOS=UMB
LASTDRIVE=E
FCBS=4,0
DOS=HIGH
DEVICEHIGH=C:\WINDOWS\IFSHLP.SYS
STACKS=9,256
DEVICEHIGH=C:\PANA\CDMKE.SYS
country=044,,c:\dos\country.sys
C:\>
```

30.4 Wild-cards (* or ?)

Wild-cards are special characters which can be used to substitute various characters in a filename. There are two wild-cards used in DOS, these are:

*   replaces any number of characters

?   replaces only one character

Sample session 30.4 shows a sample session. The user starts in the top-level directory and lists all files or directories which start with the letter 'd' (using the DIR d* command). Next the user changes the directory to \DOS. When in this directory the user lists the files which have a SYS file extension (using DIR *.SYS). Next, all files which begin with the letter 'c' are listed (using DIR C*). Finally, all files which begin with the letter 'm' have any second letter followed by the letters 'av' and with any file extension are listed (using DIR m?av.*).

30.5 Creating a text file

DOS Version 5.0/6.0 and Windows 96 have a text editor named EDIT. This editor is useful to create text files but cannot be used with software programs. For this reason an integrated development environment (IDE) editor will be used.. The example used in this section is Borland C++ Version 3.0, as shown in Figure 30.1.

Many networks allows access to the compiler through a menu option. On computers where the compiler is installed on the local hard disk, then typical set-ups are given in Table 30.1.

For the Borland products (Turbo Pascal/Borland C/Turbo C) a file is entered by simply entering text into the edit window. The main menu options are File, Edit, Run, Compile, Options, Debug. Once the file text has been entered then it is saved by using the File→Save option. As a default C files are give a C extension, C++ are given a CPP extension, and Pascal files are automatically assigned a PAS extension. To load a previously saved file the File→Open option is used.

Table 30.2   IDE packages

IDE	*Program to run*	*Typical home directory*
Borland C++ Version 4	BC.EXE	C:\BORLAND\BIN
Turbo C Version 2	TC.EXE	C:\TC
Turbo Pascal	TURBO.EXE	C:\TURBO

⌨ **Sample session 30.4**

```
C:\>dir d*
 Volume in drive C is HARD-DISK
```

```
    Directory of C:\

DOS            <DIR>        09/08/95    4:03
DOCS           <DIR>        01/01/80   23:51
DISCWLD  CD    <DIR>        23/05/95    1:12
        3 file(s)              0 bytes
                      138,100,736 bytes free
C:\>cd dos
C:\DOS>dir *.sys /w
 Volume in drive C is HARD-DISK
 Directory of C:\DOS
COUNTRY.SYS  KEYBOARD.SYS  KEYBRD2.SYS  ANSI.SYS      CHKSTATE.SYS
DISPLAY.SYS  DRVSPACE.SYS  DRIVER.SYS   HIMEM.SYS     RAMDRIVE.SYS
       10 file(s)          223,341 bytes
                      138,100,736 bytes free
C:\DOS>dir c*
 Volume in drive C is HARD-DISK
 Directory of C:\DOS
CHKDSK    EXE       12,241 31/05/94    6:22
COUNTRY   SYS       26,936 31/05/94    6:22
CURSOR    COM           91 11/12/89    7:29
CHKSTATE  SYS       41,600 31/05/94    6:22
CHOICE    COM        1,754 31/05/94    6:22
COMMAND   COM       54,645 31/05/94    6:22
        6 file(s)        137,267 bytes
                      138,100,736 bytes free
C:\DOS>dir m?av.*
 Volume in drive C is HARD-DISK
 Directory of C:\DOS
MSAV      EXE      172,198 31/05/94    6:22
MSAV      HLP       23,891 31/05/94    6:22
MWAV      EXE      142,640 31/05/94    6:22
MWAV      HLP       24,619 31/05/94    6:22
MSAV      INI          248 29/08/95    6:43
MWAV      INI           24 25/08/95    3:18
        6 file(s)        363,620 bytes
                      138,100,736 bytes free
```

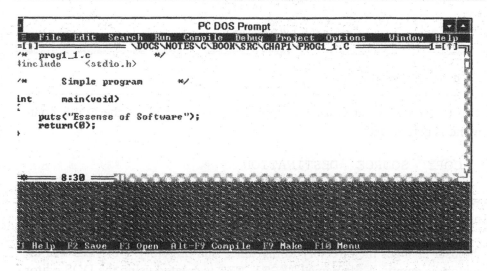

Figure 30.1 Borland C++ Version 4.0 main screen.

30.8 Deleting Files (*DEL* or *ERASE*)

The DEL and ERASE commands delete files. Wild-cards can be used to replace filenames, if required. Sample session 30.5 shows a sample session
where the file TEST3.TXT is deleted from the TEMP directory. Next, all
the files with a TXT extension are deleted.

As an extra safeguard against accidentally deleting, or selectively deleting files the /P switch can be used. Notice in Sample session 30.6 the user
is prompted to delete each of the files individually.

🖥 **Sample session 30.5**

```
C:\>cd temp
C:\TEMP>dir/w
 Volume in drive C is MS-DOS_5
 Volume Serial Number is 3B33-13D3
 Directory of C:\TEMP
[.]          [..]          TEST3.TXT    TEST2.TXT    TEST4.TXT
        5 file(s)         701 bytes
                   54509568 bytes free
C:\TEMP>erase test3.txt
C:\TEMP>dir/w
 Volume in drive C is MS-DOS_5
 Volume Serial Number is 3B33-13D3
 Directory of C:\TEMP
[.]          [..]          TEST2.TXT    TEST4.TXT
        4 file(s)         467 bytes
                   54513664 bytes free
C:\TEMP>erase *.txt
C:\TEMP>dir
 Volume in drive C is MS-DOS_5
 Volume Serial Number is 3B33-13D3
 Directory of C:\TEMP

.             <DIR>        04/10/93    18:27
..            <DIR>        04/10/93    18:27
        2 file(s)           0 bytes
                   54521856 bytes free
C:\TEMP>
```

30.9 Copying files (COPY)

The COPY command is used to copy files. The format of the COPY command is given next:

```
COPY   SOURCE   DESTINATION
```

where SOURCE is the name of the file to copy and DESTINATION is the
name of the file, or directory, to copy to.

In Sample session 30.7 the user creates a directory named TEMP. Next,
the user changes the directory to the TEMP sub-directory. Note that this directory is empty. A file named TEST1.TXT is created using the DOS editor
(EDIT). After this it is listed using the TYPE TEST1.TXT command line.

Next the directory is listed using DIR and then the COPY command is used to copy the new created file to `TEST2.TXT`. A directory listing (`DIR`) then shows that there are now two files in this directory (`TEST1.TXT` and `TEST2.TXT`). Next, the first file (`TEST1.TXT`) is deleted using `ERASE` command.

💻 **Sample session 30.6**

```
C:\TEMP>del *.txt /p
 Volume in drive C is MS-DOS_5
 Volume Serial Number is 3B33-13D3
 Directory of C:\TEMP

[.]              [..]          TEST3.TXT    TEST2.TXT    TEST4.TXT
        5 file(s)          701 bytes
                    54509568 bytes free

C:\TEMP>del *.txt /p
C:\TEMP\TEST3.TXT,      Delete (Y/N)?n
C:\TEMP\TEST2.TXT,      Delete (Y/N)?y
C:\TEMP\TEST4.TXT,      Delete (Y/N)?n

C:\TEMP>dir/w
 Volume in drive C is MS-DOS_5
 Volume Serial Number is 3B33-13D3
 Directory of C:\TEMP

[.]              [..]                TEST3.TXT          TEST4.TXT
        4 file(s)          467 bytes
                    54513664 bytes free
C:\TEMP>
```

💻 **Sample session 30.7**

```
C:\>mkdir temp
C:\>cd temp
C:\TEMP>dir
 Volume in drive C is MS-DOS_5
 Volume Serial Number is 3B33-13D3
 Directory of C:\TEMP

.              <DIR>       04/10/93    18:27
..             <DIR>       04/10/93    18:27
        2 file(s)          0 bytes

                    61628416 bytes free
C:\TEMP>edit test1.txt

C:\TEMP>type test1.txt
This is a mail message from FRED
to BERT regarding the usage of
printers on the network.

Note that the printers will be
switched off at the following times:

        5pm - 6am Mon-Fri
        All Day Sat, Sun

BERT.

C:\TEMP> dir

 Directory of C:\TEMP
```

```
.                <DIR>      04/10/93    18:27
..               <DIR>      04/10/93    18:27
TEST1    TXT         239 04/10/93    18:30
        3 file(s)          239 bytes
                      61624320 bytes free

C:\TEMP>copy test1.txt test2.txt
        1 file(s) copied

C:\TEMP>dir

 Volume in drive C is MS-DOS_5
 Volume Serial Number is 3B33-13D3
 Directory of C:\TEMP

.                <DIR>      04/10/93    18:27
..               <DIR>      04/10/93    18:27
TEST1    TXT         239 04/10/93    18:30
TEST2    TXT         239 04/10/93    18:30
        4 file(s)          478 bytes
                      61620224 bytes free
C:\TEMP>erase test1.txt

C:\TEMP>dir

 Volume in drive C is MS-DOS_5
 Volume Serial Number is 3B33-13D3
 Directory of C:\TEMP

.                <DIR>      04/10/93    18:27
..               <DIR>      04/10/93    18:27
TEST2    TXT         239 04/10/93    18:30
        3 file(s)          239 bytes
                      61624320 bytes free
C:\TEMP>
```

Worksheet 15

W15.1 Go to the top-level directory, then move around the file system looking in various directories. Try to determine the likely function of the directories.

Directory	Likely function
C:\DOS	DOS directory containing some DOS commands, help functions, system files, and so on

W15.2 Find all the files which have the DOC extension on the hard disk (hint. use the DIR /S command).

File(*.DOC)	Directory
DOS.DOC	C:\NOTES\DOCS\DOS

Find all the files on the hard disk with the BAT extension.

File (*.BAT)	Directory
AUTOEXEC.BAT	C:\

W15.3 Create a directory named TEMP on the hard disk.

Command used:

W15.4 Go into the directory you have just created.

> Command used:

W15.5 Create a file name PROG1.C (if you indent to use C) or PROG1.PAS (if you indend to use Pascal) from the IDE editor, enter the following text and save the file.

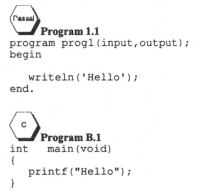
Program 1.1
```
program prog1(input,output);
begin

    writeln('Hello');
end.
```

Program B.1
```
int    main(void)
{
    printf("Hello");
}
```

W15.6 Check that the file just created exists using the DIR command

> Does it exist: YES/NO

W15.7 View the contents of this file using TYPE.

> Are the contents correct: YES/NO

W15.8 Copy this file into a file named PROG2.C (or PROG2.PAS if the file created is PROG1.PAS) using the COPY command.

> Command used:

W15.9 Verify that the COPY command has worked by listing the directory.

> Has the file been copied correctly: YES/NO

W15.10 View the contents of the newly created file using TYPE.

> Are the contents correct: YES/NO

W15.11 Insert a disk into one of the disk drives and format a floppy disk (if you have one).

Command used:	YES/NO

W15.12 Go to the floppy disk and copy the file PROG1.C (or PROG1.PAS) from the C:\TEMP directory and confirm that it has copied correctly.

Command used:

W15.13 Determine the Volume label of the hard disk you are using.

Volume label:

W15.14 Determine the names that DOS will use for the following files/directories:

Filename	DOS filename actually given
MY_FILE_42_1.TEXT	MY_FILE_.TEX
DOC$1001.D	
ACC_91_92.WK1	
BACKUP.$$1_1	
DOCS.FRED.1	
MEMO_FRED.200	

W15.15 Copy the following files that should be in the top-level directory on the hard disk.

Copy from	Copy to
AUTOEXEC.BAT	AUTOEXEC.OLD
CONFIG.SYS	CONFIG.OLD

Files copied okay:	YES/NO

W15.16 Make a directory named OLDBATS and copy the files CONFIG.OLD and AUTOEXEC.OLD into this directory.

Files copied okay:	YES/NO

PART G

C/Pascal
C++
Assembly Language
Visual Basic
HTML/Java
DOS
Windows 3.x
Windows 95
UNIX

31 Introduction

31.1 Introduction

A modern Personal Computer (PC) consists of a keyboard, a monitor, a mouse, a floppy disk drive, a hard disk drive and a system unit. An operating system allows the user to access these devices in an easy-to-use manner, as illustrated in Figure 31.1. Microsoft DOS (Disk Operating System) is a text-based system in which commands are entered via the keyboard to perform operations such as copying files and running programs.

Microsoft Windows is a program which presents an easy-to-use graphical interface to the PC hardware. It uses windows, icons, menus and pointers (WIMPs) to access application programs, disk drives, file systems, and so on. Most operations are conducted using a mouse instead of keyboard commands. The user is presented with a series of icons which represent application programs. It is far simpler to use than DOS as the information is presented graphically. A major disadvantage of DOS is that the syntax and format of the text command requires to be remembered.

Windows has gone through three major transformations, from the original Version 1.0 to the most widely available version to Version 3.1 (and Version 3.11) and now to Windows 95/NT. It is installed on almost every new PC sold and is becoming the *de-facto* standard for PC packages. This and the next chapter discuss Windows 3.1/3.11.

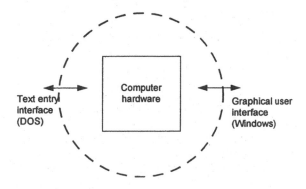

Figure 31.1 Using an operating system to access hardware

31.2 Running Windows

Windows 3.*x* is a program which is run from DOS. It will either start automatically when the PC is switched on, or can be run by typing WIN at the DOS prompt and pressing <RETURN>.

```
C:\>  WIN
```

31.3 Windows desktop

After Windows is started the screen displays the Microsoft start-up screen for a short time. This displays the version number of the software. The main Windows desktop is displayed after this. On top of the desk is the Program Manager window. It contains a number of windows and icons within it. Programs and applications are placed into groups. These groups contain icons which relate to the program. The inactive windows are displayed as group icons. There are four of these, as shown in Figure 31.2. A window is made active by double clicking on the group icon. This expands the window to its normal form.

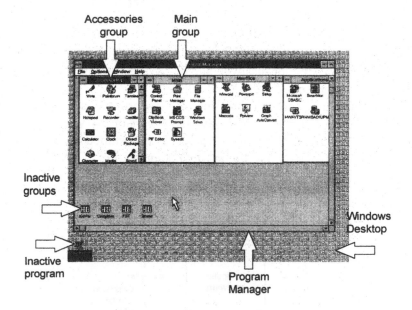

Figure 31.2 Main desktop screen

A program which has been running but is currently paused is shown as an icon at the bottom of the screen. In the example screen in Figure 31.3 there are 4 active groups, these are: **Accessories, Main, Msoffice** and **Applications**. Within the Accessories group there are 12 programs, these are: Write, Paintbrush, Terminal, Notepad, Recorder, Card-

`file`, `Calculator`, `Clock`, `Object Packager`, `Character Map-`
`per`, `Media Player` and `Sound Recorder`.

The current active window has a blue title bars whilst an inactive window has a grey title bar. Only icons within an active window can be selected.

31.4 Window items

Windows are made up to four main parts, these are:

- **Main title bar**. Every window has an identification title at the top of it, as shown in Figure 31.3.
- **Menu bar**. Some windows have menu options to choose from, as shown in Figure 31.4. Hot-keys are often underlined.
- **Scroll bar**. When only part of a window is displayed a scroll bar (often called a thumbnail) will appear. Holding down the mouse button on the small square within the scroll bar moves the contents of the window.
- **Control menu box**. As shown in Figure 31.3.

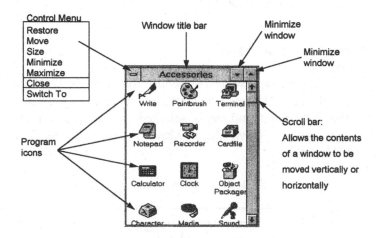

Figure 31.3 A basic group window

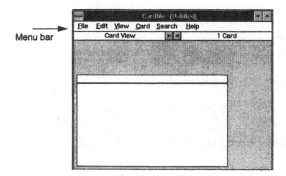

Figure 31.4 Menu bar on a window

31.5 Mouse controls

The mouse controls the pointer around the screen and the left mouse button is used either to run an application, if it is an icon, or to single click to select a window or file, as shown in Figure 31.5.

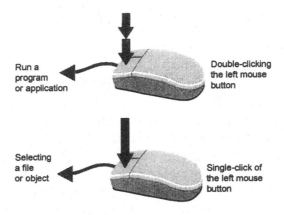

Figure 31.5 Main desktop screen

31.6 Program Manager menus

The Program Manager window has 4 menu options: File, Options, Windows and Help. Table 31.1 shows the main options and the sub-options. For example, if the File menu options is selected a sub-menu (a pull-down menu) shows the options:, New, Open, Move, Copy, Delete, Properties, Run and Exit Windows.

Table 31.1 Menu options

File	Options	Windows	Help
New	Auto Arrange	Cascade	Contents
Open	Minimize on Use	Tile	Search for Help on
Move	Save Settings on Exit	Arrange Icons	How to use Help
Copy			Windows Tutorial
Delete			About Program Manager
Properties			
Run			
Exit Windows			

31.6.1 Quitting Windows

To exit Windows, first the File option is selected from Program Manager. Next, move the mouse pointer down to the Exit Windows option and select it with a single click of the left mouse button, as shown in Figure 31.6.

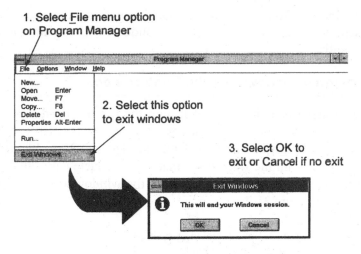

Figure 31.6   Quitting Windows

31.6.2 Getting help

The best method of getting help is to refer to the user manual. If that is not available then there is on-line help. If the Help menu option is selected on the Program Manager menu displays the options Contents, Search for Help on, How to use Help, Windows Tutorial and About Program Manager.

31.7 Moving and resizing a window

The options used to change the windows size are:

- **Maximizing a window**. A window can be made to fill the complete screen by selecting the ⬜ at the top right-hand corner of the screen.
- **Minimizing a window**. A window can be minimized to an icon using ⬜.
- **Making a window smaller**. If a window is filling the complete screen it can be made into a smaller window by selecting ⬜.
- **Moving a window**. A window can be moved by placing the mouse pointer on the blue title bar. The left mouse button is pressed down while moving the window to its required position, then the mouse button is released.
- **Resizing a window**. A window can be resized by placing the mouse pointer either at one of the corners or the sides. The pointer shape should change when it is placed over the border of the windows. To expand (or contract) the window vertically the mouse pointer is placed either at the top or bottom border. The mouse is then pressed while moving the border

of the windows. When the required position is found the mouse button is released. A similar operation can be conducted to expand (or contact) horizontally.

31.8 Closing a window with the Control menu

The Control menu is activated by selecting the top left-hand button on the window. This menu contains the options: Restore, Move, Size, Minimize, Maximize, Close, Switch to. The Restore, Move, Size, Minimize and Maximize options mimic operations that are conducted by the mouse. A useful option is the Close which exits from the window and closes it down, this is shown in Figure 31.7.

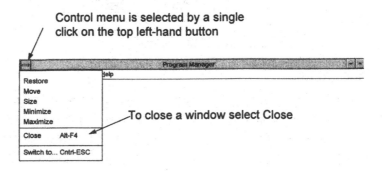

Figure 31.7   Control menu

31.9 Exercises

Conduct the following steps filling in the required information.

31.9.1   Switch PC on.   Completed successfully [   ]

31.9.2   Start Microsoft Windows and when it shows the initial start-up screen observe the version number of the software.

   Version Number: Version 3.0 [   ] Version 3.1 [   ]
   (Tick as necessary)   Version 3.11 [   ] _____

31.9.3   Identify each of the active and inactive groups within the Program Manager.

   Active groups: _____

                          _____

Inactive groups: _

                                 _

31.9.4 Identify all programs within three of the groups.

1. Group's name: _

Programs: _

                                 _

2. Group's name: _

Programs: _

                                 _

3. Group's name: _

Programs: _

                                 _

31.9.5 Resize the **Program Manager** window so that it fills the screen. Tick box if completed successfully.

Completed successfully [   ]

31.9.6 Resize the **Program Manager** window so that it is shown at its normal position on the screen.

Completed successfully [   ]

31.9.7 Iconize the **Program Manager** window and then restore it to its normal position.

Completed successfully [   ]

31.9.8 Expand the **Program Manager** window horizontally, then restore it to its normal position.

Completed successfully [   ]

31.9.9 Contract the **Program Manager** windows vertically, then restore it to its normal position.

Completed successfully [   ]

31.9.10 Find the following application programs, run them and then exit. Tick box if completed successfully.

Completed successfully [   ]

Clock

Completed successfully [   ]

Calculator

31.9.11 Exit from Windows using the <u>F</u>ile menu option in the **Program Manager**.

Operations used: _____

_____

31.9.12 Start Windows again and this time exit using the **Control** menu.

Operations used: _____

_____

31.9.13 There is an on-line help facility within Windows. To select this use the <u>H</u>elp menu option within the **Program Manager**. Use this facility to find help on the following topics.

quitting windows: _____

deleting groups: _____

starting applications: _____

deleting programs: _____

File Management

32.1 Introduction

The **Main** group contains important application programs which allow the user access to the file system, to setup a different environment, to setup different computer hardware, gain access to DOS, etc. It contains `Control Panel`, `Print Manager`, `File Manager`, `ClipBook Viewer`, `MS-DOS Prompt`, `Windows Setup`, `PIF Editor` and `Sysedit`. The icons contained within **Main** are shown in Figure 32.1.

    `Control Panel` sets up the Windows environment and can be used to change the colours of the windows, the fonts used, etc. The `Print Manager` program checks the status of files which have been printed. `File Manager` allows access to the file system and can be used to copy or move files from one directory to another or from one disk drive to another. The `MS-DOS Prompt` program gives a DOS window.

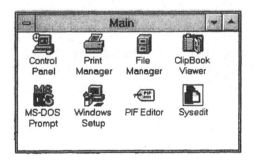

Figure 32.1   Main group

32.2 File manager

The file manager is identified by the filing cabinet icon. It is opened by double clicking the cabinet icon. Files and directories are displayed with their file name and an icon which indicates the file type. It employs a select, drag and put technique where a file (or directory), once selected, can be dragged into another directory or onto any disk drive. Note that the file manager in Windows 3.11 differs slightly from the file manager in Windows 3.1. Figure 32.2 shows the Windows 3.11 file manager. The main difference is the icons to the right of the current disk drive name.

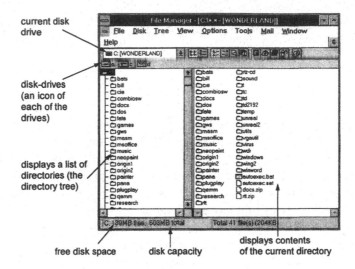

Figure 32.2 File manager

A directory is identified with the 📁 icon. Files have three main icons associated with them. Application programs are identified with a ▨, a document file by 📄 and any other file with a 🗋. Figure 32.3 shows a sample directory listing.

A document file is a file which when double clicked will start an associated application. Table 32.1 shows some typical file extensions and the application which is run when the file is double clicked. For example, a file with a doc extension will run a word processor, a bmp extension will start the paintbrush package, and so on.

Table 32.1 File associated with application

File extension	Description	Application started
wav	Sound file	Sound package
txt	Text file	Notepad
doc	Document	Word processor
wk3	Lotus 123 worksheet	Lotus 123
wri	Write document	Write
hlp	Help file	Windows help
pcx	Graphics file	Paintbrush
xls	Microsoft Excell worksheet	Microsoft Excell
ppt	Microsoft Powerpoint presentation	Microsoft Powerpoint
bmp	Bit-mapped graphics file	Paintbrush

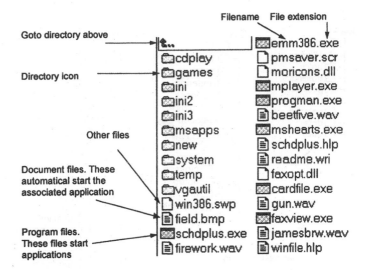

Figure 32.3 File manager

32.2.1 Selecting a drive

A disk drive is selected with a single click on one of the disk icons above the directory listing or by selecting Disk→Select drive.... In the example in Figure 32.4 there are three drives, a floppy disk on the A: drive, a hard-disk on the C: drive and a CD-ROM on the D: drive. If there are any network drives these will also be displayed.

32.2.2 Selecting a directory

A directory is selected by a single click on the directory icon (or on the directory name). The open file icon ( ) then shows which directory is currently open; the directory display window displays the contents of this directory. An example of changing the current directory is shown in Figure 32.5. In this example the user has selected to list the contents of the top-level directory on the C: drive. Next the mouse pointer is placed over the docs directory which is then opened with a single click of the left mouse button. After this the admin directory is opened. A directory can be closed by single mouse button click when it is open (another click will open it again).

32.2.3 Formatting a disk

A floppy disk must be formatted before it can store files. Some disks are pre-formatted when they are purchased, but others required to be formatted before they are used. Care must be taken when formatting a disk as the current contents of the disk will be erased.

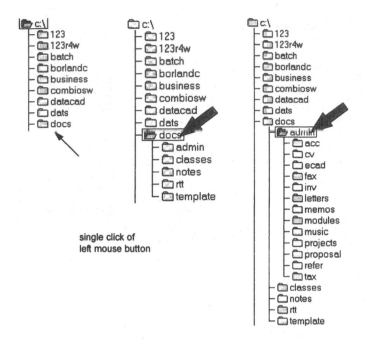

Figure 32.4 Displaying directory

To format a disk first insert it into the floppy disk drive. Next select **Disk→Format disk...** from the menu, as shown in Figure 32.5. When this is selected Windows will prompt the user for the drive which the disk has been entered and the capacity of the disk. By default this is likely to be set to A: and 1.44 MB (for a 3.5 inch floppy disk drive on the A: drive), respectively. If the drive differs from the default or its format differs then change the options by pulling down the **Disk In** or the **Capacity** options.

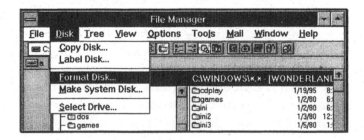

Figure 32.5 Selecting disk format option

Figure 32.6 shows the main steps that are taken to format the disk. First the disk capacity and drive name are prompted for. When these are correct the OK button is selected. Next a **Format Disk** window is displayed. Within this window the current status of the disk formatting operation is displayed (from 0 to 100% complete). When complete, a window with a message

Creating root directory will be displayed. After this the formatted disks' capacity is displayed and the user is prompted as to whether another disk is to be formatted. If no more disks are to be formatted then the No option is selected else Yes is selected. Note that the Cancel option on any of the format status windows can be selected to cancel the format process.

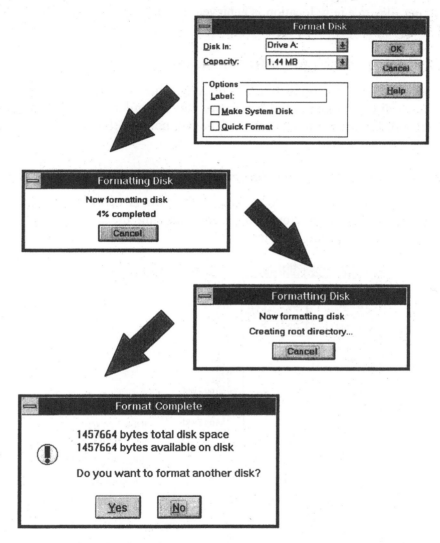

Figure 32.6 Formatting a floppy disk

32.2.4 Moving files on the same disk drive

Microsoft Windows uses a select, drag and drop procedure when copying or moving files or directories. Moving files (copying them but deleting the original file) can be achieved by moving the mouse pointer over the file and then dragging the icon, while keeping the left mouse button held down, then placing it in the desired directory. Figure 32.7 shows an example of the file acc2.wk1 being dragged into the directory C:\DOCS\BUSINESS.

32.2.5 Copying files on the same disk drive

Copying files to a new location is achieved by holding down the Ctrl key and dragging the file to the desired directory, see Figure 32.8.

32.2.6 Copying files to a different disk drive

Copying files or directories to a disk drive is achieved by simply dragging the file or directory icon to the drive icon. This is similar to moving files on the same disk drive but in this case the file will not be deleted on the source directory.

32.2.7 Moving files to a different disk drive

Moving files or directories to another disk drive is achieved by simply dragging the file or directory icon, while pressing the Shift key, and dragged it to the required drive icon.

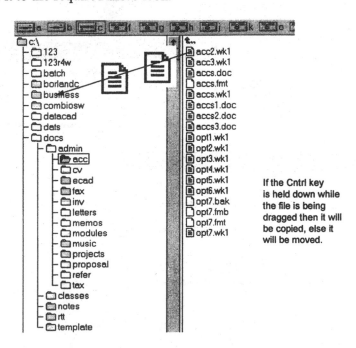

Figure 32.7   Copying or moving a file on the same disk drive

32.2.8 Deleting files

When a file is no longer required it can be deleted. This is achieved by clicking on the file to be deleted and pressing the `Del` key on the keyboard. A message asking for confirmation should appear, as shown in Figure 32.8. If the file is to be deleted then the `OK` button is selected else `Cancel`. The confirmation window for deleting a directory is different, an example is shown in Figure 32.9. After the user selects the directory to delete a confirmation is displayed with either <u>Y</u>es or `Yes to All`. If the user selects <u>Y</u>es then each file within the directory will be prompted for deletion confirmation. If `Yes to All` is selected then all files within the directory (and sub-directories) will be deleted without any confirmation.

32.2.9 Running Windows and DOS applications

Running a file is achieved by placing the mouse pointer over the application file and double clicking the left mouse button. If it is a Microsoft Windows application it will be run as a normal Windows application, else if it has an application attached to it then the associated application will be run and the selected file loaded into it. For example, if it were a document file (.DOC) then a word processor will be started with the selected document loading into it.

If a DOS-based application is selected then a DOS screen will appear which will then load the program (if it can).

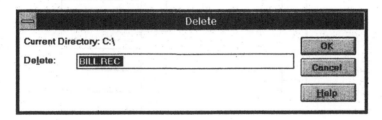

Figure 32.8   Delete confirmation

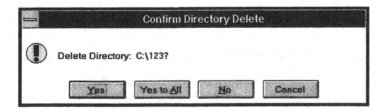

Figure 32.9   Delete directory confirmation

32.3 Exercises

Conduct the following steps filling in the required information.

32.3.1   Start Microsoft Windows.   Completed successfully [   ]

32.3.2   Identify some of the directories and sub-directories on the C: drive on the PC and complete the table given in Figure 32.10.

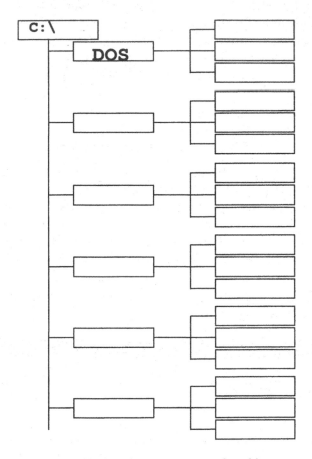

Figure 32.10   Directory structure of C: drive

32.3.3   Identify some application programs within the \dos directory.

Application programs: _____

_____

32.3.4   Identify some application programs within the \windows directory.

Application programs: _____

_____

32.3.5   Identify some files which are in the \WINDOWS directory which will run application programs

       Files: ———————————————————

                  ———————————————————

32.3.6   On the C: drive find 5 graphics files (in any directory).

       Files: ———————————————————

                  ———————————————————

32.3.7   On the C: drive find 5 document files (in any directory).

       Files: ———————————————————

                  ———————————————————

32.3.8   Insert a floppy disk in the A: drive and format it.

       Total disk space: ———————————————————

                             ———————————————————

32.3.9   Create a document file either by running Cardfile (within the **Main** group) or running Write (within **Accessories**) or using a word processor such as Word for Windows or Ami-Pro. Enter some text into the document then save it in the C:\windows directory with the name DOC1.DOC.

                                   Completed successfully [   ]

32.3.10   Verify, with **File Manager**, that this file is in the \windows directory.

                                   Completed successfully [   ]

32.3.11   Copy the file DOC1.DOC to the floppy disk in the A: drive.

                                   Completed successfully [   ]

32.3.12 Verify, with **File Manager**, that the file has been copied to the A: drive. If it is in the wrong place move it.

Completed successfully [   ]

32.3.13 Create a directory on the C: drive, named temp, below the top-level directory using the File →Create Directory menu option.

Completed successfully [   ]

32.3.14 Verify, with **File Manager**, that this directory exists. If it is in the wrong place move it.

Completed successfully [   ]

32.3.15 Copy the file DOC1.DOC to the temp directory.

Completed successfully [   ]

32.3.16 Verify, with **File Manager**, that the file is in the temp directory.

Completed successfully [   ]

32.3.17 There is an on-line help facility for the help manager within Windows. To select this use the Help menu option within the **File Manager**. Use this facility to find help on the following topics.

selecting files: _____

sorting files: _____

copying disks: _____

starting applications: _____

PART H

C/Pascal
C++
Assembly Language
Visual Basic
HTML/Java
DOS
Windows 3.x
Windows 95/NT
UNIX

33 Windows 95/NT

33.1 Introduction

Windows NT has provided an excellent network operating system. It communicates directly with many different types of networks, protocols and computer architectures. Windows NT and Windows 95 have the great advantage of other operating systems in that they have integrated network support. Operating systems now use networks to make peer-to-peer connections and also connections to servers for access to file systems and print servers. The three most widely used operating systems are MS-DOS, Microsoft Windows and UNIX. Microsoft Windows comes in many flavours; the main versions are outlined below and Table 33.1 lists some of their attributes.

- Microsoft Windows 3.*xx* – 16-bit PC-based operating system with limited multitasking. It runs from MS-DOS and thus still uses MS-DOS functionality and file system structure.
- Microsoft Windows 95 – robust 32-bit multitasking operating system (although there are some 16-bit parts in it) which can run MS-DOS applications, Microsoft Windows 3.*xx* applications and 32-bit applications.
- Microsoft Windows NT – robust 32-bit multitasking operating system with integrated networking. Networks are built with NT servers and clients. As with Microsoft Windows 95 it can run MS-DOS, Microsoft Windows 3.*x* applications and 32-bit applications.

33.2 Servers, workstations and clients

Microsoft Windows NT is a 32-bit, pre-emptive, multitasking operating system. One of the major advantages it has over UNIX is that it can run PC-based software. A Windows NT network normally consists of a server and a number of clients. The server provides file and print servers as well as powerful networking applications, such as electronic mail applications, access to local and remote peripherals, and so on.

Table 33.1   Windows comparisons

	Windows 3.1	*Windows 95*	*Windows NT*
Pre-emptive multitasking		✓	✓
32-bit operating system		✓	✓
Long file names		✓	✓
TCP/IP	✓	✓	✓
32-bit applications		✓	✓
Flat memory model		✓	✓
32-bit disk access	✓	✓	✓
32-bit file access	✓	✓	✓
Centralized configuration storage		✓	✓
OpenGL 3D graphics			✓

The Windows NT client can either:

• Operate as a stand-alone operating system.
• Connect with a peer-to-peer connection.
• Connect to a Windows NT server.

A peer-to-peer connection is when one computer logs into another computer. Windows NT provides unlimited outbound peer-to-peer connections and typically up to 10 simultaneous inbound connections.

33.3 Workgroups and domains

Windows NT assigns users to workgroups which are collections of users who are grouped together with a common purpose. This purpose might be to share resources such as file systems or printers, and each workgroup has its own unique name. With workgroups each Windows NT workstation interacts with a common group of computers on a peer-to-peer level. Each workstation then manages its own resources and user accounts. Workgroups are useful for small groups where a small number of users require to access resources on other computers.

A domain in Windows NT is a logical collection of computers sharing a common user accounts database and security policy. Thus each domain must have at least one Windows NT server.

Windows NT is designed to operate with either workgroups or domains. Figure 33.1 illustrates the difference between domains and workgroups.

Domains have the advantages that:

• Each domain forms a single administrative unit with shared security and user account information. This domain has one database containing user and group information and security policy settings.
• They segment the resources of the network so that users, by default, can view all networks for a particular domain.

User accounts are automatically validated by the domain controller. This stops invalid users from gaining access to network resources.

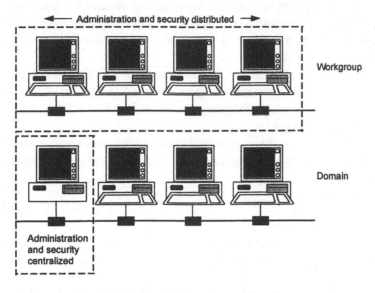

Figure 33.1   Workgroups and domains

33.4 Windows NT user and group accounts

Each user within a domain has a user account and is assigned to one or more groups. Each group is granted permissions for the file system, accessing printers, and so on. Group accounts are useful because they simplify an organization into a single administrative unit. They also provide a convenient method of controlling access for several users who will be using Windows NT to perform similar tasks. By placing multiple users in a group, the administrator can assign rights and/or permissions to the group.

Each user on a Windows NT system has the following:

- A user name (such as `fred_bloggs`).
- A password (assigned by the administrator then changed by the user).
- The groups in which the user account is a member (such as, `staff`).
- Any user rights for using the assigned computer.

Each time a user attempts to perform a particular action on a computer, Windows NT checks the user account to determine whether the user has the authority to perform that action (such as read the file, write to the file, delete the file, and so on).

Normally there are three main default user accounts: Administrator, Guest and an 'Initial User' account. The system manager uses the Administrator account to perform such tasks as installing software, adding/deleting user accounts, setting up network peripherals, installing hardware, and so on.

Guest accounts allow occasional users to log on and be granted limited rights on the local computer. The system manager must be sure that the access rights are limited so that hackers or inexperienced users cannot do damage to the local system.

The 'Initial User' account is created during installation of the Windows NT workstation. This account, assigned a name during installation, is a member of the Administrator's group and therefore has all the Administrator's rights and privileges.

After the system has been installed the Administrator can allocate new user accounts, either by creating new user accounts, or by copying existing accounts.

33.5 File systems

Windows NT supports three different types of file system:

- FAT (file allocation table) – as used by MS-DOS, OS/2 and Windows NT/95. A single volume can be up to 2 GB.
- HPFS (high performance file system) – a UNIX-style file system which is used by OS/2 and Windows NT. A single volume can be up to 8 GB. MS-DOS applications cannot access files.
- NTFS (NT file system) – as used by Windows NT. A single volume can be up to 8 GB. MS-DOS applications, themselves, cannot access the file system but they can when run with Windows NT.

The FAT file system is widely used and supported by a variety of operating systems, such as MS-DOS, Windows NT and OS/2. If a system is to use MS-DOS it must be installed with a FAT file system.

33.5.1 FAT

The standard MS-DOS FAT file and directory-naming structure allows an 8-character file name and a 3-character file extension with a dot separator (.) between them (the 8.3 file name). It is not case sensitive and the file name and extension cannot contain spaces and other reserved characters, such as:

```
 "  /  \  :  ;  |  =  ,  ^  *  ?  .
```

With Windows NT and Windows 95 the FAT file system supports long file names which can be up to 255 characters. The name can also contain multiple spaces and dot separators. File names are not case sensitive, but the case of file names is preserved (a file named FredDocument.XYz will be displayed as `FredDocument.XYz` but can be accessed with any of the characters in upper or lower case).

Each file in the FAT table has four attributes (or properties): read-only, archive, system and hidden (as shown in Figure 33.2). The FAT uses a linked list where the file's directory entry contains its beginning FAT entry number. This FAT entry in turn contains the location of the next cluster if the file is larger than one cluster, or a marker that designates this is included in the last cluster. A file which occupies 12 clusters will have 11 FAT entries and 10 FAT links.

The main disadvantage with FAT is that the disk is segmented into allocated units (or clusters). On large-capacity disks these sectors can be relatively large (typically 512 bytes/sector). Disks with a capacity of between 256 MB and 512 MB use 16 sectors per cluster (8 KB) and disks from 512 MB to 1 GB use 32 sectors per cluster (16 KB). Drives up to 2 GB use 64 sectors per cluster (32 KB). Thus if the disk has a capacity of 512 MB then each cluster will be 8 KB. A file which is only 1 KB will thus take up 8 KB of disk space (a wastage of 7 KB), and a 9 KB file will take up 16 KB (a wastage of 7 KB). Thus a file system which has many small files will be inefficient on a cluster-based system. A floppy disk normally use 1 cluster per sector (512 bytes).

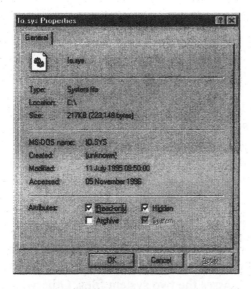

Figure 33.2 File attributes

Windows 95 and Windows NT support up to 255 characters in file names; unfortunately, MS-DOS and Windows 3.*xx* applications cannot read them. To accommodate this, every long file name has an autogenerated short file name (in the form `XXXXXXXX.YYY`). Table 33.2 shows three examples. The conversion takes the first six characters of the long name then adds a ~*number* to the name to give it a unique name. File names with the same initial six characters are identified with different *numbers*. For example, `Program Files` and `Program Directory` would be stored as `PRO-GRA~1` and `PROGRA~2`, respectively. Sample listing 33.1 shows a listing from Windows NT. The left-hand column shows the short file name and the far right-hand column shows the long file name.

Table 33.2   File name conversions

Long file name	Short file name
Program Files	PROGRA~1
Triangular.bmp	TRAING~1.BMP
Fredte~1.1	FRED.TEXT.1

🖳 **Sample listing 33.1**

```
EXAMPL~1 DOC   4,608   05/11/96   23:36 Example Document 1.doc
EXAMPL~2 DOC   4,608   05/11/96   23:36 Example Document 2.doc
EXAMPL~3 DOC   4,608   05/11/96   23:36 Example Document 3.doc
EXAMPL~4 DOC   4,608   05/11/96   23:36 Example Document 4.doc
EXAMPL~5 DOC   4,608   05/11/96   23:36 Example Document 5.doc
EXAMPL~6 DOC   4,608   05/11/96   23:36 Example Document 6.doc
EXAMPL~7 DOC   4,608   05/11/96   23:36 Example Document 7.doc
EXAMPL~8 DOC   4,608   05/11/96   23:39 Example assignment A.doc
EXAMPL~9 DOC   4,608   05/11/96   23:40 Example assignment B.doc
EXAMP~10 DOC   4,608   05/11/96   23:40 Example assignment C.doc
```

33.6 Running Windows 95 and NT

Windows 95 and NT are operating systems in their own right and thus do not require DOS. The version of DOS which is run in Window 95/NT is an emulated version.

Sometimes Windows NT/95 develops a problem when starting the computer. If this occurs then it can be started in a safe mode, which does not load network support and using the default settings (VGA monitor, no network, Microsoft mouse driver, and the minimum device drivers required to start Windows). This is achieved by pressing the F8 key when Windows is being booted. Normally the user then selects `Settings` and then `Control Panel` to change the systems settings.

Often the problem with the system start-up is to do with either `AUTOEXEC.BAT` or `CONFIG.SYS` files. If this is the case the `Command Prompt Only` operation setting can be selected when starting the computer. This option bypasses the two system files. Otherwise, the `Step-by-Step Confirmation` can be selected, which allows the user to load a specific system driver.

Another method of fixing the system is to boot the system with a bootable DOS disk. The user can then try to fix the system. Note that this can only be done if the disk has been formatted as FAT.

33.7 Basic Windows NT/95

There are many enhancements on Windows NT/95 as opposed to previous versions. The main differences are:

- **Start** **Start button and taskbar**. The task bar appears at the bottom of the screen and it contains the Start button, which can be used to quickly start a program. The taskbar also shows the programs which are currently running and it can thus be used to switch between programs.

- **games Folders**. Documents and programs are stored in folders. In DOS and previous versions of Windows these folders were called directories.

- **The desktop**. The desktop is the area of screen which appears when Windows is started. Figure 33.3 shows an example. It can be customized by adding program shortcuts, documents, printers, and the layout and style of the background. To adjust settings such as desktop colour and background, use your right mouse button to click anywhere on the desktop, and then click Properties. Figure 33.4 shows an example of the Properties window.

- **Network Neighborhood Network Neighborhood**. Windows 95/NT has extensive support for networking. If a network is present then the Network Neighborhood icon appears on your desktop. The user can then browse the entire network by double-clicking the icon.

- **My Computer My Computer**. The My Computer icon allows quick and easy access to the computer. The user can browse through the files and folders by double-clicking the My Computer icon.

- **Explorer Windows Explorer**. This is a program that allows the user to view both the hierarchy of folders on your computer and all the files and folders in each selected folder. It is typically used to copy and move files, where files can be dragged into a specific folder.

- **Properties**. Each icon which appears in Windows 95/NT has associated properties. It views the properties then the right mouse button clicks on the item, and then clicks Properties from the pop-up menu.

- **Shortcut menus**. A shortcut menu is displayed when the user clicks the right mouse button on an item. This menu allows the user to open, copy or delete the item.

- **Close, Minimize, and Maximize buttons**. Each window has a close button (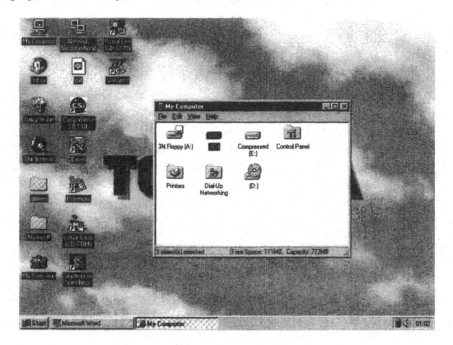) in the upper-right corner which, when clicked, will close the window and quit the program. The windows can also minimize (▬) or maximize (▢).

The current active window has a blue title bar whilst an inactive window has a grey title bar. Only icons within an active window can be selected.

Figure 33.3   Example desktop screen

Figure 33.4   Example desktop screen

33.8 Mouse controls

The mouse controls the pointer around the screen. As with previous versions of Windows, a double click of the left mouse button on an icon is used to either run an application and a single click to select a window or file.

33.9 Quitting Windows

The user must shut Windows down before switching off the computer. To do this the user selects the Shut Down... option from the Start bar. Then the user will be given a number of options, as shown in Figure 33.5. If the user is to quit and switch off then the Shut down the computer option is selected and Yes is selected.

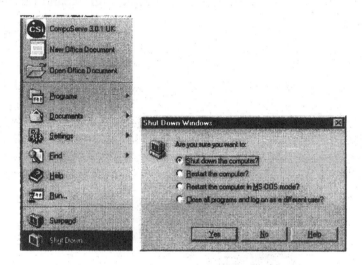

Figure 33.5   Quitting Windows

33.9.1 Getting help

The best method of getting help is to refer to the user manual. If that is not available then there is on-line help. The on-line manual is available by selecting the help manual icon.

33.10 Moving and resizing a window

The option used to change the windows size are:

- **Maximizing a window**. A window can be made to fill the complete screen by selecting the 🗗 at the top right-hand corner of the screen.
- **Minimizing a window**. A window can be minimized to an icon using 🗕.

- **Making a window smaller.** If a window is filling the complete screen it can be made into a smaller window by selecting ▣.
- **Moving a window.** A window can be moved by placing the mouse pointer on the blue title bar. The left mouse button down is pressed while moving the window to its required position, then the mouse button is released.
- **Resizing a window.** A window can be resized by placing the mouse pointer either at one of the corners or the sides. The pointer shape should change when it is placed over the border of the windows. To expand (or contract) the window vertically the mouse pointer is placed either at the top or bottom border. The mouse is then pressed while moving the border of the windows. When the required position is found the mouse button is released. A similar operation can be conducted to expand (or contract) horizontally.

33.11 Closing a window with the Control menu

The Control menu is activated by select the top left-hand button on the window. This menu contains the options: Restore, Move, Size, Minimize, Maximize and Close. The Restore, Move, Size, Minimize and Maximize options mimic operations that are conducted by the mouse. A useful option is the Close which exits from the window and closes it down; this is shown in Figure 33.6.

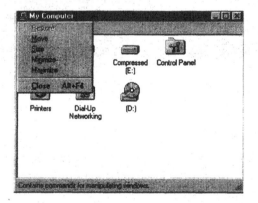

Figure 33.6 Control menu

33.12 Start

The Start button at the bottom left-hand side is used to quickly run application programs. An example Start bar is shown in Figure 33.7.

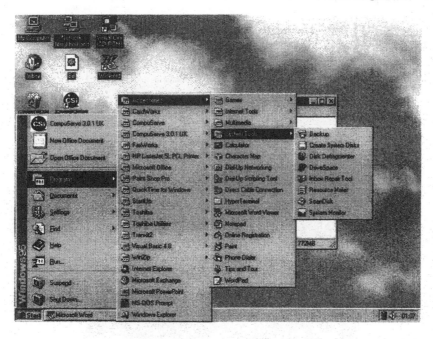

Figure 33.7 Example start options

33.13 My Computer

The My Computer window allows access to local disks, such as the floppy disk and hard disks, printers, CD-ROM, and so on. Figure 33.8 shows an example window.

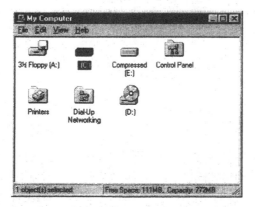

Figure 33.8 Example My Computer

33.14 Running DOS

Windows NT/95 runs an emulated version of DOS. This is typically used to run DOS-based program and also to run DOS commands. Figure 33.9 shows how a DOS window is initiated. It can either run in a window or full-screen. Figure 33.10 shows a sample DOS window.

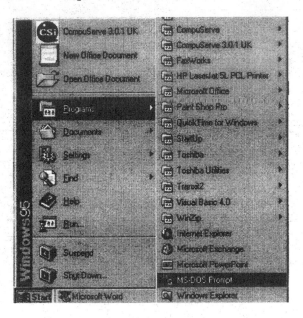

Figure 33.9 Selecting MS-DOS Prompt

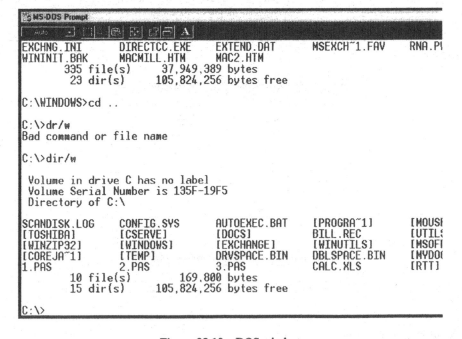

Figure 33.10 DOS window

33.15 Windows 95/NT network drives

Windows NT and 95 displays the currently mounted network drives within the group My Computer. Figure 33.11 shows drives which are either local (C: D: and E:) or mounted using NetWare (F: G: and so on). Windows NT and 95 also automatically scan the neigbouring networks to find network servers. An example is shown in Figures 1.12 and 1.13. Figure 33.13 shows the currently mounted servers (such as, EECE_1) and by selecting the Global Network icon all the other connected local servers can be shown.

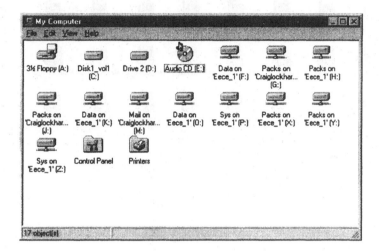

Figure 33.11   Mounted network and local drives

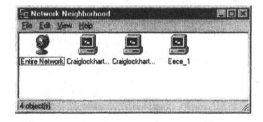

Figure 33.12   Network neighbourhood

33.16 Exercises

Conduct the following steps filling in the required information.

33.16.1   Windows version.   Windows 95 [   ] Windows NT [   ] Other [   ]

33.16.2   Networked computer. Yes [   ] No [   ]

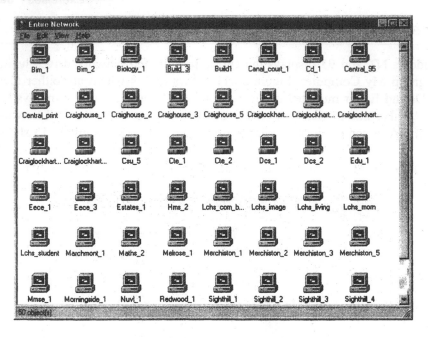

Figure 33.13   Local neighbourhood servers

33.16.3   Identify each of the icons on the desktop:

33.16.4   Select My Computer and resize it so that it fills the screen. Tic box if complete successfully.

Completed successfully [   ]

33.16.5   Resize My Computer so that it is shown at its normal position on the screen.

Completed successfully [   ]

33.16.6   Iconize My Computer and then restore it to its normal position.

Completed successfully [   ]

33.16.7   Expand My Computer horizontally, then restore it to its normal position.

Completed successfully [   ]

33.16.8   Contract My Computer window vertically, then restore it to its normal position.

Completed successfully [   ]

33.16.9 Find the following application programs, run them and then exit. Tic box if complete successfully.

Media Player   Completed successfully [   ]

Character Map   Completed successfully [   ]

Notepad   Completed successfully [   ]

Paint   Completed successfully [   ]

Calculator   Completed successfully [   ]

33.16.10 Open a DOS window and move around the file system. List some of the directories on the system:

33.16.11 Exit from Windows.

Completed successfully [   ]

33.16.12 Start Windows again.

Completed successfully [   ]

33.16.13 There is an on-line help facility within Windows. To select this, use the <u>H</u>elp menu option within the **Program Manager**. Use this facility to find help on the following topics.

Quitting windows: _____

Deleting folders: _____

Starting applications: _____

Deleting programs: _____

34 Extra Windows

34.1 Saving important set-up files

Windows 3.x and Windows 95/NT store important program configuration information in INI file. These are normally either stored in the default windows directory or in the home directory of the application. The important files for Windows configuration are:

- SYSTEM.INI. This contains information on the configuration of various programs and also programs which are initiated when Windows is started.
- WIN.INI. This contains information of the desktop, such as the groups which are displayed, their position on the desktop and their size.

It is important to backup the initialization files as these can be easily modified by mistake. Initialization files have a INI extension and are stored in the \WINDOWS directory. To make a backup of these files create a directory named INI as a subdirectory of the \WINDOWS directory can copy these INI files into it. If one of these files is modified by mistake it is a simple task to copy an older version back into the \WINDOWS directory. This procedure is shown in Test run 34.1.

💻 **Test run 34.1**

```
C:\WINDOWS>dir *.ini /w
 Volume in drive C is MS-DOS_5
 Volume Serial Number is 3B33-13D3
 Directory of C:\WINDOWS
CONTROL.INI     MOUSE.INI      MSD.INI        ORIGPROG.INI SETUP.INI
SYSTEM.INI      ORIG_WIN.INI   WINFILE.INI    DOSAPP.INI   NMINE.INI
PROGMAN.INI     MTFONTS.INI    CLOCK.INI      WINSIGHT.INI POWERPNT.INI
        32 file(s)       31849 bytes
                       4640768 bytes free
C:\WINDOWS> mkdir ini
C:\WINDOWS> copy *.ini ini
```

34.2 Running a program when Windows starts up

Programs can be initiated when Windows is started. Typical programs which can be loaded are the clock, a word processor, a spreadsheet or a calendar. This can be achieved by selecting the following folder:

Selecting Windows→Start Menu→Programs→StartUp

Then the startup item is added by either dragging it into the StartUp folder or by selecting File→New.

Figure 34.1 Adding to the StartUp folder

In Windows 3.1 it is selected by:

• The application icon can be inserted into the StartUp group so that it will automatically start when windows is started. To create a new StartUp group, select File>New from *Program Manager*, as illustrated in Figure 34.2. Next select a group and name it *StartUp*. Icons can then be dragged from another group into this one.

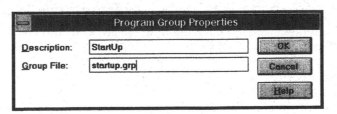

Figure 34.2 Creating a StartUp group

Once the *startup* group has been initiated an empty window should be displayed. To move a program point to it with the cursor then hold the first mouse key down and drag it into the *StartUp* group. If the program is to be copied (and not moved) then hold down the Ctrl key as the icon is dragged. Figure 34.2 shows a *StartUp* windows with the Calendar and Clock.

A program can also be automatically started by editing the WIN.INI file so that the program name is inserted after the run= directive. For example, to run the clock and Word for Windows the following can be inserted:

📄 Listing of WIN.INI

```
[windows]
spooler=yes
run=clock.exe
Beep=yes
device=PostScript Printer,pscript,LPT2:
```

To load a program which is minimized (that is, it will appear as an icon at the bottom of the screen) the `load=` directive is used. For example, to start the programs Word for Windows 6.0 and `cardfile` the following can be inserted into the `WIN.INI` file

📄 Listing of WIN.INI

```
[windows]
spooler=yes
run=
load=c:\word6\winword.exe clock.exe
Programs=com exe bat pif
```

34.3 Capturing Windows screen

There are two hot keys set-up in Windows to capture screens:

- Print Screen key captures a complete screen.
- Alt-Print Screen captures only the current window.

These captured screens are put into the clipboard. Figure 34.3 shows the Clipboard Viewer screen after a screen has been captured from Microsoft Word for Windows 2.0. The contents of the clipboard can be imported into a document or other application by pasting. This can either be done by selecting the Paste from a menu or Ctrl-V.

34.4 Swapping between applications

The Alt-TAB hot key can be used to switch between one application and another. For example:

1. Run an application.
2. Press Alt-TAB to cycle through the Windows applications on the desktop. The name of the application appears at the top of the screen.

Figure 34.4 shows an example of swapping between a Windows screen and another application.

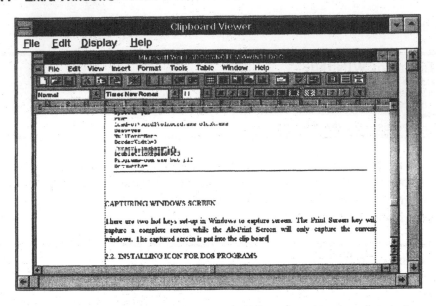

Figure 34.3   Contents of Clipboard Viewer after a screen has been captured

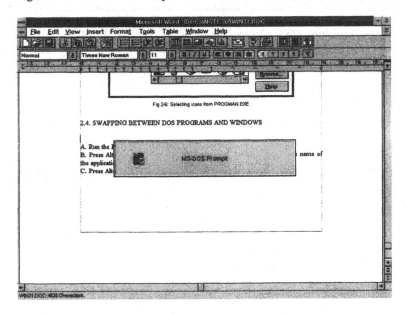

Figure 34.4   Swapping between applications

34.5 Terminating programs which have crashed

If a program has crashed it can normally be terminated using the Ctrl-Alt-Del keystrokes. If Windows is still running it will display a blue screen which asks the user to either press <RETURN> if the program is to be terminated to use Ctrl-Alt-Del to reboot the system. Normally it is only required to terminate the crashed program so press <RETURN>.

34.6 Windows file extensions

Windows uses a large number of file types. Table 34.1 gives the basic file type extensions.

Table 34.1 File extensions

Extension	Description	Extension	Description
BMP	Bitmapped image	HLP	Help data file
CRD	Cardfile data	INI	Initialization file
DLL	Dynamic Linked Library	PIF	Program Information File
DRV	Device driver	REC	Recorder macro file
EXE	Executable file	SYS	Device driver
FON	Font	TXT	Text file
GRP	Program Manager group	WRI	Write document

34.7 Exercises

34.7.1 Start the calculator program and then press Ctrl-Alt-Del. Confirm that a similar screen to the one shown in Figure 34.5. Next shut down the program.

Completed successfully [   ]

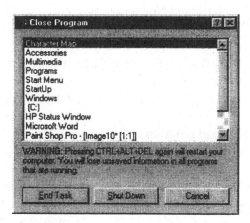

Figure 34.5 Closing an application

34.7.2 Identify some INI files and the application program that they are likely to be linked to.

Completed successfully [   ]

34.7.3 List some of the WIN.INI and SYSTEM.INI files. Identify some of the lines in these files.

Completed successfully [   ]

34.7.4   Run the Calculator and Notepad programs (from the Accessories group) and use the Alt-TAB keystroke to swap between these applications.

Completed successfully [   ]

34.7.5   With the Calculator program, determine the following:

(i) 54×32
(ii) $10432 \div 54$
(iii) $\sqrt{163}$

Completed successfully [   ]

34.7.6   With the Notepad program, create a file called MYFILE.TXT and add the following text:

```
This is a text file that has been created
from Windows.

Freddy Bloggs.
```

Completed successfully [   ]

34.7.7   Quit the Calculator and Notepad programs.

Completed successfully [   ]

34.7.8   Locate the MYFILE.TXT file and open it to see that it contains the entered text.

Completed successfully [   ]

34.7.9   Close the Notepad program.

Completed successfully [   ]

34.7.10   From the operating system, change the name of MYFILE.TXT to MYFILE2.TXT.

Completed successfully [   ]

34.7.11   From the operating system, copy the MYFILE2.TXT file into MYFILE3.TXT file.

Completed successfully [   ]

34.7.12   Format a floppy disk and copy the MYFILE2.TXT file onto it.

Completed successfully [   ]

PART I

C/Pascal
C++
Assembly Language
Visual Basic
HTML/Java
DOS
Windows 3.x
Windows 95/NT
UNIX

35 Introduction to UNIX

35.1 Introduction

UNIX is an extremely popular operating system and dominates in the high-powered, multi-tasking workstation market. It is relatively simple to use and to administer, and also has a high degree of security. UNIX computers use TCP/IP communications to mount disk resources from one machine onto another. Its main characteristics are:

- Multi-user.
- Pre-emptive multitasking.
- Multi-processing.
- Multi-threaded applications.
- Memory management with paging (organizing programs so that the program is loaded into pages of memory) and swapping (which involves swapping the contents of memory to disk storage).

The two main families of UNIX are UNIX System V and BSD (Berkeley Software Distribution) Version 4.4. System V is the operating system most often used and has descended from a system developed by the Bell Laboratories and was recently sold to SCO (Santa Cruz Operation). Popular UNIX systems are:

- AIX (on IBM workstations and mainframes).
- HP-UX (on HP workstations).
- Linux (on PC-based systems).
- OSF/1 (on DEC workstations).
- Solaris (on Sun workstations).

An initiative by several software vendors has resulted in a common standard for the user interface and the operation of UNIX. The user interface standard is defined by the common desktop environment (CDE). This allows software vendors to write calls to a standard CDE API (application specific interface). The common UNIX standard has been defined as Spec 1170 APIs. Compliance with the CDE and Spec 1170 API is certified by X/Open, which is a UNIX standard organization.

35.2 Login into the system

In order to connect to the system a valid user ID and a password are required. These are assigned initially by the system manager. In Sample session 35.1 the user bill_b logs in with the correct password. If the user ID and password are valid then system messages are then displayed in a start-up screen (such as system shut-downs, holiday arrangements, and so on) and the command line prompt is displayed (in this case [51: miranda : / net/ castor_win/ local_user/ bill_b] %). The computer is now ready for a command.

Sample session 35.1

```
HP-UX miranda A.09.01 A 9000/720 (ttys0)

login:      bill_b
Password:   ********

(c)Copyright 1983-1992 Hewlett-Packard Co.,   All Rights Reserved.
    ::::::
(c)Copyright 1988 Carnegie Mellon

                    RESTRICTED RIGHTS LEGEND
Use, duplication, or disclosure by the U.S. Government is subject to
restrictions as set forth in sub-paragraph (c)(1)(ii) of the Rights in
Technical Data and Computer Software clause in DFARS 252.227-7013.

                    Hewlett-Packard Company
                    3000 Hanover Street
                    Palo Alto, CA 94304 U.S.A.

Rights for non-DOD U.S. Government Departments and Agencies are as set
forth in FAR 52.227-19(c)(1,2).

[51:miranda :/user/bill_b ] %
```

35.3 Directory structure

Files are used to store permanent information which are used by programs. The information could be schematics, text files, documents, and so on. In order to facilitate the recovery of files they are arranged into directories in a structure that is similar to an office filing system. The UNIX file directory structure takes the form of a tree with the root directory at the highest level. This top level, or root directory, is given the name /. An example of a directory system is shown in Figure 35.1. In this case there are 5 sub-directories below the root level, these are bin, usr, etc, dev and user. Below the usr directory there are 3 sub-directories, these are lib, adm and bin. In this case, all the users of the system have been assigned to a sub-directory below the users directory, that is, bill_b, fred_a and alan_g.

The full pathname of the bill_b directory is /users/bill_b and the full pathname of the adm directory is /usr/admin. Files can be stored within a directory structure. Figure 35.2 shows an example structure.

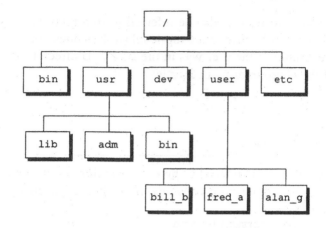

Figure 35.1 Basic directory structure

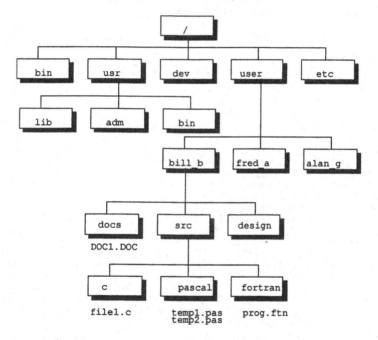

Figure 35.2 Basic directory structure showing files within directories

In this case, the full pathname of the FORTRAN file prog.ftn is:

/user/bill_b/src/fortran/prog.ftn

and the full pathname of the c directory is:

/user/bill_b/src/c

Files and sub-directories can also be referred to in a relative manner, where the directory is not reference to the top-level (it thus does not have a preceding /). For example, if the user was in the `bill_b` directory then the relative path for the C program `file1.c` is:

```
src/c/file1.c
```

36.4 On-line manual

Unix provides an on-line manual to give information on all the UNIX commands. To get information man *command-name* is used, such as:

```
man command-name
```

Examples are `man ls`, `man cd`, and so on. Sample session 35.2 shows an example of the help manual for the `ls` command.

Sample session 35.2
```
[1:miranda :/user/bill_b ] % man ls

ls(1)

NAME
     ls, l, ll, lsf, lsr, lsx - list contents of directories

SYNOPSIS
     ls [-abcdfgilmnopqrstuxACFHLR1] [names]
     l [ls_options] [names]
     ll [ls_options] [names]
     lsf [ls_options] [names]
     lsr [ls_options] [names]
     lsx [ls_options] [names]

DESCRIPTION
For each directory argument, ls lists the contents of the directory.
For each file argument, ls repeats its name and any other information
requested.  The output is sorted in ascending collation order by
default (see Environment Variables below).  When no argument is given
the current directory is listed.  When several arguments are given,
the arguments are first sorted appropriately, but file arguments
appear before directories and their contents.
--More--
```

35.5 Changing directory

The `pwd` command can be used to determine the present working directory. and the `cd` command can be used to change the current working directory. When changing directory either the full pathname or the relative pathname is given. If a / precedes the directory name then it is a full pathname, else it is a relative path. Some special character sequences are used to represent other directory, such as the directory above the current directory is specified by a double dot (. .).

Thus to move to the directory above the command `cd ..` can be used. If the `cd` command is used without any preceding directory specifier then the directory will be changed to the user's home directory. Some example command sessions are given next.

`cd ..`	move to the directory above
`cd /`	move to the top-level directory
`cd /user/bill_b/fortran`	move to the directory `/user/bill_b/fortran`
`cd src/c`	move to the sub-directory `c` which is below the `src` sub-directory
`cd`	move to the user's home directory

35.6 Listing directories

The `ls` command lists the contents of a directory. If no directory-name is given then it lists the contents of the current directory. Sample session 35.3 shows a typical session. In [5] the user moves to the directory above and in [11] the user move back to the home directory.

💻 Sample session 35.3

```
[2:miranda :/user/bill_b ] % ls
compress  design   docs      mentor    research  shells    spwfiles
[3:miranda :/user/bill_b ] % cd design
[4:/user/bill_b/design ] % ls
analogue  digital  ic        pcb       vhdl
[5:/user/bill_b/design ] % cd ..
[6:/user/bill_b ] % ls
compress  design   docs      mentor    research  shells    spwfiles
[7:/user/bill_b ] % cd ..
[8:/user ] % ls
bill_b    george_r
[9:/user ] % cd ..
[10:/ ] % ls
TEX.install         local_user          spwsys
TEX.new             lost+found          syn_sample_lib
backup_log          mentor_bdf          syn_sample_lib.8.2_1
caedata             mentor_fonts        system
caesys              mentor_fonts_2      temp
castor_win          mentor_fonts_new    tmp
designer-2.0        methdata            ue
espresso            mf                  user
espresso.ex.tar.Z   planb               users
[11:/ ] % cd
[12:/user/bill_b ] % ls
compress  design   docs      mentor    research  shells    spwfiles
[13:/user/bill_b ] %
```

The basic directory listing gives no information about the size of files, if it is a directory, and so on. To list more information the `-l` option is used. An example session is shown in Sample session 35.5. In [15] the user requests extended information on the files.

```
🖳  Sample session 35.4
[14:/user/bill_b ] % ls
compress  design    docs      mentor    research  shells    spwfiles
[15:/user/bill_b ] % ls -l
total 14
drwxr-xr-x    2 bill_b   10           1024 Nov  1 17:09 compress
drwxr-xr-x    7 bill_b   10           1024 Nov  7 10:11 design
drwxr-xr-x    2 bill_b   10             24 Oct 31 09:52 docs
drwxr-xr-x    3 bill_b   10           1024 Sep 14 13:48 mentor
drwxr-xr-x    3 bill_b   10           1024 Oct 31 09:38 research
drwxr-xr-x    2 bill_b   10           1024 Nov  7 10:21 shells
drwxr-xr-x    2 bill_b   10           1024 Jun 20 18:12 spwfiles
[16:/user/bill_b ] % cd shells
[17:/user/bill_b/shells ] % ls -l
total 6
-rw-r--r--    1 bill_b   10            988 Nov  7 10:20 Cshrc
-rw-r--r--    1 bill_b   10             43 Nov  7 10:20 Login
-rwxr-xr-x    1 bill_b   10             28 May 12  1993 gopc
[18:/user/bill_b/shells ] %
```

Other options can be used with `ls`; to get a full list use the on-line manual. In Sample session 35.1 it was seen that other possible extensions are a b c d f g i l m n o p q r s t u x A C F H L R 1.

Examples of usage are:

`ls -d`	lists only directories
`ls -r`	reverse alphabetic order
`ls -t`	order in time last modified

It is also possible to specify more than one extension, such as `ls -dr` which lists only directories in reverse order.

35.7 File attributes

UNIX provides system security by assigning each file and directory with a set of attributes. These give the privileges on the file usage and their settings are displayed with the `ls -l` command. In the case of [**15**] in Sample session 35.4, the user uses `ls -l` to get extended information. The directory listing gives six main fields; these are:

- File attributes.
- Owner of the files. Person (user ID) who owns the file.
- Group information. The group name defines the name of the group the group the attributes concern.
- Size of file. The size of the file in bytes (8 bits).
- Date and time created or last modified. This gives the date and time the file was last modified. If it was modified in a different year then the year and date are given, but no time information is given.
- Filename.

Figure 35.3 defines the format of the extended file listing. The file attributes contain the letters r, w, x which denote read, write and executable. If the attribute exists then the associated letter is placed at a fixed position, else a – appears. The definition of these attributes are as follows:

- Read (r). This file can be copied, viewed, and so on, but it cannot be modified.
- Write (w). This file can be copied, viewed and changed, if required.

Executable (x). It is possible to execute this program.

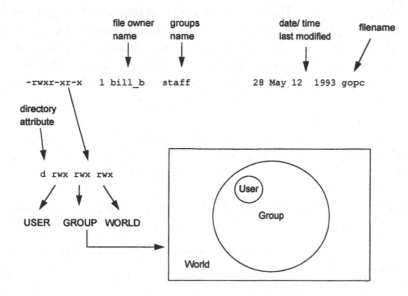

Figure 35.3 Extended file listing

The file attributes split into four main sections. The first position identifies if it is a directory or a file. A d character identifies a directory, else it is a file. Positions 2–4 are the owner attributes, positions 5–7 are the groups attributes and positions 8–10 are the rest of the world's attributes. Table 35.1 lists some examples.

Table 35.1 Example file attributes

Attributes	Description
-r-x--x---	This file can be executed by the owner and his group (e.g. staff, students, admin, research, system, and so on). It can be viewed by the owner but no-one else. No other privileges exist.
drwxr-xr-x	This directory can be cannot be written to by the members of group and others. All other privileges exist.
-rwxrwxrwx	This file can be read and written to by everyone and it can also be executed by everyone (beware of this).

35.7.1 Changing attributes of a file

As we have seen, files have certain attributes such a read/write privileges, and so on.

The chmod command can be used by the owner of the file to change any of the file attributes. A general format is given next.

chmod *settings filename*

where *settings* define how the attributes are to be changed and which part of the attribute to change. The characters which define which part to modify are u (user), g (group), o (others), or a (all). The characters for the file attributes are a sign (+, − or =) followed by the characters r, w, x. A '+', specify that the attribute is to be added, a '−' specifies that the attribute is to be taken away. The '=' defines the actual attributes. Some examples are given next:

chmod u+x prog1.c	owner has executable rights added
chmod a=rwx prog2.c	sets read, write, execute for all
chmod g-r cprogs	resets read option for group

In Sample session 35.5 [20] the owner of the file changes the execute attribute for the user.

🖥 **Sample session 35.5**
```
[19:/user/bill_b/shells ] % ls -l
total 6
-rw-r--r--    1 bill_b    10           988 Nov   7 10:20 Cshrc
-rw-r--r--    1 bill_b    10            43 Nov   7 10:20 Login
-rwxr-xr-x    1 bill_b    10            28 May 12  1993 gopc
[20:/user/bill_b/shells ] % chmod u+x Login
[21:/user/bill_b/shells ] % ls -l
total 6
-rw-r--r--    1 bill_b    10           988 Nov   7 10:20 Cshrc
-rwxr--r--    1 bill_b    10            43 Nov   7 10:20 Login
-rwxr-xr-x    1 bill_b    10            28 May 12  1993 gopc
[22:/user/bill_b/shells ] %
```

35.8 Special characters (*, ? and [])

There are several special characters which aid access to files. These are stated in Table 35.2.

Table 35.2 Special characters

Char	Description
?	matches any single character in a filename.
*	matches zero or more characters in a filename.
[]	matches one character at a time, these characters are contained in the squared brackets.

Sample session 35.6 shows a few sample uses of wildcards. In [**24**] the user lists all the files which begin with the letter 'm'. In [**26**] all two letter file-names beginning with 'c' are listed. Then in [**27**] the files which begin with the letters 'a', 'b' or 'c' are listed.

💻 **Sample session 35.6**
```
[23:/user/bill_b ] % cd /bin
[24:/bin ] % ls m*
mail    make    mesg    mkdir   mkfifo  mktemp  model   mstm    mt
m25:/bin ] % ls c*
c89     cd      chmod   cmp     cp      crypt
cat     chacl   chown   cnodes  cpio    csh
cc      chgrp   cksum   command cps     cstm
[26:/bin ] % ls c?
cc  cd  cp
[27 :/bin ] % ls [a-c]*
alias     basename  cat     chacl   chown   cnodes   cpio  csh
ar        bg        cc      chgrp   cksum   command  cps   cstm
as        c89       cd      chmod   cmp     cp       crypt
[28 :/bin ] % ls [asz]*
alias   as      sh      sleep   strip   su    sync
ar      sed     size    sort    stty    sum   sysdiag
[29 :/bin ] %
```

Note that [a–c] represents [abc] and [1–9] represents [123456789].

35.9 Listing contents of a file

The command to list the contents of a file is cat, its form is:

cat *filename*

Sample session 35.7 shows how it is used. If a file is larger than one screen full it is possible to stop the text from scrolling by using Cntl-S (^S) to stop the text and Cntl-Q (^Q) to start.

💻 **Sample session 35.7**
```
[30:/user/bill_b ] % cd shells
[31:/user/bill_b/shells ] % ls
Cshrc  Login  gopc
[32:/user/bill_b/shells ] % cat gopc
setenv DISPLAY pc9:0
xterm
[33:/user/bill_b/shells ] %
```

35.10 Making and removing directories

To make a directory the mkdir command is used, its form is:

mkdir   directory-name

to remove a directory the rmdir command is used, its form is:

```
rmdir   directory-name
```

In both cases the full pathname or relative pathname can be given.

35.11 Copying, renaming and removing

The command to copy files is the cp command. The standard format is:

```
cp source destination
```

and the command to rename (or move) is called mv. Its standard format is:

```
mv source destination
```

Files are erased, or deleted, using the rm command. Sample session 35.8 shows some uses of these commands.

💻 **Sample session 35.8**
```
[34 :/user/bill_b ] % mkdir src
[35 :/user/bill_b ] % ls
compress dirlist   mentor    shells    src
design   docs      research  spwfiles  temp
[36 :/user/bill_b ] % cd src
[37 :/user/bill_b/src ] % mkdir cprogs
[38:/user/bill_b/src ] % cd cprogs
[39 :/user/bill_b/src/cprogs ] % ls
[40 :/user/bill_b/src/cprogs ] % cp ../../docs/*.c .
[41 :/user/bill_b/src/cprogs ] % ls
file.c
[42 :/user/bill_b/src/cprogs ] % cp file.c file1.c
[43 :/user/bill_b/src/cprogs ] % ls
file.c   file1.c
[44 :/user/bill_b/src/cprogs ] % mv file1.c file2.c
[45 :/user/bill_b/src/cprogs ] % ls
file.c   file2.c
[46 :/user/bill_b/src/cprogs ] % mkdir list
[47 :/user/bill_b/src/cprogs ] % cp *.c list
[48 :/user/bill_b/src/cprogs ] % ls
file.c   file2.c list
[49 :/user/bill_b/src/cprogs ] % ls -l
total 10
drwxr-xr-x   3 bill_b   10           1024 Nov 14 10:40 .
drwxr-xr-x   3 bill_b   10           1024 Nov 14 10:38 ..
-rw-r--r--   1 bill_b   10            213 Nov 14 10:39 file.c
-rw-r--r--   1 bill_b   10            213 Nov 14 10:39 file2.c
drwxr-xr-x   2 bill_b   10           1024 Nov 14 10:40 list
[50 :/user/bill_b/src/cprogs ] % cat file.c file2.c > src.lis
[51 :/user/bill_b/src/cprogs ] % ls
file.c   file2.c list      src.lis
[52 :/user/bill_b/src/cprogs ] % ls -l
total 12
drwxr-xr-x   3 bill_b   10           1024 Nov 14 10:43 .
drwxr-xr-x   3 bill_b   10           1024 Nov 14 10:38 ..
-rw-r--r--   1 bill_b   10            213 Nov 14 10:39 file.c
-rw-r--r--   1 bill_b   10            213 Nov 14 10:39 file2.c
drwxr-xr-x   2 bill_b   10           1024 Nov 14 10:40 list
-rw-r--r--   1 bill_b   10            426 Nov 14 10:43 src.lis
[53 :/user/bill_b/src/cprogs ] %
```

35.12 Standard input and output

The standard input device for a program is via the keyboard and the standard output is to the monitor. In UNIX all input output devices are communicated with through a device file, which are stored in the /dev directory. For example, each connected keyboard to the system (including remote computers) has a different device name. To determine which is your terminal pathname use the tty command as shown in Sample session 35.9.

Sample session 35.9
```
[54  :/user/bill_b ] % tty
/dev/ttys0
[55  :/user/bill_b ] %
```

35.12.1 Redirection

It is possible to direct the input and/or output from a program to another file or device.

Redirecting output

The redirection symbol (>) redirects the output of a program to a given file (or output device). This output will not appear on the monitor (unless it is redirected to it). Sample session 35.10 shows how the output from a directory listing can be send to a file named dirlist (see **[57]**). The contents of this file are then listed in **[58]**.

Sample session 35.10
```
[56  :/user/bill_b ] % ls
compress design    docs      mentor    research  shells    spwfiles
[57:/user/bill_b ] % ls > dirlist
[58  :/user/bill_b ] % cat dirlist
compress design    dirlist   docs      mentor    research  shells
spwfiles
[59  :/user/bill_b ] %
```

The redirection of output is particularly useful when a process is running and an output to the screen is not required. Another advantage of redirection is that it is possible to keep a permanent copy of a program's execution.

To create a text file the cat command is used with the redirection, as shown next. The file is closed when the user uses the CNTRL-D keystroke, as shown in Sample run 35.11.

If the user does not want the output to appear on the screen then it can be redirected to the file called /dev/null. This is the systems wastepaper basket and it will automatically be deleted.

Redirecting input

The input can be redirected with the redirect symbol (<). The file given after

560 Introduction to UNIX

the symbol is taken as the input. For example:

```
% prog1 < inputfile
```

In this case the program prog1 takes its input from the file inputfile, and not from the keyboard.

💻 **Sample session 35.11**
```
[60  :/user/bill_b ] % cat > file
This is an example of a
file created by cat
^D
[61  :/user/bill_b ] % ls
compress  dirlist   file      research  spwfiles
design    docs      mentor    shells
[62  :/user/bill_b ] % cat file
This is an example of a
file created by cat
```

35.12.2 Pipes

The pipe allows the output of one program to be sent to another as the input. The symbol used is the vertical bar (|) and its standard form is:

```
program_a [arguments]   | program_b [arguments]
```

This notation means that the output of program_a is used as the input to program_b. The pipe helps in commands where a temporary file/s needs to be created. For example, to find out who is logged into the system the who -a command can be used. If names in a file need to be sorted alphabetically the sort command can be used. Thus to sort the users on the system alphabetically we can use:

💻 **Sample session 35.12**
```
[63  :/user/bill_b ] % who -a > temp
[64  :/user/bill_b ] % sort temp
aed_9     ttyp8    Nov  9 13:51  old    14496  id=  p8 term=0   exit=0
aed_9     ttyp9    Nov  9 13:51  old    14497  id=  p9 term=0   exit=0
bill_b    ttyp1    Nov  4 16:05  old     6288  id=  p1 term=0   exit=0
bill_b    ttypa    Nov  4 16:02  old     9567  id=  pa term=0   exit=0
bill_b    ttypb    Nov  4 16:05  old     9582  id=  pb term=0   exit=0
bill_b    ttys0    Nov 14 08:57   .      18715  ees10
julian_m ttyp6     Nov  9 13:59  old    14190  id=  p6 term=0   exit=0
peter_t ttyp4      Nov 10 11:10  old    15169  id=  p4 term=0   exit=0
root      ttys1    Nov  8 09:50  old    11292  id=  s1 term=0   exit=0
steve_w ttyp3      Nov  9 10:02  old    13205  id=  p3 term=0   exit=0
steve_w ttyp5      Nov  9 11:17  old    13493  id=  p5 term=0   exit=0
steve_w ttyp7      Nov  9 10:49  old    13570  id=  p7 term=0   exit=0
xia       ttys2    Nov  7 12:07  old     9961  id=  s2 term=0   exit=0
[65 :/user/bill_b ] %
```

It is possible to achieve this with one line using pipes.

🖳 **Sample session 35.13**
```
[66  :/user/bill_b ] % who -a | sort
aed_9      ttyp8      Nov   9 13:51   old      14496   id=  p8 term=0      exit=0
aed_9      ttyp9      Nov   9 13:51   old      14497   id=  p9 term=0      exit=0
bill_b     ttyp1      Nov   4 16:05   old       6288   id=  p1 term=0      exit=0
bill_b     ttypa      Nov   4 16:02   old       9567   id=  pa term=0      exit=0
bill_b     ttypb      Nov   4 16:05   old       9582   id=  pb term=0      exit=0
bill_b     ttys0      Nov  14 08:57   .        18715   ees10
julian_m   ttyp6      Nov   9 13:59   old      14190   id=  p6 term=0      exit=0
peter_t    ttyp4      Nov  10 11:10   old      15169   id=  p4 term=0      exit=0
root       ttys1      Nov   8 09:50   old      11292   id=  s1 term=0      exit=0
steve_w    ttyp3      Nov   9 10:02   old      13205   id=  p3 term=0      exit=0
steve_w    ttyp5      Nov   9 11:17   old      13493   id=  p5 term=0      exit=0
steve_w    ttyp7      Nov   9 10:49   old      13570   id=  p7 term=0      exit=0
xia        ttys2      Nov   7 12:07   old       9961   id=  s2 term=0      exit=0
[67 :/user/bill_b ] %
```

35.13 Compiling C programs

The cc command is used to compile and link a program. If there are no errors in the C source code then, by default, a file named a.out will be produced. This can then be executed. If another executable filename is required then the −o filename option can be used. Sample session 35.14 shows how the C file file.c is compiled and linked to an executable file named file.

🖳 **Sample session 35.14**
```
[68 :/user/bill_b/src/cprogs ] % cat file.c
#include <stdio.h>

int     main(void)
{
int     value;

        printf("Enter a decimal value (0-127)>> ");
        scanf("%d",&value);
        printf("Decimal %d Hex %x Octal %o Character <%c>\n",
                      value,value,value,value);
        return(0);
}
[69 :/user/bill_b/src/cprogs ] % cc -o file file.c
[70 :/user/bill_b/src/cprogs ] % file
Enter a decimal value (0-127)>> 53
Decimal 53 Hex 35 Octal 65 Character <5>
[71 :/user/bill_b/src/cprogs ] %
```

35.14 Displaying the time and date

The command to display the date is simply date. This is shown in Sample session 35.15.

🖳 **Sample session 35.15**
```
[72 :/user/bill_b] % date
Mon Nov 14 11:30:08 GMT 1994
[73 :/user/bill_b] %
```

35.15 Where to find things

UNIX has various default directories which store standard command programs and configuration files. Some of these are defined in the following sections.

35.15.1 /bin

The /bin directory contains most of the standard commands, such as compilers, UNIX commands, program development tools, and so on. Examples are:

- FORTRAN and C compilers, f77 and cc.
- commands such as ar, cat, man utilities.

35.15.2 /dev

The /dev directory contains special files for external devices, terminals, consoles, line printers, disk drives, and so on.

- /dev/console console terminal.
- /dev/null   system wastebasket.
- /dev/tty*   terminals (such as /dev/tty1 /dev/tty2).

35.15.3 /etc

The /etc directory contains restricted system data and system utility programs which are normally used by the system manager. These include password file, login, and so on. Examples are:

- /etc/groups   system group tables.
- /etc/hosts   list of system hosts.
- /etc/passwd   list of passwords and users.
- /etc/termcap table of terminal devices.
- /etc/ttys   terminal initialization information.
- /etc/ttytype table of connected terminals.
- /etc/utmp   table of users logged in.

35.15.4 /lib

The /lib directory contains system utilities and FORTRAN and C runtime support, system calls and input/output routines.

35.15.5 /tmp

Temporary (scratch) files used by various utilities such as editors, compilers, assemblers. These are normally stored in the /tmp directory.

35.15.6 /usr/adm

The /usr/adm stores various administrators files, such as:

/usr/adm/lastlog table of recent logins

35.15.7 /usr/bin

The /usr/bin contains less used utility programs, such as:

```
/usr/bin/at
/usr/bin/bc
/usr/bin/cal
```

35.15.8 /usr/include

The /usr/include directory contains C #include header files, such as:

```
/usr/include/stdio.h
/usr/include/math.h
```

35.15.9 /usr/lib

The /usr/lib directory contains library routines and setup files, such as:

```
/usr/lib/Cshrc
/usr/lib/Login
/usr/lib/Logout
/usr/lib/Exrc
/usr/lib/calendar
```

35.15.10 /usr/man

The /usr/man contains the manual pages, such as

```
/usr/man/cat[1-8]
/usr/man/man[1-8]
```

35.16 Exercises

35.17.1 Go to your home directory.

DIRECTORY NAME: . . . .

35.17.2 Display the current system date.

TIME/DATE: . . . .

35.17.3 Display the on-line help manual for the following commands:

```
ls
cp
rm
```

Completed successfully: YES/NO

35.17.4 List all the users currently logged into the system.

```
USERS:.................................
```

35.17.5 Create a directory named `src` in your home directory.

35.17.6 Go into the directory `src` and create a sub-directory named `cprogs`.

35.17.7 Go into the `cprogs` directory and create a file named `file1.c` using `cat` (or the system editor). Enter the following text and save the file.

```
/*    file1.c                                      */
/*    Program to determine binary equivalent of an 8-bit*/
/*    integer                                       */
int   main(void)
{
int   val, bit;
   printf("Enter value (0-255) >>");
   scanf("%d",&val);
   printf("Binary value is ");
   for (bit=0x80;bit>0;bit>>=1)
   {
       if (bit & val)    printf("1");
       else              printf("0");
   }
   return(0);
}
```

35.17.8 List the file you have just created.

Completed successfully: YES/NO

35.17.9 Change the name of this file to file2.c.

Completed successfully: YES/NO

35.17.10 Make a copy of this file and name it `file3.c`.

Completed successfully: YES/NO

35.17.11 Change the file attributes of this file so that the public cannot read from or write to this file.

35.17.12 Create another named `file2.c` and enter the following source code.

```
/*    file2.c                          */
/*    Program to print ASCII characters    */
#include <stdio.h>
int   main(void)
{
int   i,start,end;

    printf("Enter start and end for ASCII characters >>");
    scanf("%d %d",&start,&end);
    puts("INTEGER   HEX      ASCII");

    for (i=start;i<=end;i++)
        printf("%5d   %5x   %5c\n",i,i,i);
    return(0);
}
```

35.17.13 Concatenate the two files together. Call the resultant file `src.lis`.

Completed successfully: YES/NO

35.17.14 Compile and run the two C programs just created.

Completed successfully: YES/NO

35.17.15 Remove the file `src.lis`

Completed successfully: YES/NO

35.17.16 Investigate the following commands/ utilities:

```
cal
ftp
more
mail
```

35.17.17 Locate the following utility files and determine their usage.

```
spell
csh
sort
cc
wc
sleep
touch
ed
ps
find
```

35.17.18 Sketch a rough outline of the directory structure of the system.

35.17 Summary

Table 35.3 lists some standard UNIX commands.

Table 35.3   Commands summary

Command	Description
man subject	on-line manual on subject
ls	list of a directory (-l long listing; -a all; -r reverse listing.
cd	change directory
mv	move file
cp	copy file
mkdir	make directory
pwd	present working directory
cat	concatenate a file/s
> file	redirect output to file
< file	redirect input from file

36 UNIX Commands

36.1 Process control

UNIX is a multitasking, multiuser operating system, where many tasks can be running at a certain time. Typically there are several processes which are started when the computer is rebooted; these are named daemon processes and they run even when there is no user logged into the system.

36.1.1 ps (process status)

The `ps` command prints information about the current process status. The basic `ps` list gives a list of the current jobs of the user. An example is given in Sample session 36.1.

Sample session 36.1

```
% ps
  PID  TTY  TIME  CMD
   43   01  0:15  csh
   51   01  0:03  ls -R /
  100   01  0:01  ps
%
```

The information provided gives the process identification number (PID), the terminal at which the process was started (TTY), the amount of CPU time used and the command line that was entered (CMD). A process can be stopped using the `kill` command.

A long listing is achieved using the `-l` option and for a complete listing of all processes on the system the `-a` option is used, as shown in Sample session 36.2.

Sample session 36.2

```
% ps -al
F S UID PID PPID CPU PRI NICE ADDR SZ WCHAN TTY TIME CMD
1 S 101 43    1   3  30   20   3211 12 33400 01  0:15 csh
1 S 104 44    2   2  27   20   4430 12 51400 04  0:08 sh
1 S 104 76   32   3  30   20   3223 12 33400 04  0:03 vi tmp
1 S 104 89    1   3  30   20  10324 02 44103 04  0:01 ls
1 R 101 99   55  43  52   20   4432 12 33423 01  0:01 ps
```

The column headed S gives the state of the process. An S identifies that the process is sleeping (the system is doing something else), W specifies that the system is waiting for another process to stop and R specifies that the process is currently running. In summary:

- R process is running.
- T process has stooped.
- D process in disk wait.
- S process is sleeping (that is, less than 20 secs.).
- I process is idle (that is, longer than 20 secs.).

UID is the user identity number. PRI is the priority of the process. A high number means a low priority.

36.1.2 kill (send a signal to a process, or terminate a process)

The kill command sends a signal to a process. The general format is:

$$\text{kill } \textit{-sig processid}$$

The *processid* is the number given to the process by the computer. This can be found by using the ps command. The *sig* value defines the amount of strength that is given to the kill process and the strongest value is a −9. The owner of a process can kill his own, but only the super-user can kill any process. Sample session 36.3 gives an example session.

Sample session 36.3

```
% ps
   PID  TTY  TIME  CMD
   112  01   1:15  csh
   145  01   0:05  lpr temp.c
   146  01   0:01  ps
% kill -9 145
% ps
   PID  TTY  TIME  CMD
   112  01   1:15  csh
   146  01   0:01  ps
% find / -name "*.c" -print > listing &
% ps
   PID  TTY  TIME  CMD
   112  01   1:15  csh
   177  01   0:03  find -nam
   179  01   0:01  ps
% kill -9 177
%
```

36.1.3 at (execute commands at later date)

The at command when used in conjunction with another command will execute that command at some later time. The standard format is:

at *time* [*date*] [*week*]

where *time* is given using from 1 to 4 digits, followed by either 'a', 'p', 'n' or 'm' for am, pm, noon or midnight, respectively. If no letters are given then a 24-hour clock is assumed. A 1- or 2-digit time is assumed to be given in hours, whereas a 3- or 4-digit time is assumed to be hours and minutes. A colon may also be included to separate the hours from the minutes.

The *Date* can be specified by the month followed by the day-of-the-month number, such as Mar 31. A *Week* can be given instead of the day and month.

For example, to compile a program at quarter past eight at night:

🖥 **Sample session 36.4**
```
%  at 20:15
     cc - test test.c
     ^D
     520776201.a at Tue May 26 20:15:00 1997
```

---

and to send fred a message at 14:00:

🖥 **Sample session 36.5**
```
%  at 14:00
     echo "Time for a tea-break" | mail fred
     ^D
     520777201.a at Mon Jun 4 14:00:00 1989
```

To remove all files with the .o extensions from the current directory on September 9th at 1 noon.

🖥 **Sample session 36.6**
```
% at 1n sep 9
     rm *.o
     ^D
     520778201.a at Sat Sep 9 13:00:00 1989
```

To list all jobs that are waiting to executed at some later time use the −1 option.

🖥 **Sample session 36.7**
```
% at -1
     520776201.a  Mon Jun 4 20:15:00 1989
     520778201.a  Sat Sep 9 13:00:00 1989
     520777201.a  Mon Jun 4 14:00:00 1989
```

To remove jobs from the schedule the −r option can be used and giving the job number.

💻   **Sample session 36.8**
```
% at -r 520777201.a
```

36.1.4 nice (run a command at a low priority)

The `nice` command runs a command at a low priority. The standard format is as follows:

$$\texttt{nice} \; -number \; command \; [arguments]$$

The lowest priority is –20 and the default is –10. Sample session 36.9 gives a sample session.

💻   **Sample session 36.9**
```
% nice -15 ls -al
% nice -20 find / -name "*.c" -print > Clistings &
```

36.2 Compilers

The two main compilers on UNIX systems are C (cc) and FORTRAN (f77).

36.2.1 cc (C compiler)

The `cc` command is the C language compiler and is used to convert a C program into binary code. It can also link several binary modules together to make an executable file.

Files with the extension .c are taken as C programs and when a C program is compiled the resulting binary module (the object code) is placed in the current directory. If a `-c` option is not used the binary code will be called `a.out`. If the `-c` option is used the binary code will take the name given, the modules produced will have a .o extension. Options are:

`-c`   produce single binary module (object code) with the extension .o.
`-o` *outfil* this option is used to give the name for the output file.
`-g`   produce debug information for use with a debugger (such as dbx).

In Sample session 36.11, the first example of the C program prog1.c is compiled; the resulting file produced is called `a.out`. This file can be executed as required. The second type of compilation (`cc -c *.c`) compiles all the C programs and produces an object module with a .o extension. These object modules can be linked together using the compiler. The last example (`cc -o prog prog1.o prog2.o`) links together the object modules for `prog1.c` and `prog2.c`. The executable program, in this case, will be called `prog`.

📟 **Sample session 36.10**
```
%  cc   prog1.c
%  ls
    a.out   prog1.c   prog2.c
%  rm a.out
%  cc -c *.c
%  ls
   prog1.c   prog1.o   prog2.c   prog2.o
%  cc -o prog   prog1.o   prog2.o
%  ls
   prog   prog1.c   prog1.o   prog2.c   prog2.o
```

36.2.2 f77 (FORTRAN compiler)

The FORTRAN compiler f77 is used to compile and/or link FORTRAN programs. The source file is assumed to have the .f extension. If the −c option is chosen the compiled module will have the .o extension; this is an object module. Sample session 36.11 gives some examples.

📟 **Sample session 36.11**
```
%   f77 -c *.f
%   f77 -o prog *.o
```

36.2.3 dbx (debugger for C, FORTRAN and Pascal)

The dbx utility is a source-level debugger for C, FORTRAN and Pascal. It allows the user to trace a program's execution and list variables. The general format is:

dbx *filename*

Some commands for dbx are:

run	runs the current program.
step	step through the program execution one line at a time.
cont	continue program execution where it stopped.
print	print contents of a variable such as print input_value.
list	list next ten lines of code if no arguments are given; to list from linestart to lineend use list linestart,lineend.
help	get help information.
trace	trace program execution.
quit	exit dbx.

36.2.4 lint (C program verifier)

The lint utility is used to verify C programs. It attempts to detect bugs, non-portable warnings and any wasteful use of C code. The basic C compiler, cc, does little run time checking or inefficiency checking. To run the lint program:

lint *filename*

Sample session 36.12 gives a sample session.

📟 **Sample session 36.12**
```
% lint prog1.c
   function returns value is always ignored
      printf  strcmp
```

36.3 File manipulation commands

UNIX has a number of file manipulation commands, some of which were defined in the previous chapter and many others are given in this section.

36.3.1 cp (copy files)

The cp command will copy a given file or directory to a given file or directory. There are several options that can be used:

-i   uses interactive mode, the user is prompted whenever a file is to be overwritten. The user answers 'y' or 'n'.

-r   uses recursive mode, the cp command all files including each subdirectory it encounters.

The following example will copy a file called file1 into a file called file2. Note that if file2 were a directory then file2 would be copied into that directory.

📟 **Sample session 36.13**
```
% cp file1 file2
```

If you want to copy a whole directory and all subdirectories the -r option is used. In the next example the whole directory structure of /usr/staff/bill is copied into the directory /usr/staff/fred:

📟 **Sample session 36.14**
```
% cp -r /usr/staff/bill /usr/staff/fred
```

To copy a file into a directory:

🖳 **Sample session 36.15**
```
% cp type.c   cprogs
```

36.3.2 ls (list directory)

The ls command lists the contents of a directory. A summary of the various options are given below:

-a   lists all entries including files that begin with a . (dot)

-l   lists files in the long format. Information given includes size, ownership, group and time last modified.

-r   lists in reverse alphabetic order.

-t   lists by time last modified (latest first) instead of name.

-1   lists one entry per line.

-F   marks directories with a trailing slash (/), executables with a trailing star (*).

-R   recursively lists subdirectories encountered.

Sample session 36.16 gives some examples.

36.3.3 file (determine file type)

The file command is used to test a file for its type such as a C program, text file, binary file, and so on. Sample session 36.17 gives a typical session. Typical file types include:

- mc68020 demand paged executable.
- C program text.
- ASCII text.
- Empty.
- Archive random library.
- Symbolic link.
- Executable shell script.

36.3.4 du (disk usage)

The du command lists the size of a directory and its subdirectories. If no directory name is given the current directory is assumed. Two options are given next:

```
-a          all file sizes are listed
-s          summary only
```

💻 **Sample session 36.16**
```
% ls -al
-rw-rw----  4 bill   staff   1102 Jun  4 12:05 .temp
drwxr-xr-x  2 bill   staff     52 Jun  4 14:20 cprogs
r   r   r   1 fred   staff  10320 Jan 29 15:11 data_file
-rw-rw----  4 bill   staff    102 Jun  1 11:13 file.txt
-rw-rw----  4 bill   staff    102 Jun  1 11:13 file1.f
-rwx--x--x  1 root   staff     20 May 31  9:02 list
-rwxrwx---  4 bill   staff   9102 Feb  4  1988 runfile
dr-xr-xr-x  1 joe    staff    100 Jan 14 13:11 temp
% ls -l
drwxr-xr-x  2 bill   staff     52 Jun  4 14:20 cprogs
-r--r--r--  1 fred   staff  10320 Jan 29 15:11 data_file
-rw-rw----  4 bill   staff    102 Jun  1 11:13 file.txt
-rw-rw----  4 bill   staff    102 Jun  1 11:13 file1.f
-rwx--x--x  1 root   staff     20 May 31  9:02 list
-rwxrwx---  4 bill   staff   9102 Feb  4  1988 runfile
dr-xr-xr-x  1 joe    staff    100 Jan 14 13:11 temp
% ls -r
    temp  runfile  list  file1.f  file.txt  data_file
    cprogs
% ls -1
    cprogs
    data_file
    file.txt
    file1.f
    list
    runfile
    temp
% ls -t
    cprogs  file.txt  file1.f  list  runfile  data_file
    temp
% ls -F
    cprogs/  data_file  file.txt  file1.f  list*  runfile*
    temp/
```

💻 **Sample session 36.17**
```
% file *
    prog1.c:  C program text
    test :    executable shell script
    fred_dir: symbolic link
    docs:     ascii text
```

36.3.5 cat (concatenate and display)

The cat command concatenates and displays the given files to the standard output, which is normally the screen (although this output can be changed using the redirection symbol). For example, to list a file to the screen:

💻 **Sample session 36.18**
```
% cat file.c
    This is the contents of file.c
```

To concatenate two files together (`file1.txt` and `file2.txt`) and put them into a file called `file3.txt`:

💻 **Sample session 36.19**
```
% cat file1.txt file2.txt  > file3.txt
```

If no filename is given then the input is taken from the standard input, normally the keyboard. If a redirect symbol is used then this input will be sent to the given file. The end of the input is given by a ^D (a control-D). For example:

💻 **Sample session 36.20**
```
% cat > newfile.txt
   Mary did not have a little lamb
   She had a fox instead
   ^D
% cat newfile.txt
   Mary did not have a little lamb
   She had a fox instead
```

36.3.6 compress, uncompress (compress and expand files)

The compress command uses the adaptive Lempel-Ziv coding to reduce the size of a file. A .Z is added onto the filename so that it can be identified. Sample session 36.21 shows an example.

💻 **Sample session 36.21**
```
% ls -al
-rw-rw----  4 bill   staff   102 Jun  1 11:13 file.c
-rw-rw----  4 bill   staff  1102 Jun  1 11:15 file.o
-rw-rw----  3 bill   staff   102 Jun  1 11:13 file1.f
-rw-rw----  4 bill   staff   102 Jun  1 11:13 file1.o
-rwxrw----  4 bill   staff 10010 Mar  2 15:23 runfile
% compress *
% ls -al
-rw-rw----  4 bill   staff    62 Jun  1 11:13 file.c.Z
-rw-rw----  4 bill   staff   542 Jun  1 11:15 file.o.Z
-rw-rw----  3 bill   staff    50 Jun  1 11:13 file1.f.Z
-rw-rw----  4 bill   staff    50 Jun  1 11:13 file1.o.Z
-rwxrw----  4 bill   staff  5005 Mar  2 15:23 runfile.Z
```

The contents of the compressed files are in a coded form so that they cannot be viewed by a text editor. The uncompress command can be used to uncompress a compress file. Only files with the extension .Z can be uncompressed.

💻 **Sample session 36.22**
```
%  uncompress *
```

36.3.7 chmod (change mode)

Files in UNIX have various attributes; these attributes can be set or reset with the chmod command. The attributes are:

Owner	Group	Public
r w x	r w x	r w x

The owner is the person who created the file and the group is a collection of several users, such as research, development, production, admin groups, and so on. The public is for everyone else on the system.

The r attribute stands for read privilege and if it is set then the file can be read (that is, listed or copied) by that type of user. The w attribute stands for write privilege and if it is set then the file can be written to (that is, listed, copied or modified) by that type of user. The x attribute stands for execute privilege and if it is set then the file can be executed (that is, it can be run) by that type of user.

For example -rw-r--r-- is a file that the owner can read or write but the group and the public can only read this file. Another example is -r-x--x--x; with these attributes the owner can only read the file. No-one else can read the file. No-one can change the contents of the file. Everyone can execute the file. The ls -al listing gives the file attributes.

💻　**Sample session 36.23**
```
% ls -al
-rw-rw---- 4 bill staff  1102 Jun  4 12:05 Run_prog
drwxr-xr-x 2 bill staff    52 Jun  4 14:20 cprogs
-r--r--r-- 1 fred staff 10320 Jan 29 15:11 data_file
-rw-rw---- 4 bill staff   102 Jun  1 11:13 file.txt
-rw-rw---- 4 bill staff   102 Jun  1 11:13 file1.f
-rwx--x--x 1 root staff    20 May 31  9:02 list
-rwxrwx--- 4 bill staff  9102 Feb  4  1988 runfile
```

There are several ways to change the attributes. The general format is given below:

chmod *<permission>* *<file>*

The permission can be set by using the octal system. If an attribute exists a 1 is set, if not it is set to a 0. For example, rw-r--r-- translates to 110 100 100, which is 644 in octal. For example:

to set to	rwx--x---	use 710
to set to	r-x------	use 500
to set to	rwxrwxrwx	use 777
to set to	--x------	use 100
to set to	rw-rw-rw-	use 666

To make the file file.txt into `rw-rw-r--` :

📟 **Sample session 36.24**
```
%  chmod 664  file.txt
```

To make the file `Run_prog` into `--x--x---` :

📟 **Sample session 36.25**
```
% chmod 110  Run_prog
% ls -al
---x--x--- 4 bill staff  1102 Jun  4 12:05 Run_prog
drw-r--r-- 2 bill staff    52 Jun  4 14:20 cprogs
-r--r--r-- 1 fred staff 10320 Jan 29 15:11 data_file
-rw-rw-r-- 4 bill staff   102 Jun  1 11:13 file.txt
-rw-rw---- 4 bill staff   102 Jun  1 11:13 file1.f
-rwx--x--x 1 root staff    20 May 31  9:02 list
-rwxrwx--- 4 bill staff  9102 Feb  4 1988 runfile
dr--r--r-- 1 joe  staff   100 Jan 14 13:11 temp
```

The other method used is symbolic notation, these are listed next:

u user permission   g group permission
o others (public) permission   a all of user, group and other permissions
= assign a permission   + add a permission
- take away permission   r read attribute
w write attribute   x execute attribute

For example, to change the file `Run_list` so that it can be executed by all use:

📟 **Sample session 36.26**
```
% chmod +x  Run_list
```

To take away read privilege for all files with the .c extension for the group use:

📟 **Sample session 36.27**
```
% chmod g-r *.c
```

36.3.8 df (disk space)

The df command allows you to list the usage of the disk and all other mounted disk drives. Sample session 36.28 gives some examples.

📟 **Sample session 36.28**
```
% df
Filesystem kbytes  used   avail   capacity  Mounted on
/dev/nst0  200000  50003  159997  25%       /
/dev/nst1  5000    100    4900    2%        /temp
```

36.3.9 diff (differences between files)

The `diff` command shows the difference between two files or two directories.

$$\text{diff } \textit{file1 file2}$$

There are various options used and these are summarized next:

-i   ignores the case of letters (such as 'b' is same as 'B').

-w   ignore all blanks (such as fred = 16.2 is same as fred=16.2).

Some examples are next. In the output listing the < character refers to the first file given and the > character refers to the second file given. A c refers to a change, a d to a line deleted and an a refers to text that has been appended. The line numbers of the first file always appear first.

🖳 **Sample session 36.29**

```
%  cat oldfile
    This is the contents of the old
    ***
    file. It will be modified and
    a diff will be done.
% cat newfile
    This is the contents of the new
    file. It will be modified and
    a diff will be DONE.
% diff -i oldfile  newfile
    1c1
    < This is the contents of the old
    ---
    > This is the contents of the new
    2d1
    < ***
```

36.3.10 mv (move files)

The mv command moves files or directories around the file system. The standard formats are:

$$\text{mv } [\textit{-i}] \ [\textit{-f}] \ \textit{filename1 filename2}$$
$$\text{mv } [\textit{-i}] \ [\textit{-f}] \ \textit{directory1 directory2}$$
$$\text{mv } [\textit{-i}] \ [\textit{-f}] \ \textit{filename directory}$$

There are three different permutations. These are:

- Moving a file into another file (like renaming).
- Moving a directory into another directory (like renaming).
- Moving a file into another directory.

The options that can be used are:

-i   files are moved interactively. When a file is to be moved the system will prompt the user as to whether he wants to move the file. If the answer is 'y' the file will be moved, else it will not be moved.

-f   force mode; override any restrictions.

Sample session 36.30 shows examples.

🖥 **Sample session 36.30**
```
% ls
   fortran  prog1.c  prog2.c  prog3.c  prog.f
% mv prog.f fortran
% ls
   fortran  prog1.c  prog2.c  prog3.c
% ls fortan
   prog.f
% mv fortran fortran_new
% ls
   fortran_new  prog1.c  prog2.c  prog3.c
```

36.3.11 more (page a file)

The more command allows the user to print one page of text at a time to the standard output. It pauses at the end of the page with the prompt '--More--' at the bottom of the page. Sample session 36.31 shows some examples.

🖥 **Sample session 36.31**
```
% more doc.txt
   fsdfsd dfsfs ddfsdfs d
   plpfd fdfsfdf fdfsfpl
      etc
      : :
   --more--
   dfsfsdf dfsfdf dfsffgf
   dfsdf fdfdfhgf
% cat file | more
   fsdf dfd fghfg fgfg
   lk;lk;l fdf poper
%
```

36.3.12 rm (remove files or directories)

The rm command removes files or directories. To remove a directory which is empty the rmdir command can be used. There are various options, such as:

-f   force mode; remove files without asking questions
-r   recursive mode; delete content of a directory, its subdirectories and the directory itself;
-i   interactive mode; asks question as to whether each file is to be deleted.

Sample session 36.32 gives some examples.

💻 **Sample session 36.32**
```
% ls
   fortran  pascal  text.1  text.2  text.3
% ls fortran
   progs1  progs2
% rm -r fortran/progs2
% ls fortran
   progs1
% rm text.*
% ls
   fortran  pascal
```

36.3.13 ln (make links)

The ln command makes a soft link to a file or directory. When the
linkname is used the system will go to the place indicated by the link. The
general format is:

$$\text{ln } \textit{-s filename [linkname]}$$

Sample session 36.33 shows some example.

💻 **Sample session 36.33**
```
% ls
  fred1  fred2  fred3
% ln -s /usr/staff/bill/prog.txt prog
% ls
  fred1  fred2  fred3  program
% ls -al
drw-r--r-- 2 bill staff  52 Jun  4 14:20 fred1
dr--r--r-- 1 fred staff  10 Jan 29 15:11 fred2
drw-rw---- 4 bill staff 102 Jun  1 11:13 fred3
lrw-rw---- 4 bill staff 102 Jun  4 13:13 prog->/usr/bill/prog.txt
% cat prog
    This is the contents of the
    prog.txt file.
```

36.3.14 find (find file)

The find command searches recursively through a directory structure to
find files that match certain criteria. It uses a pathname from where to start
seaching; this is the first argument given after find. The name of the file is
specified after the -name argument and if the user wants the files found
printed to the standard output the -print is specified at the end. Sample
session 36.34 gives an example of finding a file called fred.f, starting
from the current directory.

💻 **Sample session 36.34**
```
% find . -name fred.f -print
   dir1/fred.f
   fortran/progs/fred.f
```

Sample session 36.35 shows a search of the file pas swd, starting from the top-level directory.

```
% find / -name passwd -passwd
    /etc/passwd
```

The wild-card character can be used in the name but this must be inserted in inverted commas (" "). Sample session 36.36 gives an example of search for all C files starting with the /usr/staff/bill directory.

```
% find /usr/staff/bill -name "*.c" -print
    /usr/staff/bill/prog1.c
    /usr/staff/bill/cprogs/prog2.c
    /usr/staff/bill/cprogs/prog3.c
```

Other extensions can be used such as -atime which defines the time of last access. The argument following -atime is the number of days since it has been accessed. Sample session 36.37 gives an example of searching for all .o files that have not been used within 10 days.

```
% find . -name "*.o" -atime +10 -print
```

36.3.15 grep (search a file for a pattern)

The grep command will search through given files looking for a given string pattern. There are various options which can be used, a summary of these is given below:

-v   print lines that do not match;
-x   display only lines that match exactly;
-c   display count of matching lines;
-i   ignore case.

Sample session 36.38 gives some examples and the standard format is:

grep [*-v*][*-c*][*-x*][*-i*] *expression* [*file*]

36.3.16 head (displays first few lines of a file)

The head command prints the top n lines of a file. The default number is 10 lines and Sample 2.39 gives examples. The format is as follows:

head *-n filename*

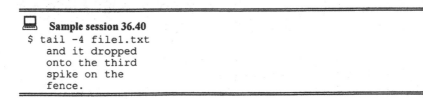
Sample session 36.38
```
%  grep function *.c
      prog1.c: function add(a,b)
      prog1.c: function subtract(a,b)
      prog3.c: function xxx

% grep -i function *.f
      man.f:          FUNCTION ON_LINE
      man.f:C  This is the function that prints

% grep -v fred listnames
      bert baxter
      sim pointer
      al gutter
```

Sample session 36.39
```
% head -3 diary.txt
      June 5th 1989
      Dear Diary,
      Today I got my head stuck inside a

% head -2 *.doc
    ==>first.doc<==
      This is the first
      document
      ==>second.doc<==
      And this is the
      second document.
%
```

36.3.17 tail (display last part of file)

The tail command displays the last part of a file, where the first argument defines the number of lines to be displayed. For example, to display the last 4 lines of the file file1.txt:

Sample session 36.40
```
$ tail -4 file1.txt
    and it dropped
    onto the third
    spike on the
    fence.
```

36.3.18 wc (word count)

The wc utility counts the words, characters and lines in a file. If several files are given then the sum total of the files will be given. There are three options that can be used; these are c, w and 1, which stand for characters, words and lines.

wc -*options filename*

If no filename is given then the keyboard is taken as the input and a control-D ends the file input. Examples are:

⌨ **Sample session 36.41**

```
%  cat file1
      This is the contents
      of the file to be
      used as an example.
% wc -l file1
      3 file1
% wc -lc file1
      3   46   file1
% wc -w file1
      13  file1
% wc file*
      3 13   46   file1
      5 32  103   file2
     10 44  294   file4
     18 89  433   total
```

36.4 Other commands

36.4.1 ar (archive and library maintainer)

The `ar` command is used to create and maintain libraries and archives. A library is a file which contains compiled programs. These files are called object modules and are self-contained modules. The object modules have the extension .o. The libraries are used in the linking process. There are certain standard libraries on the system; these include the standard maths library and standard input/output library (normally found in /usr/lib).

>   `ar` *options filename*

There are various options used with `ar`; these are given in Table 36.1. For example, to create an archive

```
% ar cr newarc xxx.c
```

Then to list the archive:

```
% ar tv newarc
   xxx.c
```

Table 36.1   `ar` options

Option	Description
d	delete the named files from the archive
r	replace the named files from archive
t	list the contents of an archive
x	extract a file from an archive
c	create an archive
v	extended listing of file

and to extract a file from the archive:

```
% ar xv newarc xxx.c
   x- xxx.c
```

Then adding all files with .c extension:

```
% ar rv newarc *.c
   a- prog1.c
   a- prog2.c
   a- zzz.c
```

and listing the archive again:

```
% ar tv newarc
   a- prog1.c
   a- prog2.c
   a- xxx.c
   a- zzz.c
```

To extract a file:

```
% ar xv newarc zzz.c
   x- zzz.c
```

and to create a library with .o object modules:

```
% ar cva Newlib *.o
   a- prog1.o
   a- prog2.o
   a- xxx.o
   a- zzz.o
```

Archives are useful for keeping many files together in the one file. This makes them useful for printing, storing on tape, and so on. These are standard libraries on the UNIX system, such as:

```
/lib/libc.a        which is the standard C library
/usr/lib/libF77.a  which is the standard FORTRAN library
```

36.4.2 banner (make posters)

The banner command prints the words given on the screen in large letters. A sample session is shown in Sample session 36.42. The maximum letters on one line is ten. Its standard form is:

banner *line_of_text*

36.4.3 cal (calendar)

The cal command displays a calendar for a given year and/or month. The year may vary from 1 to 9999 and if no month or year is given the current month will be printed. Sample session 36.43 shows an example of displaying the current month and Sample session 36.44 displays a whole year.

💻 **Sample session 36.42**
```
% banner hello
   *   *   ****   *        *        ****
   *   *   *      *        *        *  *
   ****  ***    *        *        *  *
   *   *   *      *        *        *  *
   *   *   ****   ****   ****   ****
```

💻 **Sample session 36.43**
```
% cal 3 1989
      March 1989
    M  T  W  T  F  S  S
             1  2  3  4  5
    6  7  8  9 10 11 12
   13 14 15 16 17 18 19
   20 21 22 23 24 25 26
   27 28 29 30 31
```

💻 **Sample session 36.44**
```
% cal 1966
    : :
```

36.4.4 chpass (change password)

The chpass command is used to change the current login password. Sample session 36.45 shows an example.

💻 **Sample session 36.45**
```
$ chpass
 Enter old password: scooby
 Enter new password: scrappy
 Re-enter new password: scrappy
 Password modified.
```

36.4.5 date (display or set date)

The date command either gives the current date and time or can be used to set the current date and time (although only the superuser can do this). If no arguments are given then the current date and time is printed. If a valid argument is given then the current time is changed. The general format is given below:

$$\text{date } [yymmddhhmm[.ss]]$$

where *yy* is the year, *mm* is the month, *dd* is the day of month, *hh* is the hour

(24 hour clock), *mm* is the minute and *ss* is the seconds (optional).

It is possible to leave out various field such as the year and the seconds. Sample session 36.46 gives some examples.

Sample session 36.46
```
% date
   Mon Jun 4 09:24:01 PST 1989
% date   06050133
% date
   Tue Jun 5 01:33:10 PST 1989
```

36.4.6 echo (echo arguments)

The echo command writes its argument to the standard output. A C-like standard is used for printing. Sample session 36.47 shows a few examples and the characters used are given below:

\b	backspace	\c	print line without new-line
\f	form-feed	\n	new-line
\r	carriage return	\t	tab
\v	vertical tab	\\	backslash
\0xx	8-bit ASCII octal code		

Sample session 36.47
```
% echo "Mary had \na littla\be lamb"
   Mary had
   a little lamb
% echo "Time to logout" | mail fred
% echo "Hello fred" > /dev/tty2
```

36.4.7 finger (display information on users)

The finger command gives information on a given user. This information includes:

- Login name.
- Full name.
- Terminal name.
- Idle time.
- Login time.
- Office location.
- Phone number, if known.

The files used are:

/etc/utmp for information about who is logged in;
/etc/passwd for user names;

36.4.8 ftp (file transfer program)

The ftp utility is the standard File Transfer Protocol. It is used to transfer files to and from a remote network. The format is:

> ftp *hostname*

36.4.9 lpr (print file)

The lpr command spools a file to the printer.

36.4.10 mail (interactive mail)

The mail utility is used to send and receive messages from/to other users. To send mail the user's name is simply specified after the program name. For example, to send mail to a user fred:

Sample session 36.48
```
% mail fred
    This is a message to you
    fred. It is time for a coffee
    ^D
```

Whenever the control-D (^D) keystroke is used then the message is sent to a specified recipient. Other users can be sent the same mail by simply giving their user name as one of the arguments, such as:

Sample session 36.49
```
% mail fred jack bill farquar
    Hello to you
    ^D
%
```

To send a file which has mail the redirect symbol (<) is used, such as:

Sample session 36.50
```
% mail fred tom < todays_message
%
```

When reading mail the system will typically tell the user when there is mail waiting to be read, such as:

Sample session 36.51
```
% You have mail
```

Then to read the mail the mail program is run, such as:

Then to read the mail the mail program is run, such as:

📟 **Sample session 36.52**

```
% mail
   From fred Tue Jun 4 11:12:00 1989
   I have had enough of this computer, HELP***
   ?
```

The mail response with a question mark (?) is used to indicate that the mail program is waiting on the user to enter a command. Possible options are:

?	display help information
q	quit
x	exit without changing mail
p	print
s *file*	save to file
d	delete mail
+	next (no delete)
m user	mail user.

36.4.11 passwd (change login password)

The passwd command is used to change the current password. Each user's password is stored in a coded form in /etc/passwd. Each new password's must be at least 5 characters long if they include upper and lower-case, or 6 if they are monocase. Sample session 36.53 shows an example.

📟 **Sample session 36.53**

```
% passwd
   New password : Freddy
   Retype new password: Freddy
```

36.4.12 pwd (present working directory)

The pwd command prints the pathname of the current working directory. Examples are:

📟 **Sample session 36.54**

```
% cd /usr/staff/fred
% pwd
   /usr/staff/fred
% cd fortran
% pwd
   /usr/staff/fred/fortran
% cd ..
% pwd
   /usr/staff/fred
% cd ../graham
% pwd
   /usr/staff/graham
%
```

36.4.13 rlogin (remote login)

The rlogin command allows the user to log into a remote machine. The format is:

> `rlogin` *rhost*

36.4.14 sleep (suspend execution for an interval)

The `sleep` command is used to suspend to command for a given time. The standard format is:

> `sleep` *time*

where the *time* is in seconds. Sample session 36.55 shows an example.

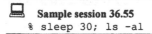 **Sample session 36.55**
```
% sleep 30; ls -al
```

36.4.15 spell (find spelling errors)

The spell utility finds spelling errors. The UNIX dictionary resides in the file `/usr/dict/words`. The format is given below.

> `spell` *option filename*

There are various *options* used:

-b British spelling (such as colour instead of color).

-v verbose, full listing.

-x lists every plausible stem for doubtful words.

Sample session 36.56 shows an example.

Sample session 36.56
```
% spell -b file1
    expact
    conder
%
```

36.4.16 stty (set terminal options)

The stty command sets up the terminal characteristics. There are two main types of options; these are toggles and settings. The standard format is given below.

> `stty` *option*

Various settings are:

`even`	allow even parity
`odd`	allow odd parity
`echo`	echo every character to screen
`lcase`	convert capital letters to lowercase
`tabs`	preserve tabs
`ek`	set erase and shell characters to # and @
`erase c`	set erase character to c (usually B/S)
`kill c`	set kill character to c (usually @)
`300, 1200,` and so on	set baud rate

Sample session 36.57 gives examples.

🖥 **Sample session 36.57**
```
% stty
   speed 9600 baud
   erase = '#'  kill = '@'

% stty erase ^H

% stty kill ^F
```

36.4.17 tar (tape archiver)

The `tar` utility allows files to be saved, recovered or listed from a tape drive. Its standard form is:

> `tar` *options filenames*

Options used are:

`c`	to create tape archive
`t`	to list tape archive
`r`	to recover tape archive
`v`	complete listing of file names

Sample session 36.58 shows examples.

🖥 **Sample session 36.58**
```
% tar cv /usr/staff/fred
% tar tv
   /usr/staff/fred/doc/doc1
   /usr/staff/fred/doc/doc2
   /usr/staff/fred/doc/doc3
   /usr/staff/fred/prog/prog1.c
   /usr/staff/fred/prog/prog2.c
% tar xv /usr/staff/fred/doc/doc2
```

36.4.18 who (users logged in)

The who command lists all users on the system and which terminal they are using. This information is found in the /etc/utmp file. If the arguments given are am i (that is, who am i) then the system prints details about the user. For example:

💻 **Sample session 36.59**
```
% who
    fred   tty2   Jun 5 10:10
    bill   tty3   Jun 5 13:05
% who am i
    bill tty3 Jun 5 13:05
%
```

The information given is the user name, terminal logged in on and date that the user logged in.

36.4.19 write (sends messages to users)

The write utility sends a message to a specified user. A control-D ends the message input and sends it. Examples are:

💻 **Sample session 36.60**
```
% who
    fred   tty2   Jun 5 10:10
    bill   tty3   Jun 5 13:05

% write fred
    Time for a break
    I think
    ^D
%
```

36.5 Exercises

36.5.1 Login in the UNIX system and determine the current processes that the user is currently running.

36.5.2 Determine all the current processes that are running and the user which initiated the process. Identify the status of the processes.

36.5.3 Use the cat > newfile command and then the Ctrl-Z keystroke to suspend the process. Show that the process has been suspended.

36.5.4 Kill the process which has just been suspended.

36.5.5 Using the at command, run an ls -l at a time one minute from the current time.

36.5.6 From the current directory, determine the file types of the files.

36.5.7 From the current directory, determine the size of all the directories.

36.5.8 Determine the present working directory.

36.5.9 Determine the present date.

36.5.10 List the current directory so that directories have a trailing slash.

36.5.11 Recursively list the current directory, showing all subdirectories.

36.5.12 Determine the users who are currently logged into the computer.

36.5.13 If there is a user logged into the computer then send an electronic message to them.

36.5.14 Using `stty`, determine the terminal characteristics.

36.5.15 For the top-level directory, find all the C program files.

37 Editing and Text Processing

37.1 Introduction

UNIX has several command programs which can be used to edit and process text. These are defined in Table 37.1.

Table 37.1 Editing and text processing programs

Text editors	Text processing	Text manipulation
vi , the visual editor	nroff, format document	grep, looks for a pattern
ex, line editor	troff, print formated document	sort, sorts lines of text
ed, line editor	tbl, formatting tables	spell, finds spelling errors
sed, stream text editor	pic, graphics language for typesetting	tr, translates characters
awk, pattern scanning and processing	eqn, typesetting mathematics	wc, counts characters, words and lines in a file.

37.2 Visual editor

Most UNIX systems have their own text editor which is selected from the Graphical User Interface. The vi editor (pronounced *vee-eye*) is a text-based editor and is standard on all UNIX systems. It is not the easiest editor to use but it is available on all UNIX-based system. The editor is called with:

> vi *filename*

37.2.1 Making changes

The vi editor involves using special characters to identify editing commands. The user enters text only when the editor is in insert mode. All characters entered this mode are added to the file. When not in insert mode the characters are interpreted as editor commands. For example, an 'x' keystroke tells the editor to delete a single character and a 'd' followed by a 'w' deletes a word.

Insert mode

To insert text the user must be in the insert mode. A summary of the main insertion and delete options is given below. To insert text, place the cursor on the character you want to insert before then press 'i'; this will put you into insert mode. The text can then be entered with as many lines as required. To exit from insert mode, the ESC key is pressed, as illustrated in Figure 37.1. This then puts the editor back into normal mode. Other methods of insertion are similar, such as an 'a' puts inserted text after the current cursor position. While 'O' starts a new line above the line the user is currently on and 'o' puts it after the current line. The insert mode characters are listed in Table 37.2.

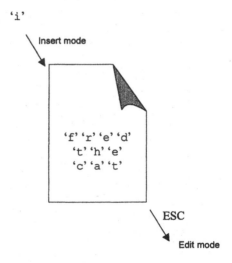

Figure 37.1 Edit and insert mode

37.2.2 Deleting text

Deleting text is performed in the edit mode. The various methods are:

- **Delete lines**. A single line is deleted with two 'd' characters, that is 'd' 'd'. To delete several lines the number of lines to be deleted is entered first followed by two 'd' characters, such as 6 'd' 'd' deletes 6 lines.
- **Delete words**. A single word (a sequence of characters delimited by spaces) is deleted with a 'd' followed by 'w'. As with delete lines, several words are deleted by entering the number of words to delete followed by the 'd' and 'w' characters.

Table 37.2 Editing and text processing programs

Option	Description
i	Enter insert text mode before cursor. The ESC key is used to exit from insert mode.
O	Enter insert mode and create a new text line above line you are on.
o	Enter insert mode below current line.
a	Enter insert mode after cursor (that is, append).
dd	Delete whole line.
x	Delete one character.
dw	Delete word.
db	Delete word before.
ns	Substitutes n characters and enters into insert mode.

- **Delete characters**. A single character is deleted with the 'x' character. Again, if this character is preceded by a number then a number of characters are deleted. For example, 5 followed by 'x' deletes 5 characters.

37.2.3 Substituting

The 's' character identifies a substitution of characters. First the number of characters to be substituted is entered and the editor marks the end of the text to be substituted with a $. The editor is then put into insert mode and the new text is entered followed by the ESC key. For example, to change fred to tom the cursor is moved to the f of fred and then the user types 4stom <ESC>.

37.2.4 Cut and paste (yank and put)

The 'y' 'y' keystroke is used to copy a line of text (yank) and the 'p' keystroke (put) is then used to put this text to a defined part of a file. This is equivalent to copy then paste operation. If the yank option is preceded by a number then that number of lines are yanked. For example 7 'y' 'y' stores 7 lines of text. Note that the 'p' character puts lines after the present line and 'P' after the present line.

37.2.5 Undoing

The vi editor allows the last change to be undone. The keystroke for the last operation to be undone is 'u' and to restore a whole line the 'U' keystroke is used.

37.2.6 Moving around

There are various methods of moving around these are outlined in Table 37.3; and are defined as:

- **Moving cursor**. The cursor can be used with the arrow key (←,→, and so on) if they are supported by the terminal being used. If the keyboard has no arrow keys (or they are not supported) then the characters used are 'h' (left), 'j' (down), 'k' (up) and 'l' (right) keys
- **Moving by page**. The page can be moved back and forward using the PgUp and PgDn keys, if they are supported. Else, to move backwards by one page the ^b (press Cntrl and b at the same time) and to move one page forward then ^f. Λ ^u and ^d allows the screen to be scrolled up or down, respectively.
- **Going to a line**. In many situations the user requires to go to a specific line number (especially when compiling a program). To do this the 'G' character is followed by the line number. For example, 'G' 1234 goes to line number 1234.

Table 37.3 Editing and text processing programs

Option	Description
l or →	move cursor forward one cursor position
h or ←	move cursor back one position
^b	backwards one page
^f	forward one page
^d	scroll down file
^u	scroll up file
k or ↓	next line
j or ↑	previous line
G n	go to line number n
W	forward one word
b	back one word
e	end of current word
w	word after this word

37.2.7 Searching for strings

The '/' and '?' characters are used to find a string. A '/' searches forwards for the next occurrence of a string, whereas, '?' searches back for the occurrence of the string. Thus:

```
/ string        looks for next occurrence of string

? string        looks for previous occurrence of string
```

37.2.8 Getting out of the editor and saving

There are various options that can be used (depending on what you require). These are specified in Table 37.4. The user's changes are not stored to the file until the user uses either 'Z' 'Z', ':' 'w' 'q' or ':' 'w'. These differ in that the ':' 'w' 'q' and 'Z' 'Z' sequence causes the changes to be saved to the file and quit from the editor, whereas ':' 'w' saves changes and the editor stays active.

A user can abandon the changes that have been made since the last save by using either ':' 'q' '!' or ':' 'e' '!'.

Table 37.4   Methods of exiting

Option	Description
Z Z	Exits from editor and saves all changes.
:wq <CR>	Same as above (write and quit).
:w <CR>	Write changes but do not quit.
:w *filename* <CR>	Write changes to filename, no quit.
:q! <CR>	Quit from editor, no changes.
:e! <CR>	Re-edit, discarding changes.

Note that <CR> stands for pressing the return key.

37.2.9   Reading-in and writing-to files

The command to read-in files is :r *filename* and the command to write the file to a specified file is :w *filename*. For example, :r fred reads in a file called fred and :w fred2.txt writes to a file called fred2.txt.

37.2.10   Useful advanced editing facilities

There are many powerful text editing facilities available, some useful examples are:

```
:s/name1/name2/g 10
```

this changes all occurrences of name1 to name2 for 10 lines:

```
:5,18w fred1.txt
```

writes lines 5 to 18 to file called fred.txt.

37.3   Example of text editing

This section contains a practical example of text editing. Start the editor with:

```
vi lion.txt
```

then when the editor has started enter 'i' to put the editor into insert mode. Enter the text given below and make the changes required. If mistakes are made as the text is being typed-in then the back-space key will delete the previous character, else wait until all the text has been entered and then make the required modifications. The text is:

```
            THE LOIN AND THE UNICORN
The next moment soldiers came running through the wood,
at first in twos and  threes, then ten or twenty
together, and at last in such crowds that they   seemed
to fill the whole forest. Alice got behind a tree, for
fear of being  run over, and watched them go by.
    She thought that in all her life she had never seen
soldiers so uncertain on   their feet: they were always
tripping over something or other, and whenever  one went
down, several more always fell over him, so that the
ground was soon covered with little sheeps of men.
    Then  came  horses. Having  four  feet,  these managed
rather  better  that  the   foot-soldiers; but  even  they
stumbled and then; and it seemed as a regular   rule
that, whenever a horse stumbled, the rider fell of
instantly. The  confusion got worse every moment, and
Alice was very glad to get out of the  wood into an open
place,  where  she  found  the  White  King  seated  on  the
ground,   busily writing in his memorandum-book.
```

Once the text has been entered then make the following changes/operations:

(a) **Change LOIN in the title to LION.** This can be done in a number of ways, such as:

- Move cursor to the L of LOIN press 4x (or xxxx) and then press i to enter insert mode and enter correct word LION. Press ESC button to exit from insert mode.
- Move cursor to the L of LOIN press 4s (substitute 4 letters) the $ sign shows the end of the word to be substituted. You will be in insert mode automatically. Enter new text LION and press ESC to return to normal mode;
- Move cursor to the L of LION press R (overwrite mode and replace) this will automatically put you into insert mode. Next enter correct LION and press ESC.
- For advanced editing go to start of text then type: :s/LOIN/LION/g 100 <CR>. This substitutes globally LOIN for LION for 100 lines.

(b) **Copy the first 4 lines of the second paragraph and put these after the third paragraph.** To do this first yank (to get lines) and the put. First, position the cursor at any position on the line which begins the second paragraph. Next, get the lines by using 4yy (which gets 4 lines or use 4Y) then move the cursor to the last line and press p to put the lines below last line.

(c) **Delete these new lines.** This can be done by moving the cursor to the first line of the new lines and then using 4dd (or dddd) to delete these 4 new lines.

(d) **Search for the word sheeps and change this to heaps.** To find the word sheeps use /sheeps <CR>, the editor will then find the first occurrence after the current cursor position. If the word lies backwards from the cursor then use ?sheeps <CR>. Once found replace it with heaps.

(e) **Search for the word stumbled.** How many occurrences are there?

(f) **Write modified text to the file lion2.txt.** This is done by using :w lion2.txt <CR>.

(g) **Read this file in at the end of the file.** Move the cursor to the end of file then use :r lion2.txt. This will read the file in.

(h) **Delete these new lines.** Position the cursor on the first new line then use dd to delete a line. Repeat this deletion with the repeat option until all the required lines are deleted.

(i) **Find the word confusion and change it to unconfusion, then undo this press u (the undo option).**

(j) **Save the text.** Use :wq <CR>.

37.4 sed editor

The sed is a non-iterative stream editor. It is useful for editing large files and performing global edits. The following text can be entered and operated on in the examples given; call the file Walrus.

```
The sun was shining on the sea,
Shining with all her might
He did her very best to make
The billows smooth and bright,
And ther was odd, because it was
The middle of the night.
```

37.4.1 Global changing

To globally change *string1* to *string2* the following is used:

```
sed s/string1/string2/g filename
```

For example, if we use the example given previously to change his to her, the following can be used:

```
% sed s/his/her/g Walrus > Walrus1
% cat Walrus1

    The sun was shining on the sea,
    Shining with all her might
    He did her very best to make
    The billows smooth and bright,
    And ther was odd, because it was
    The middle of the night.
```

We can see that on the fifth line this has been replace by ther. To undo this the following could be used:

```
% sed s/ther/this/g Walrus1 >Walrus2
% cat Walrus2

    The sun was shining on the sea,
    Shining with all her might
    He did her very best to make
    The billows smooth and bright,
    And this was odd, because it was
    The middle of the night.
```

37.5 grep

The grep command searches through given files for a given string pattern. Its standard format is:

grep [-v][-c][-x][-i] *expression* [*file*]

where the *options* which can be used are:

-v   print lines that do not match
-x   display only lines that match exactly
-c   display count of matching lines
-i   ignore case

Examples are:

```
%  grep function *.c
    prog1.c: function add(a,b)
    prog1.c: function subtract(a,b)
    prog3.c: function xxx
% grep -i function *.f
    man.f:          FUNCTION ON_LINE
    man.f:C  This is the function that prints
% grep -v fred listnames
    bert baxter
    sim pointer
    al gutter
```

The grep is typically used with a pipe to search the output from a command. For example, to find every occurrence of the word set in the manual on the C shell:

```
% man csh | grep set
```

which is equivalent to:

```
% man csh > temp
% grep set temp
```

37.6 sort

The sort utility sorts lines in the given files and writes the output to the standard output. It always uses the characters to sort the text and ignores the values of numerical values. The standard format is:

sort +*pos1* –*pos2 filename*

As an example, create the following file and call it league. Its function is to list a number of individuals, the number of games they have played and the number of points scored. Each field is separated by spaces or tabs.

```
Bert Ripon        3   3
Fred Green        6   11
James Fortie      5   3
Simon Ripon       1   1
Alan Doo          6   4
Frank Doobie      3   2
Charles Zebidee   4   6
Paul McDonald     7   5
```

To sort the first field, that is the first name, then the sort is used with no options.

```
% sort league
Alan Doo          6   4
Bert Ripon        3   3
Charles Zebidee   4   6
Frank Doobie      3   2
Fred Green        6   11
James Fortie      5   3
Paul McDonald     7   5
Simon Ripon       1   1
```

Next, the surnames can be sorted by adding the +1 option, which causes the sort to skip to the first field.

```
% sort +1 league
  Alan Doo          6    4
  Frank Doobie      3    2
  Fred Green        6   11
  James Fortie      5    3
  Paul McDonald     7    5
  Simon Ripon       1    1
  Bert Ripon        3    3
  Charles Zebidee   4    6
```

This is an incorrect sort as Simon Ripon comes before Bert Ripon. The reason that it has not sorted these names properly is that it has tested only the surname. To test the first field after the second the −2 option is added. Thus the sort +1 −2 league command is interpreted as 'skip the first field and then sort but stop sorting after the second field and go back to the start of the line'.

```
% sort +1 -2 league
  Alan Doo          6    4
  Frank Doobie      3    2
  Fred Green        6   11
  James Fortie      5    3
  Paul McDonald     7    5
  Bert Ripon        3    3
  Simon Ripon       1    1
  Charles Zebidee   4    6
```

To sort the third field (that is, the games played) then the +2n option is used. Thus sort +2n +1 League is interpreted as 'skip two fields then sort numerically, then they are sorted by skipping the first field'.

```
% sort +2n +1 League
  Simon Ripon       1    1
  Frank Doobie      3    2
  Bert Ripon        3    3
  Charles Zebidee   4    6
  James Fortie      5    3
  Fred Green        6   11
  Alan Doo          6    4
  Paul McDonald     7    5
```

To sort in reverse the r option is used.

```
% sort +2nr +1 League
  Simon Ripon       1    1
  Bert Ripon        3    3
  Frank Doobie      3    2
  Charles Zebidee   4    6
  James Fortie      5    3
  Fred Green        6   11
  Alan Doo          6    4
  Paul McDonald     7    5
```

To sort with the fourth field (that is, points):

```
% sort +3nr +1 League
  Fred Green        6  11
  Charles Zebidee   4   6
  Paul McDonald     7   5
  Alan Doo          6   4
  James Fortie      5   3
  Bert Ripon        3   3
  Frank Doobie      3   2
  Simon Ripon       1   1
```

37.7 Exercises

37.7.1 Write a command using ls, | and the grep that will determine all files, in the current working directory, which contain the letter r.

37.7.2 Using who, | and grep, determine all the users on teletype terminals (that is, *ttyxx*).

37.7.3 Using the ls -al and grep, show any file which has the attributes rwx-r--r--.

37.7.4 Determine any usage of the string ">>" in the manual on csh and sh.

37.7.5 The following are lists of teams and their games played, goals for, goals against and points. Using the sort utility, determine:

    (i)   The list alphabetically of teams.
    (ii)   The teams sorted by decending goals for.
    (iii)   The teams sorted by ascending goals against.
    (iv)   The teams sorted by points then by goals for.

```
Hippo Town      4   2   3   3
Lion City       7  14  12   6
Tiger United    1   0   0   0
Horse Head      5  11   0   9
Hippo City      2   7   3   3
Fox Albion     10   2   2   0
Zebra United    5  12  17   3
```

37.7.6 Using sort and ls -al, list the files in the current directory in order of decreasing size.

38 Csh (C Shell)

38.1 Introduction

The C shell (csh) is one of the interfaces between the user and the UNIX operating system; others include:

- Bourne shell (sh).
- Korne shell (ksh).

The C shell includes a command interpreter and a high-level programming language. It has several advantages over the Bourne shell, these include:

- Variables are more versatile.
- It supports arrays of numbers and strings.
- Improved protection over accidental logout.
- Improved protection over accidental file overwrite.
- Alias mechanism.
- History mechanism to recall past events.

38.2 Entering the C Shell

The Bourne shell and the C shell can be used at any time but only one can be used at a time. To call-up the C shell:

```
%   csh
```

and the Bourne shell is called-up using sh. Normally, the shell at which a user is put into initially is setup in the /etc/passwd file. This file can only be change by the system administrator. To list the passwd file:

```
%   more /etc/passwd
```

this gives a paged list and the shells are given defined by either /bin/sh or /bin/csh. An excerpt from a passwd file is given next:

```
root:FDEc6.32:1:0:Super unser:/user:/bin/csh
fred:jt.06hLdiSDaA:2:4:Fred Bloggs:/user/fred:/bin/csh
fred2:jtY067SdiSFaA:3:4:Fred Smith:/user/fred2:/bin/csh
```

In this case, the passwd file has 3 users defined, these are root, fred and fred2. The encrypted password is given in the second field (between the first and second colon). The third field is a unique number that defines the user (in this case fred is 2 and fred2 is 3). The fourth field, in this case defines the group number. The fifth field is simply a comment field and, in this case, contains the user's names. In the next field the user's home directory is defined and the final field contains the initial UNIX shell (in this case it is the C shell).

When entering the C shell the .cshrc file is executed. This file sets up the C shell to the needs of the user and is found in the user's home directory. It can be found by typing ls -al in the user's home directory. It can be listed as follows:

```
% cd
% cat .cshrc
```

The .cshrc file will be discussed later.

38.3 Leaving the C shell

An annoying feature of the Bourne shell is that if the user presses the CONTROL-D keystroke by mistake then they will be logged-out. In the C shell if the ignoreeof variable is set then the CONTROL-D (^D) log out process is inactive. A user can exit from the C shell by typing exit. To set the ignoreeof variable just type:

```
%   set ignoreeof
```

This is usually done automatically when entering the shell. An example of entering and leaving the C shell is given below.

```
%   csh
%   ls
    fred1    fred2    fred3
%   exit
```

Note: If the ignoreeof variable is not set then ^D will exit the C shell.

38.4 History

The history mechanism maintains a list of previous command lines, which are also called events. To set the number of events the set history command is used. For example, to set the number of stored events to 10:

```
% set history = 10
```

and to recall the previous events the history command is typed:

```
% history
    3 cat fred1
    4 ls
    5 mkdir temp
    6 cd temp
    7 ls -al
    8 cd ..
    9 ls
   10 cat fred2
   11 vi fred2
   12 history
```

38.5 Reexecuting events

It is possible to reexecute any event in the history list. The reference to an event is made by starting the line with a !.

38.5.1 Event numbers

An event can be reexecuted using the event number. This is achieved by a ! followed by the event number.

```
% history
    1 ls
    2 cat temp
    3 vi temp
    4 history
% !1
   fred1   fred2   fred3
% !2
   this is the contents
   of the temp file.
```

A negative number after the exclamation mark refers to an event prior to the current event.

```
% history
    1 ls
    2 cat temp
    3 vi temp
    4 history
% !-2
   this is the contents
   of the temp file.
% history
    1 ls

    2 cat temp

    3 vi temp
    4 history
    5 cat temp
```

38.5.2 Reexecuting previous event

Two exclamation marks (! !) are used to reexecute the previous event. For example:

```
% ls -al
-rw-rw-r-- 2 fred    5 Mar 11 14:31 file1
% !!
-rw-rw-r-- 2 fred    5 Mar 11 14:31 file1
```

38.5.3 Events using text

To execute a previous event with text then an exclamation mark followed by text at the beginning of the event can be used. For example:

```
% cat temp
  This is the content of
  the temp file.
% ls
   fred1 fred2 fred3
% vi temp
% !ca
  This is the content of
  the temp file.
% !l
   fred1 fred2 fred3
```

38.5.4 Changing last event

Typically a user makes a mistake in a command line and wants to replace a part of the command line with another. To change a string of text in the last event then the caret (^) symbol is used. For example ^fred^temp reexecutes the last event but changes fred for temp. For example:

```
% cat fred
  file not found
% ^fred^temp
  This is the contents
  of the temp file.
% history
   1 cat fred
   2 cat temp
   3 history
```

38.5.5 Changing previous events

To modify a previous event it is possible to use the : s/old/new to change a given event. The event is specified by the methods given previously. For example:

```
% !13:s/temp/fred
   cat fred |lpr
% !cat:s/xxx/yyy
   more yyy
```

The first example changes event 13 but substitutes temp for fred. The second example executes the last event beginning with cat and substitutes yyy for xxx.

38.6 Alias

The alias mechanism allows a user to substitute a given command for a single word (or alias). The following are common aliases:

```
% alias h history
% alias dir ls -al
% alias w who
% alias ^L clear
% alias zap kill -9
```

The first example executes the history command whenever h is entered at the command line. The second executes an ls -al when dir is entered.

The user can list the currently defined aliases by simply typing alias, such as:

```
% alias
    h        history
    dir      ls -al
    w        who
% h
    5 dir
    6 cat file1
    7 vi file1
    8 h
% dir
    -rw-rw-r-- 2 fred    5 Mar 11 14:31 file1
    -rw-rw-r-- 2 fred    5 Mar 11 14:31 file2
```

The unalias command is used to remove an alias, such as:

```
% alias goodbye exit
% alias
    h        history
    dir      ls -al
    w        who
    goodbye  exit
% unalias goodbye
% alias
    h        history
    dir      ls -al
    w        who
```

38.7 Variables

The C shell only uses strings and does not use numbers. Thus, for example, 5 is interpreted as the character 5 and not as the number 5. To operate on numbers the @ is used.

38.7.1 Variable substitution

The commands that are used to manipulate variables are:

- `setenv` which operates on global variables (that is, system variables).
- `set` and `@` which operate on local variables to the current shell.

A variable is recognized by a preceding $ followed by the variable name.
Examples are $fruit, $numbers, and so on.

38.7.2 String variables

The C shell treats variables as strings. To declare a variable the `set` command is used. For example:

```
% set myname = Bill
% echo $myname
   Bill
% set
   argv()
   home       /usr/staff/eeea15
   Shell      /bin/csh
   myname     Bill
```

This will declare the variable `myname` which contains the string "`Bill`".
Note that a space on either side of the = is necessary.

```
% set chess1 = king
% set chess2 = $chess1 + pawn
% echo $chess1 and $chess2.
   king and king + pawn.
```

38.7.3 Arrays of strings

Before an array member is used it must be declared, that is, set. It is used to declare every member of the array.

```
% set people = ( fred bert frank farquar angus )
% echo $people
   fred bert frank farquar angus
% echo $people[4]
    farquar
% echo $people[1-3]
    fred bert frank
% set sysusers = ( '' '' '' '' '' '' '' )
% echo $sysusers
% set sysusers[4] = $people[2]
% echo $sysusers
   bert
```

38.7.4 Numerical variables

The `@` command is used to assign a numeric variable. The format of the `@` command is:

@ variable-name operator expression

The operator can be any of the C operators. These are:

Arithmetic:
%   remainder   / divide   + add   *   multiply   – subtract

Shift operators:
<<   shift left   >> shift right

Relationship:
> greater than   < less than   >= greater than or equal to
<= less than or equal to   ! = not equal to   == equal to

Bitwise:
& AND   ^ Exclusive OR | Inclusive OR

Logical:
&& AND | | OR

An example is given next:

```
%   @ number = 4
 %   echo $number
    4
 %   @ number = ( 5 * 2 )
 %   echo $number
   10
 %   @ test = ( $number > 8 )
 %   echo $test
    0
```

38.7.5 Files

Files can be tested for their status using the *–n* option, the standard form is:

    –n filename

where *n* can be:

    d   file is a directory
    e   file exists
    f   file is a plain file
    o   user owns the file
    r   user has read access to file
    w   user has write access to file

x user has execute access to file
z file is empty

For example

```
if (-z prog1.c) echo "File does not exist"
```

38.8 Special forms ($#, $? and ~)

The number of elements in an array is determined using the $# operator. The general format is:

```
$#variable-name
```

To determine if a variable has been declared the $? can be used. The general format is:

```
$?variable-name
```

The value returned will be zero if it has not been declared and 1 if it has been declared.

```
% set cars = ( fiat ford vauxhall nissan rover )
% echo The number of car types are $#cars
 The number of car types are 5
% echo $?cars
 1
% unset cars
% echo cars
% echo $?cars
 0
```

The ~ (tilde) operator is used to substitute for the full pathname of a given user. For example:

```
%  cd ~fred
%  cp ~eeea15/doc/*.txt .
%  rmdir ~bill/fortran
```

Thus ~fred stands for user fred's home directory, and so on.

38.9 Shell variables

These variables can be set by either the shell or the user. There are two basic types: the first is given values, whereas the second is either set or unset.

38.9.1 System variables that take a value

The standard variables that take on values are:

- $argv – The argv variable contains all the command line arguments. argv[0] holds the name of the calling program, argv[1] contains the first argument, and so on. For example, if the command line is:

  ```
  % find abc def xde
  ```

  Then the argv variables will be:

  ```
  argv[0] contains find    { name of program }
  argv[1] contains abc     { first argument}
  argv[2] contains def
  argv[3] contains xde
  ```

- $#argv –contains the number of command line arguments. In the previous example this will be 4.

- $cdpath – The variable is used by the cd command. When a cd command with a simple filename is given, then the system will search the current working directory. If the file is not in the current working directory then the system will search in the directories given by the cdpath variable. If a file is found in any of the search paths then the system changes to that directory. It is usually set in the .login file, for example:

  ```
  set cdpath = ( /usr/staff/fred/bin   /usr/staff/fred/docs )
  ```

- $cwd – full pathname of current directory.

- $history – defines the size of the history list. This value is usually set in the .cshrc file, such as:

  ```
  set history = 30
  echo $history
  ```

- $home – defines the home directory of the user.

  ```
  set home = /usr/staff/eeea15
  echo $home
  ```

- $mail – files where the shell checks for mail.

- $path – gives the directories in which the system searches for executables. Usually the search path is '.' , '/bin' and '/usr/bin'. For example:

```
set path = ( . /bin   /usr/bin )
echo $path
```

- $prompt – A string that is printed before each command is read. If a '!'
  appears in the string the current event number is substituted. The default
  prompt is a %. An example prompt setting is:

```
set prompt = " {`whoami`} $cwd: [\!] %"
```

which will print as a prompt:

```
{bill_b} /usr/staff/bill_b/docs: [15] %
```

if the current directory is /usr/staff/bill_b/docs, the current
event was 15 and the user's name is bill_b.

- $savehist – contains the number of history events that are stored
  when the user logs out. These events are stored in the ~/.history file.
  When the user next logs in then the current history list will be as the
  saved event history.

38.9.2 System variables are set

The standard variables that are set are:

- $filec – enables file name completion. If it is set then the filename is
  completed when the <ESC> key is pressed and the file is found.

```
set filec
```

```
% vi tem<ESC>
```

may become:

```
% vi temporary_file.ftn
```

The <Cntl-D> (^D) character is used to list any files which match the
current input. For example:

```
% cd FOR<control-D>
  FORTRAN FOR_DO FORever
% cd FOR
```

In this example the files which match the first three characters FOR are
listed.

- $ignoreeof – if set the shell ignores the <Cntrl-D> logout command

and `exit` is used to leave the shell. It is set with:

```
set ignoreeof
```

- `$noclobber` – when set there are restrictions on the redirection to files. If a > is given and the file exists it will not be destroyed.

```
set noclobber
```

38.10 Shell scripts

A shell script is a file that contains C shell commands. To execute a file the mode must be changed so that it can be executed. For example:

```
% cat Shell1
   #simple C shell
   echo This is a simple shell script
% chmod +x Shell1
% Shell1
   This is a simple shell script
```

Note that the first nonblank character in a C shell script must be a #. This tells the shell that it is a C shell script.

38.11 Control Structures

The C shell control statements are `if`, `foreach`, `while` and `switch`.

38.11.1 If

The general format of the if control structure is given below.

    `if` (*expression*) *statement*

An example of a C shell script using the `if` control structure is:

```
# simple routine to print number of command line
# arguments 0 to 2

if ($#argv == 0) echo "No arguments given"
if ($#argv == 1) echo "One argument"
if ($#argv > 1) echo "More than one argument"
```

If the script's filename were `argtest` then a typical run session may be:

```
% vi argtest
% chmod +x argtest
% argtest acb ddd xvf dddddd
  More than one argument
% argtest
  No arguments given
```

38.11.2 If-then else

There are three different types of if-then else structures:

Type 1	*Type 2*	*Type 3*

`if (expression) then`
`  commands`
`endif`
	`if (expression) then`
	`  commands`
	`else`
	`  commands`
	`endif`
		`if (expression) then`
		`  commands`
		`else if (expression) then`
		`  .`
		`  .`
		`  .`
		`endif`

An example of a Shell script using if-then is:

```
#  C shell script to determine a grade for a given
# mark
set grade
set mark = $argv[1]
#
if ( 0 <= $mark && $mark <= 50 ) then
   grade = fail
else if ( 50 < $mark && $mark <= 60 ) then
   set grade = C
else if ( 60 < $mark && $mark <= 70 ) then
   set grade = B
else if ( 70 < $mark && $mark <= 100 ) then
   set grade = A
else
   set grade = Invalid
endif
#
echo "Mark is $mark , the grade is $grade ."
```

If the file is called `grade` then a typical run session may be:

```
% vi grade
% chmod +x grade
% grade 55
  Mark is 55, the grade is C.
% grade -100
  Mark is -100, the grade is Invalid
```

38.11.3 Foreach

The `foreach` structure reads each argument of an argument list in turn. The standard format is as follows:

```
foreach loopindex (argument-list)
    commands
end
```

When the loop is first executed then *loopindex* will be assigned the first value in the *argument-list*. The commands are then executed and the loop returns to the start and the *loopindex* then takes on the second value in the *argument-list*, and so on.

38.11.4 While

The general format of the while loop is:

```
while (expression)
  commands
end
```

The following shell script is an example which determines the factorial of a number.

```
#
echo "C Shell program to determine factorial"
if ($#argv == 0) goto error
set fact = 1
set number = $argv[1]
set count = 1
#
while ( $count <= $number )
   @ fact *= $count
   echo "Count is " $count
   @ count++
end
#
echo "Answer is " $fact
exit
error:
echo "Incorrect arguments given"
echo "Format <fact number>"
echo "For example fact 4 will give 24"
```

38.11.5 Switch

The switch statement can choose between several different strings. The general format is:

```
switch (string)
   case pattern1:
     commands
   breaksw

   case pattern2:
     commands
   breaksw

   default:
     commands
   breaksw
endsw
```

An example of the switch structure which implements a basic calculator is given next:

```
# C shell script to act as a calculator

if ($#argv != 3) then
   echo "Incorrect arguments use <calc 2 - 3> for example"
```

```
else
  set result
  set value1 = $argv[1]
  set value2 = $argv[3]
  set operator = $argv[2]

  switch ($operator)

    case +:
     @ result = $value1 + $value2
     breaksw

    case -:
     @ result = $value1 - $value2
     breaksw

    case x:
     @ result = $value1 * $value2
     breaksw

    case /:
     @ result = $value1 / $value2
     breaksw

    default:
     echo "Operator not recognized use +, -, x or /"
     exit
     breaksw

  endsw
  echo "Answer is " $result
endif
```

38.12 Automatically executed C shell scripts (.login .cshrc)

When someone logs on to the system, or starts a C shell, three C shell scripts are run at various times.

38.12.1 The .login file

The .login file is executed when the user logs onto the system. Usually this file contains terminal and keyboard commands. An example of a typical .login file is shown below. In this case, the setenv command is used to set a variable for all shells, that is csh and sh. The terminal type setup is a vt100.

```
# .login file
echo "Terminal set up is a vt100"
echo "erase is ^H and kill is ^?"

setenv TERM vt100
stty erase ^H kill ^?
```

38.12.2 The .cshrc file

The .cshrc file is executed whenever a C shell is run. Again it is found in the home directory of the user. It is typically used to set up parameters used by the C shell, such as the path, prompt, and so on. A quick way of listing the .cshrc file wherever you are is:

```
% cat ~/.cshrc
```

A typical .cshrc file is:

```
# .cshrc file
set noclobber
set ignoreeof
set history=50
set savehist =50
set prompt = " ! %"
set path = (. /bin ~bill/bin /usr/bin /usr/ucb)

alias h history
alias m more
alias dir ls -al
```

38.12.3 The .logout file

The .logout file is automatically executed when the user logs out of the system. It typically contains commands which tidy up the user's work area and also to write messages.

38.13 Exercises

38.13.1 Using an alias, define a command named rename which is equivalent to:

```
mv -i file1 file2
```

38.13.2 Using the date command, set user prompt to give the date.

38.13.3 Write a C shell script that calculates the sum of numbers up-to and including the number given.

38.13.4 Write a C shell script named add_dot_C.which will add a .c to any given file name. For example:

```
add_dot_C fred
```

would cause the file fred to become fred.c.

38.13.5 The command for finding a given file on the system is:

```
find / -name filename -print
```

which searches from the top-level directory for the file called `filename`. For example, to find `prog1.c` the following can be used:

```
find / -name prog1.c -print
```

Write a C shell script which is called `findfile` in which the argument passed to it is the filename. For example,

```
findfile prog1.c
```

38.13.6 The stream editor `sed` can be used to replace a given word with another word. The format of this command is given below:

```
sed s/word_old/word_new/g filename
```

This changes `wordold` for `wordnew` globally in file called `filename`. Write a shell script called `replace` which gives three arguments. The first two are for the words and the third for the filename. For example, to replace fred for terry in the file name_file:

```
replace fred terry name_file
```

38.13.7 Write a C shell script which will give information about a given file. The information should include access rights, whether it is of zero length, file type, and so on. Name the shell script `testfile`.

A Java Classes

java/
java/lang/
java/lang/Object.class
java/lang/Exception.class
java/lang/Integer.class
java/lang/NumberFormatException.class
java/lang/Throwable.class
java/lang/Class.class
java/lang/IllegalAccessException.class
java/lang/StringBuffer.class
java/lang/ClassNotFoundException.class
java/lang/IllegalArgumentException.class
java/lang/Number.class
java/lang/InterruptedException.class
java/lang/String.class
java/lang/RuntimeException.class
java/lang/InternalError.class
java/lang/Long.class
java/lang/Character.class
java/lang/CloneNotSupportedException.class
java/lang/InstantiationException.class
java/lang/VirtualMachineError.class
java/lang/Double.class
java/lang/Error.class
java/lang/NullPointerException.class
java/lang/Cloneable.class
java/lang/System.class
java/lang/ClassLoader.class
java/lang/Math.class
java/lang/Float.class
java/lang/Runtime.class
java/lang/
    StringIndexOutOfBoundsException.class
java/lang/IndexOutOfBoundsException.class
java/lang/SecurityException.class
java/lang/LinkageError.class
java/lang/Runnable.class
java/lang/Process.class
java/lang/SecurityManager.class
java/lang/Thread.class
java/lang/UnsatisfiedLinkError.class
java/lang/IncompatibleClassChangeError.class
java/lang/NoSuchMethodError.class
java/lang/IllegalThreadStateException.class
java/lang/ThreadGroup.class
java/lang/ThreadDeath.class
java/lang/
    ArrayIndexOutOfBoundsException.class

java/lang/Boolean.class
java/lang/Compiler.class
java/lang/NoSuchMethodException.class
java/lang/ArithmeticException.class
java/lang/ArrayStoreException.class
java/lang/ClassCastException.class
java/lang/NegativeArraySizeException.class
java/lang/IllegalMonitorStateException.class
java/lang/ClassCircularityError.class
java/lang/ClassFormatError.class
java/lang/AbstractMethodError.class
java/lang/IllegalAccessError.class
java/lang/InstantiationError.class
java/lang/NoSuchFieldError.class
java/lang/NoClassDefFoundError.class
java/lang/VerifyError.class
java/lang/OutOfMemoryError.class
java/lang/StackOverflowError.class
java/lang/UnknownError.class
java/lang/Win32Process.class
java/io/
java/io/FilterOutputStream.class
java/io/OutputStream.class
java/io/IOException.class
java/io/PrintStream.class
java/io/FileInputStream.class
java/io/InterruptedIOException.class
java/io/File.class
java/io/InputStream.class
java/io/BufferedInputStream.class
java/io/FileOutputStream.class
java/io/FileNotFoundException.class
java/io/BufferedOutputStream.class
java/io/FileDescriptor.class
java/io/FilenameFilter.class
java/io/FilterInputStream.class
java/io/PipedInputStream.class
java/io/PipedOutputStream.class
java/io/EOFException.class
java/io/UTFDataFormatException.class
java/io/DataInput.class
java/io/DataOutput.class
java/io/DataInputStream.class
java/io/PushbackInputStream.class
java/io/ByteArrayInputStream.class
java/io/SequenceInputStream.class
java/io/StringBufferInputStream.class
java/io/LineNumberInputStream.class

java/io/DataOutputStream.class
java/io/ByteArrayOutputStream.class
java/io/RandomAccessFile.class
java/io/StreamTokenizer.class
java/util/
java/util/Hashtable.class
java/util/Enumeration.class
java/util/HashtableEnumerator.class
java/util/Properties.class
java/util/HashtableEntry.class
java/util/Dictionary.class
java/util/Date.class
java/util/NoSuchElementException.class
java/util/StringTokenizer.class
java/util/Random.class
java/util/VectorEnumerator.class
java/util/Vector.class
java/util/BitSet.class
java/util/EmptyStackException.class
java/util/Observable.class
java/util/Observer.class
java/util/ObserverList.class
java/util/Stack.class
java/awt/
java/awt/Toolkit.class
java/awt/peer/
java/awt/peer/WindowPeer.class
java/awt/peer/TextFieldPeer.class
java/awt/peer/ContainerPeer.class
java/awt/peer/PanelPeer.class
java/awt/peer/CanvasPeer.class
java/awt/peer/FramePeer.class
java/awt/peer/ChoicePeer.class
java/awt/peer/CheckboxMenuItemPeer.class
java/awt/peer/TextAreaPeer.class
java/awt/peer/FileDialogPeer.class
java/awt/peer/TextComponentPeer.class
java/awt/peer/ScrollbarPeer.class
java/awt/peer/ButtonPeer.class
java/awt/peer/ComponentPeer.class
java/awt/peer/MenuComponentPeer.class
java/awt/peer/MenuItemPeer.class
java/awt/peer/CheckboxPeer.class
java/awt/peer/MenuPeer.class
java/awt/peer/ListPeer.class
java/awt/peer/MenuBarPeer.class
java/awt/peer/LabelPeer.class
java/awt/peer/DialogPeer.class
java/awt/Image.class
java/awt/MenuItem.class
java/awt/MenuComponent.class
java/awt/image/
java/awt/image/ImageProducer.class
java/awt/image/ColorModel.class
java/awt/image/DirectColorModel.class
java/awt/image/ImageConsumer.class
java/awt/image/ImageObserver.class
java/awt/image/CropImageFilter.class
java/awt/image/ImageFilter.class
java/awt/image/FilteredImageSource.class
java/awt/image/IndexColorModel.class
java/awt/image/MemoryImageSource.class
java/awt/image/PixelGrabber.class
java/awt/image/RGBImageFilter.class
java/awt/FontMetrics.class
java/awt/Checkbox.class
java/awt/CheckboxGroup.class
java/awt/MenuContainer.class
java/awt/Menu.class
java/awt/Insets.class
java/awt/MenuBar.class
java/awt/List.class
java/awt/Label.class
java/awt/Component.class
java/awt/TextField.class
java/awt/TextComponent.class
java/awt/Dialog.class
java/awt/Font.class
java/awt/Window.class
java/awt/FocusManager.class
java/awt/Panel.class
java/awt/Container.class
java/awt/Graphics.class
java/awt/CheckboxMenuItem.class
java/awt/Canvas.class
java/awt/Frame.class
java/awt/Choice.class
java/awt/Event.class
java/awt/TextArea.class
java/awt/AWTError.class
java/awt/Polygon.class
java/awt/FlowLayout.class
java/awt/Point.class
java/awt/FileDialog.class
java/awt/Scrollbar.class
java/awt/Dimension.class
java/awt/Color.class
java/awt/Button.class
java/awt/LayoutManager.class
java/awt/Rectangle.class
java/awt/BorderLayout.class
java/awt/GridLayout.class
java/awt/GridBagConstraints.class
java/awt/GridBagLayout.class
java/awt/GridBagLayoutInfo.class
java/awt/CardLayout.class
java/awt/MediaTracker.class
java/awt/MediaEntry.class
java/awt/ImageMediaEntry.class
java/awt/AWTException.class
java/net/
java/net/URL.class
java/net/URLStreamHandlerFactory.class
java/net/InetAddress.class
java/net/UnknownContentHandler.class
java/net/UnknownHostException.class
java/net/URLStreamHandler.class
java/net/URLConnection.class
java/net/MalformedURLException.class
java/net/ContentHandlerFactory.class
java/net/ContentHandler.class
java/net/UnknownServiceException.class

java/net/ServerSocket.class
java/net/PlainSocketImpl.class
java/net/SocketImpl.class
java/net/ProtocolException.class
java/net/SocketException.class
java/net/SocketInputStream.class
java/net/Socket.class
java/net/SocketImplFactory.class
java/net/SocketOutputStream.class
java/net/DatagramPacket.class
java/net/DatagramSocket.class
java/net/URLEncoder.class
java/applet/
java/applet/Applet.class
java/applet/AppletContext.class
java/applet/AudioClip.class
java/applet/AppletStub.class
sun/
sun/tools/
sun/tools/debug/
sun/tools/debug/BreakpointQueue.class
sun/tools/debug/DebuggerCallback.class
sun/tools/debug/RemoteThread.class
sun/tools/debug/StackFrame.class
sun/tools/debug/RemoteAgent.class
sun/tools/debug/AgentConstants.class
sun/tools/debug/AgentIn.class
sun/tools/debug/RemoteObject.class
sun/tools/debug/RemoteStackVariable.class
sun/tools/debug/RemoteValue.class
sun/tools/debug/RemoteClass.class
sun/tools/debug/Agent.class
sun/tools/debug/RemoteBoolean.class
sun/tools/debug/RemoteChar.class
sun/tools/debug/RemoteString.class
sun/tools/debug/NoSessionException.class
sun/tools/debug/Field.class
sun/tools/debug/
  NoSuchLineNumberException.class
sun/tools/debug/RemoteShort.class
sun/tools/debug/RemoteThreadGroup.class
sun/tools/debug/RemoteInt.class
sun/tools/debug/ResponseStream.class
sun/tools/debug/RemoteDouble.class
sun/tools/debug/LocalVariable.class
sun/tools/debug/BreakpointSet.class
sun/tools/debug/RemoteStackFrame.class
sun/tools/debug/MainThread.class
sun/tools/debug/BreakpointHandler.class
sun/tools/debug/AgentOutputStream.class
sun/tools/debug/RemoteLong.class
sun/tools/debug/RemoteFloat.class
sun/tools/debug/RemoteArray.class
sun/tools/debug/InvalidPCException.class
sun/tools/debug/LineNumber.class
sun/tools/debug/RemoteField.class
sun/tools/debug/NoSuchFieldException.class
sun/tools/debug/RemoteByte.class
sun/tools/debug/EmptyApp.class
sun/tools/debug/RemoteDebugger.class
sun/tools/java/

sun/tools/java/RuntimeConstants.class
sun/tools/java/Constants.class
sun/tools/java/Environment.class
sun/tools/java/ClassPath.class
sun/tools/java/ClassDeclaration.class
sun/tools/java/FieldDefinition.class
sun/tools/java/Type.class
sun/tools/java/ClassNotFound.class
sun/tools/java/ClassType.class
sun/tools/java/ClassDefinition.class
sun/tools/java/Parser.class
sun/tools/java/ClassPathEntry.class
sun/tools/java/CompilerError.class
sun/tools/java/Identifier.class
sun/tools/java/Package.class
sun/tools/java/ClassFile.class
sun/tools/java/Imports.class
sun/tools/java/ArrayType.class
sun/tools/java/AmbiguousField.class
sun/tools/java/MethodType.class
sun/tools/java/Scanner.class
sun/tools/java/SyntaxError.class
sun/tools/java/BinaryClass.class
sun/tools/java/BinaryField.class
sun/tools/java/AmbiguousClass.class
sun/tools/java/BinaryConstantPool.class
sun/tools/java/ScannerInputStream.class
sun/tools/java/BinaryAttribute.class
sun/tools/java/BinaryCode.class
sun/tools/java/BinaryExceptionHandler.class
sun/tools/javac/
sun/tools/javac/Main.class
sun/tools/javac/SourceClass.class
sun/tools/javac/CompilerField.class
sun/tools/javac/SourceField.class
sun/tools/javac/BatchEnvironment.class
sun/tools/javac/ErrorConsumer.class
sun/tools/javac/ErrorMessage.class
sun/tools/javac/BatchParser.class
sun/tools/zip/
sun/tools/zip/ZipFile.class
sun/tools/zip/ZipEntry.class
sun/tools/zip/ZipFileInputStream.class
sun/tools/zip/ZipConstants.class
sun/tools/zip/ZipFormatException.class
sun/tools/zip/ZipReaderInputStream.class
sun/tools/zip/ZipReader.class
sun/tools/tree/
sun/tools/tree/ConstantExpression.class
sun/tools/tree/LocalField.class
sun/tools/tree/Expression.class
sun/tools/tree/IncDecExpression.class
sun/tools/tree/SuperExpression.class
sun/tools/tree/NaryExpression.class
sun/tools/tree/StringExpression.class
sun/tools/tree/UnaryExpression.class
sun/tools/tree/Context.class
sun/tools/tree/ExpressionStatement.class
sun/tools/tree/ConditionVars.class
sun/tools/tree/Node.class
sun/tools/tree/CharExpression.class

sun/tools/tree/CaseStatement.class
sun/tools/tree/LessExpression.class
sun/tools/tree/IntegerExpression.class
sun/tools/tree/SubtractExpression.class
sun/tools/tree/ArrayAccessExpression.class
sun/tools/tree/TryStatement.class
sun/tools/tree/BinaryEqualityExpression.class
sun/tools/tree/Statement.class
sun/tools/tree/AssignSubtractExpression.class
sun/tools/tree/FinallyStatement.class
sun/tools/tree/ForStatement.class
sun/tools/tree/DivRemExpression.class
sun/tools/tree/BinaryExpression.class
sun/tools/tree/ShiftRightExpression.class
sun/tools/tree/AssignMultiplyExpression.class
sun/tools/tree/BooleanExpression.class
sun/tools/tree/BinaryArithmeticExpression.class
sun/tools/tree/ThrowStatement.class
sun/tools/tree/AssignDivideExpression.class
sun/tools/tree/AssignShiftLeftExpression.class
sun/tools/tree/NewArrayExpression.class
sun/tools/tree/AndExpression.class
sun/tools/tree/AssignBitOrExpression.class
sun/tools/tree/BreakStatement.class
sun/tools/tree/SynchronizedStatement.class
sun/tools/tree/PreDecExpression.class
sun/tools/tree/CompoundStatement.class
sun/tools/tree/DoubleExpression.class
sun/tools/tree/ConvertExpression.class
sun/tools/tree/NullExpression.class
sun/tools/tree/LessOrEqualExpression.class
sun/tools/tree/IdentifierExpression.class
sun/tools/tree/ReturnStatement.class
sun/tools/tree/BitNotExpression.class
sun/tools/tree/LongExpression.class
sun/tools/tree/VarDeclarationStatement.class
sun/tools/tree/MethodExpression.class
sun/tools/tree/ThisExpression.class
sun/tools/tree/BitOrExpression.class
sun/tools/tree/PositiveExpression.class
sun/tools/tree/IfStatement.class
sun/tools/tree/FloatExpression.class
sun/tools/tree/NotEqualExpression.class
sun/tools/tree/InstanceOfExpression.class
sun/tools/tree/NotExpression.class
sun/tools/tree/BitAndExpression.class
sun/tools/tree/DivideExpression.class
sun/tools/tree/ShortExpression.class
sun/tools/tree/RemainderExpression.class
sun/tools/tree/NewInstanceExpression.class
sun/tools/tree/SwitchStatement.class
sun/tools/tree/AddExpression.class
sun/tools/tree/AssignOpExpression.class
sun/tools/tree/EqualExpression.class
sun/tools/tree/PostIncExpression.class
sun/tools/tree/GreaterExpression.class
sun/tools/tree/PostDecExpression.class
sun/tools/tree/AssignExpression.class
sun/tools/tree/WhileStatement.class
sun/tools/tree/ContinueStatement.class
sun/tools/tree/ConditionalExpression.class

sun/tools/tree/AssignAddExpression.class
sun/tools/tree/BinaryBitExpression.class
sun/tools/tree/CastExpression.class
sun/tools/tree/AssignBitXorExpression.class
sun/tools/tree/ArrayExpression.class
sun/tools/tree/InlineMethodExpression.class
sun/tools/tree/InlineNewInstanceExpression.class
sun/tools/tree/CodeContext.class
sun/tools/tree/AssignShiftRightExpression.class
sun/tools/tree/UnsignedShiftRightExpression.class
sun/tools/tree/AssignBitAndExpression.class
sun/tools/tree/ShiftLeftExpression.class
sun/tools/tree/CatchStatement.class
sun/tools/tree/IntExpression.class
sun/tools/tree/TypeExpression.class
sun/tools/tree/CommaExpression.class
sun/tools/tree/AssignUnsignedShiftRightExpression.class
sun/tools/tree/ExprExpression.class
sun/tools/tree/AssignRemainderExpression.class
sun/tools/tree/ByteExpression.class
sun/tools/tree/BinaryAssignExpression.class
sun/tools/tree/DoStatement.class
sun/tools/tree/DeclarationStatement.class
sun/tools/tree/MultiplyExpression.class
sun/tools/tree/InlineReturnStatement.class
sun/tools/tree/BitXorExpression.class
sun/tools/tree/BinaryCompareExpression.class
sun/tools/tree/BinaryShiftExpression.class
sun/tools/tree/CheckContext.class
sun/tools/tree/PreIncExpression.class
sun/tools/tree/GreaterOrEqualExpression.class
sun/tools/tree/FieldExpression.class
sun/tools/tree/OrExpression.class
sun/tools/tree/BinaryLogicalExpression.class
sun/tools/tree/NegativeExpression.class
sun/tools/tree/LengthExpression.class
sun/tools/asm/
sun/tools/asm/Assembler.class
sun/tools/asm/Instruction.class
sun/tools/asm/LocalVariable.class
sun/tools/asm/ArrayData.class
sun/tools/asm/LocalVariableTable.class
sun/tools/asm/SwitchDataEnumeration.class
sun/tools/asm/ConstantPool.class
sun/tools/asm/ConstantPoolData.class
sun/tools/asm/NameAndTypeConstantData.class
sun/tools/asm/NumberConstantData.class
sun/tools/asm/FieldConstantData.class
sun/tools/asm/TryData.class
sun/tools/asm/Label.class
sun/tools/asm/SwitchData.class
sun/tools/asm/CatchData.class
sun/tools/asm/StringExpressionConstantData.class
sun/tools/asm/NameAndTypeData.class
sun/tools/asm/StringConstantData.class
sun/tools/asm/ClassConstantData.class
sun/tools/ttydebug/
sun/tools/ttydebug/TTY.class
sun/tools/javadoc/
sun/tools/javadoc/Main.class

sun/tools/javadoc/DocumentationGenerator.class
sun/tools/javadoc/
    HTMLDocumentationGenerator.class
sun/tools/javadoc/
    MIFDocumentationGenerator.class
sun/tools/javadoc/MIFPrintStream.class
sun/net/
sun/net/MulticastSocket.class
sun/net/URLCanonicalizer.class
sun/net/NetworkClient.class
sun/net/NetworkServer.class
sun/net/ProgressData.class
sun/net/ProgressEntry.class
sun/net/TelnetInputStream.class
sun/net/TelnetProtocolException.class
sun/net/TelnetOutputStream.class
sun/net/TransferProtocolClient.class
sun/net/ftp/
sun/net/ftp/FtpInputStream.class
sun/net/ftp/FtpClient.class
sun/net/ftp/FtpLoginException.class
sun/net/ftp/FtpProtocolException.class
sun/net/ftp/IftpClient.class
sun/net/nntp/
sun/net/nntp/NewsgroupInfo.class
sun/net/nntp/NntpClient.class
sun/net/nntp/UnknownNewsgroupException.class
sun/net/nntp/NntpProtocolException.class
sun/net/nntp/NntpInputStream.class
sun/net/smtp/
sun/net/smtp/SmtpPrintStream.class
sun/net/smtp/SmtpClient.class
sun/net/smtp/SmtpProtocolException.class
sun/net/www/
sun/net/www/auth/
sun/net/www/auth/Authenticator.class
sun/net/www/auth/basic.class
sun/net/www/content/
sun/net/www/content/text/
sun/net/www/content/text/Generic.class
sun/net/www/content/text/plain.class
sun/net/www/content/image/
sun/net/www/content/image/gif.class
sun/net/www/content/image/jpeg.class
sun/net/www/content/image/x_xbitmap.class
sun/net/www/content/image/x_xpixmap.class
sun/net/www/FormatException.class
sun/net/www/MessageHeader.class
sun/net/www/MeteredStream.class
sun/net/www/ProgressReport.class
sun/net/www/MimeEntry.class
sun/net/www/MimeLauncher.class
sun/net/www/MimeTable.class
sun/net/www/URLConnection.class
sun/net/www/UnknownContentException.class
sun/net/www/UnknownContentHandler.class
sun/net/www/protocol/
sun/net/www/protocol/file/
sun/net/www/protocol/file/Handler.class
sun/net/www/protocol/file/
    FileURLConnection.class

sun/net/www/protocol/http/
sun/net/www/protocol/http/Handler.class
sun/net/www/protocol/http/
    HttpURLConnection.class
sun/net/www/protocol/http/
    HttpPostBufferStream.class
sun/net/www/protocol/doc/
sun/net/www/protocol/doc/Handler.class
sun/net/www/protocol/verbatim/
sun/net/www/protocol/verbatim/Handler.class
sun/net/www/protocol/verbatim/
    VerbatimConnection.class
sun/nct/www/protocol/gopher/
sun/net/www/protocol/gopher/GopherClient.class
sun/net/www/protocol/gopher/
    GopherInputStream.class
sun/net/www/http/
sun/net/www/http/
    UnauthorizedHttpRequestException.class
sun/net/www/http/HttpClient.class
sun/net/www/http/AuthenticationInfo.class
sun/awt/
sun/awt/HorizBagLayout.class
sun/awt/VerticalBagLayout.class
sun/awt/VariableGridLayout.class
sun/awt/FocusingTextField.class
sun/awt/win32/
sun/awt/win32/MToolkit.class
sun/awt/win32/MMenuBarPeer.class
sun/awt/win32/MButtonPeer.class
sun/awt/win32/Win32Image.class
sun/awt/win32/MScrollbarPeer.class
sun/awt/win32/MDialogPeer.class
sun/awt/win32/MCheckboxMenuItemPeer.class
sun/awt/win32/Win32Graphics.class
sun/awt/win32/MListPeer.class
sun/awt/win32/MWindowPeer.class
sun/awt/win32/MMenuItemPeer.class
sun/awt/win32/ModalThread.class
sun/awt/win32/MCanvasPeer.class
sun/awt/win32/MFileDialogPeer.class
sun/awt/win32/MTextAreaPeer.class
sun/awt/win32/MPanelPeer.class
sun/awt/win32/MComponentPeer.class
sun/awt/win32/MCheckboxPeer.class
sun/awt/win32/MLabelPeer.class
sun/awt/win32/Win32FontMetrics.class
sun/awt/win32/MFramePeer.class
sun/awt/win32/MMenuPeer.class
sun/awt/win32/MChoicePeer.class
sun/awt/win32/MTextFieldPeer.class
sun/awt/win32/Win32PrintJob.class
sun/awt/image/
sun/awt/image/URLImageSource.class
sun/awt/image/ImageWatched.class
sun/awt/image/InputStreamImageSource.class
sun/awt/image/ConsumerQueue.class
sun/awt/image/ImageDecoder.class
sun/awt/image/ImageRepresentation.class
sun/awt/image/ImageInfoGrabber.class
sun/awt/image/XbmImageDecoder.class

sun/awt/image/GifImageDecoder.class
sun/awt/image/ImageFetcher.class
sun/awt/image/PixelStore.class
sun/awt/image/JPEGImageDecoder.class
sun/awt/image/PixelStore8.class
sun/awt/image/ImageFetchable.class
sun/awt/image/OffScreenImageSource.class
sun/awt/image/PixelStore32.class
sun/awt/image/ImageFormatException.class
sun/awt/image/FileImageSource.class
sun/awt/image/Image.class
sun/awt/UpdateClient.class
sun/awt/ScreenUpdaterEntry.class
sun/awt/ScreenUpdater.class
sun/misc/
sun/misc/Ref.class
sun/misc/MessageUtils.class
sun/misc/Cache.class
sun/misc/CacheEntry.class
sun/misc/CacheEnumerator.class
sun/misc/CEFormatException.class
sun/misc/CEStreamExhausted.class
sun/misc/CRC16.class
sun/misc/CharacterDecoder.class
sun/misc/BASE64Decoder.class
sun/misc/UCDecoder.class
sun/misc/UUDecoder.class
sun/misc/CharacterEncoder.class
sun/misc/BASE64Encoder.class
sun/misc/HexDumpEncoder.class
sun/misc/UCEncoder.class
sun/misc/UUEncoder.class

sun/misc/Timeable.class
sun/misc/TimerTickThread.class
sun/misc/Timer.class
sun/misc/TimerThread.class
sun/misc/ConditionLock.class
sun/misc/Lock.class
sun/audio/
sun/audio/AudioDataStream.class
sun/audio/AudioData.class
sun/audio/AudioDevice.class
sun/audio/AudioPlayer.class
sun/audio/AudioStream.class
sun/audio/NativeAudioStream.class
sun/audio/InvalidAudioFormatException.class
sun/audio/AudioTranslatorStream.class
sun/audio/AudioStreamSequence.class
sun/audio/ContinuousAudioDataStream.class
sun/applet/
sun/applet/StdAppletViewerFactory.class
sun/applet/TextFrame.class
sun/applet/AppletViewerFactory.class
sun/applet/AppletViewer.class
sun/applet/AppletCopyright.class
sun/applet/AppletAudioClip.class
sun/applet/AppletSecurity.class
sun/applet/AppletThreadGroup.class
sun/applet/AppletClassLoader.class
sun/applet/AppletPanel.class
sun/applet/AppletViewerPanel.class
sun/applet/AppletProps.class
sun/applet/AppletSecurityException.class
sun/applet/AppletZipClassLoader.class

B ANSI-C Functions

Table B.1 Time functions

Conversion functions	Header file	Description
`char *asctime( struct tm *ttt);`	time.h	Function: Converts date and time to string. The time passed as a pointer to a `tm` structure. Return: A pointer to the date string.
`char *ctime( time_t *ttt);`	time.h	Function: Converts date and time to string. The time passed as a pointer to by `ttt`. Return: A pointer to the date string.
`double difftime( time_t time2, time_t time1);`	time.h	Function: To determine the number of seconds between `time2` and `time1`. Return: Difference in time returned as a `double`.
`struct tm *gmtime( time_t *ttt);`	time.h	Function: Converts time into Greenwich Mean Time. The time is passed as a pointer to `ttt` and the result is put into a `tm` structure. Return: A pointer to the `tm` structure.
`int localtime( time_t *tt);`	time.h	Function: Converts time into local time. The time is passed as a pointer to `ttt` and the result is put into a `tm` structure. Return: A pointer to the `tm` structure.
`time_t time( time *ttt);`	time.h	Function: To determine the time of day. The time is passed as a pointer to `ttt` and the result gives the number of seconds that have passed since 00:00:00 GMT January 1970. This value is returned through the pointer `ttt`. Return: The number of seconds that have passed since January 1970.

Table B.2 Classification functions

Classification function	Header file	Description
`int isalnum(int ch);`	ctype.h	Function: To determine if character ch is a digit ('0'-'9') or a letter ('a'-'z' or 'A'- 'Z'). Return: A non-zero value if the character is a digit or letter.
`int isalpha(int ch);`	ctype.h	Function: To determine if character ch is a letter ('a'-'z', 'A'-'Z'). Return: A non-zero value if the character is a letter.
`int iscntrl(int ch);`	ctype.h	Function: To determine if character ch is a control character, i.e. ASCII 0-31 or 127 (DEL). Return: A non-zero value if the character is a control character.
`int isdigit(int ch);`	ctype.h	Function: To determine if character ch is a digit ('0'-'9'). Return: A non-zero value if the character is a digit.
`int isgraph(int ch);`	ctype.h	Function: To determine if character ch is a printing character (the space character is excluded). Return: A non-zero value if the character is printable (excluding the space character).
`int islower(int ch);`	ctype.h	Function: To determine if character ch is a lowercase letter ('a'- 'z'). Return: A non-zero value if the character is a lowercase letter.
`int ispunct(int ch);`	ctype.h	Function: To determine if character ch is a punctuation character. Return: A non-zero value if the character is a punctuation character.
`int isprint(int ch);`	ctype.h	Function: To determine if character ch is a printing character. Return: A non-zero value if the character is printable (including the space character).
`int isspace(int ch);`	ctype.h	Function: To determine if character ch is either a space, horizontal tab, carriage return, new line, vertical tab or form-feed. Return: A non-zero value if the character is either a space, horizontal tab, carriage return, new line, vertical tab or form-feed.

`int isupper(int ch);`	ctype.h	Function: To determine if character `ch` is an uppercase letter ('A'- 'Z'). Return: A non-zero value if the character is an uppercase letter.
`int isxdigit(int ch);`	ctype.h	Function: To determine if character `ch` is a hexadecimal digit i.e. '0'-'9','a'-'f', 'A'-'F'. Return: A non-zero value if the character is a hexadecimal digit.

Table B.3   String functions

String function	*Header file*	*Description*

`int strcmp(`
`  char *str1,`
`  char *str2);`
	string.h	Function: Compares two strings `str1` and `str2`.
		Return: A 0 (zero) is returned if the strings are identical, a negative value if `str1` is less than `str2`, or a positive value if `str1` is greater than `str2`.

`int strlen(` `  char *str);`	string.h	Function: Determines the number of characters in `str`. Return: Number of characters in `str`.

`char *strcat(`
`  char *str1,`
`  char *str2);`
	string.h	Function: Appends `str2` onto `str1`. The resultant string `str1` will contain `str1` and `str2`.
		Return: A pointer to the resultant string.

`char *strlwr(` `  char *str1);`	string.h	Function: Converts uppercase letters in a string to lowercase Return: A pointer to the resultant string
`char *strupr(` `  char *str1);`	string.h	Function: Converts lowercase letters in a string to uppercase. Return: A pointer to the resultant string.

`char *strcpy(`
`  char *str1,`
`  char *str2);`
	string.h	Function: Copies `str2` into `str1`.
		Return: A pointer to the resultant string.

`int sprintf(`
`  char *str,`
`  char *format_str,`
`  arg1,....);`
	stdio.h	Function: Similar to `printf()` but output goes into string `str`.
		Return: Number of characters output.

`int sscanf(`
`  char *str,`
`  char *format_str,`
`  arg1,...);`
	stdio.h	Function: Similar to `scanf()` but input is from string `str`.
		Return: Number of fields successfully scanned.

Table B.4   Conversion functions

Conversion functions	Header file	Description
`double atof( char *str);`	stdlib.h	Function: Converts a string `str` into a floating-point number. Return: Converted floating-point value. If value cannot be converted the return value is 0.
`int atoi(char *str);`	stdlib.h	Function: Converts a string `str` into an integer. Return: Converted integer value. If value cannot be converted the return value is 0.
`long atol(char *str);`	stdlib.h	Function: Converts a string `str` into a long integer. Return: Converted integer value. If value cannot be converted the return value is 0.
`char *itoa( int val, char *str, int radix);`	stdlib.h	Function: Converts `val` into a string `str`. The number base used is defined by `radix`. Return: A pointer to `str`.
`char *ltoa( long val, char *str, int radix);`	stdlib.h	Function: Converts `val` into a string `str`. The number base used is defined by `radix`. Return: A pointer to `str`.
`int _tolower( int ch);`	ctype.h	Function: Converts character `ch` to lowercase. Return: Converted value lowercase character. Note, `ch` must be in uppercase when called.
`int tolower( int ch);`	ctype.h	Function: Converts character `ch` to lowercase. Return: Converted value lowercase character. Return value will be `ch` unless `ch` is in uppercase.
`int _toupper( int ch);`	ctype.h	Function: Converts character `ch` to uppercase. Return: Converted value lowercase character. Note, `ch` must be in lowercase when called.
`int toupper( int ch);`	ctype.h	Function: Converts character `ch` to uppercase. Return: Converted value lowercase character. Return value will be `ch` unless `ch` is in lowercase.

Table B.5   Input/output functions

Input/output functions	Header file	Description
`void clearerr( FILE *fptr);`	stdio.h	Function: Resets file error or end-of-file indicator on a file. Return: None.
`int fclose( FILE *fptr);`	stdio.h	Function: Closes a file currently pointed to by file pointer `fptr`. Return: A 0 on success, otherwise, `EOF` if any errors are encountered.
`int  feof(FILE *fptr);`	stdio.h	Function: Detects the end-of-file. Return: A non-zero if at the end of a file, otherwise a 0.
`int  ferror(FILE *fptr);`	stdio.h	Function: Detects if there has been an error when reading from or writing to a file. Return: A non-zero if an error has occurred, otherwise, a 0 if no error.
`int fflush(FILE *fptr);`	stdio.h	Function: Flushes a currently open file. Return: A 0 on success, otherwise, `EOF` if any errors are encountered.
`int  fgetc(FILE *fptr);`	stdio.h	Function: Gets a character from a file. Return: The character is read. On an error it returns `EOF`.
`char *fgets(char *str, int n, FILE *fptr);`	stdio.h	Function: To read a string from the file pointed to by `fptr` into string `str` with n characters or until a new-line character is read (whichever is first). Return: On success, the return value points to string `str`, otherwise a `NULL` on an error or end-of-file.
`FILE *fopen( char *fname, char *mode);`	stdio.h	Function: Opens a file named `fname` with attributes given by `mode`. Attributes include "r" for read-only access, "w" for read/write access to an existing file, "w+" to create a new file for read/write access, "a" for append, "a+" for append and create file is it does not exist and "b" for binary files. Return: If successful a file pointer, otherwise a `NULL` is returned.
`int fprintf( FILE *fptr, char *fmt,arg1...);`	stdio.h	Function: Writes formatted data to a file. Return: The number of bytes outputted. On an error the return is `EOF`.

`int fputc(`
`  int ch,`
`  FILE *fptr);`
	stdio.h	Function: Writes a character ch to a file.
		Return: The character written, otherwise on an error it returns EOF.

`int fread(`
`  void *buff,`
`  size_t size,`
`  size_t n,`
`  FILE *fptr);`
	stdio.h	Function: Reads binary data from a file. It reads n items of data, each of length size bytes into the block specified by buff.
		Return: The number of items read. In the event of an error the return will be less than the specified number (n).

`int fscanf(`
`  FILE *fptr,`
`  char *format,`
`  &arg1...);`
	stdio.h	Function: Scans and formats input from a file in a format specified by format.
		Return: The number of fields successfully scanned. In the event of a reading from an end-of-file the return is EOF.

`int fseek(`
`  FILE *fptr,`
`  long offset,`
`  int whence);`
	stdio.h	Function: The file pointer fptr is positioned at an offset specified by offset beyond the location specified by whence. This location can be either to SEEK_SET (the start of the file), SEEK_CUR (the current file position) or SEEK_END (the end-of-file).
		Return: A 0 on success; otherwise, a non-zero value if any errors are encountered.

`int fwrite(`
`  void *buff,`
`  size_t size,`
`  size_t n,`
`  FILE *fptr);`
	stdio.h	Function: Writes binary data to a file. It writes n items of data, each of length size bytes from the block specified by buff.
		Return: The number of items written. On the event of an error the return will be less than the specified number (n).

`int getc(FILE *fptr);`	stdio.h	Function: Gets a character ch from a file. Return: The character read, or in the event of an error it returns EOF.
`int getchar(void);`	stdio.h	Function: Gets a character from the standard input (normally the keyboard). Return: The character read, or on an error it returns EOF.
`char *gets(` `  char *str);`	stdio.h	Function: Gets a string str from the standard input (normally the keyboard). String input is terminated by a carriage return (and not with spaces or tabs, as with scanf()). Return: On success, the return value points to string str, otherwise a NULL on an error.

`int printf(`
`  char *fmt,`
`  arg1....);`
	stdio.h	Function: Writes formatted data to the standard output (normally the display).
		Return: The number of bytes output. On an error the return is EOF.

`int putc(`
`  int ch,`
`  FILE *fptr);`
	stdio.h	Function: Puts a character ch to a file.
		Return: The character written, else in the event of an error it returns EOF.

`int putchar(` `  int ch);`	stdio.h	Function: Puts a character ch to the standard output (normally the display). Return: The character written, else on an error it returns EOF.
`int puts(` `  char *str);`	stdio.h	Function: Puts a string str to the standard output (normally the display). The string is appended with a new-line character. Return: The character written, else on an error it returns EOF.
`void rewind(` `  FILE *fptr);`	stdio.h	Function: Repositions a file pointer to the start of a file. Any file errors will be automatically cleared. Return: None

`int scanf(`
`  char *format,`
`  &arg1...);`
	stdio.h	Function: Scans and formats input from the standard input (normally the keyboard) in a format specified by format.
		Return: The number of fields successfully scanned. In the event of a reading from an end-of-file the return is EOF.

Table B.6   Other standard functions

Conversion functions	Header file	Description
`void exit(` `  int status);`	stdlib.h	Function: To terminate the program. The value passed status indicates the termination status. Typically, a 0 indicates a normal exit and any other value indicates an error. Return: None.
`void free(` `  void *block);`	stdlib.h	Function: To free an area of memory allocated to block. Return: None.
`void *malloc(` `  size_t size);`	stdlib.h	Function: To allocate an area of memory with size bytes. Return: If there is enough memory a pointer to an area of memory is returned, otherwise a NULL is returned.
`int system(` `  char *cmd);`	stdlib.h	Function: Issues a system command given by cmd. Return: A 0 on success, otherwise a −1.

Table B.7   Math functions

Math function	Header file	Description
`int abs(int val);`	math.h, stdlib.h	Function: To determine the absolute value of `val`. Return: Absolute value.
`double acos(` `  double val);`	math.h	Function: To determine the inverse cosine of `val`. Return: Inverse cosine in radians. If the range of `val` is invalid then `errno` is set to `EDOM` (domain error).
`double asin(` `  double val);`	math.h	Function: To determine the inverse sine of `val`. Return: Inverse sine in radians. If the range of `val` is invalid then `errno` is set to `EDOM` (domain error).
`double atan(` `  double val);`	math.h	Function: To determine the inverse tangent of `val`. Return: Inverse tangent in radians.

`double atan2(`
`  double val1,`
`  double val2);`
	math.h	Function: To determine the inverse tangent of `val1/val2`.
		Return: Inverse tangent in radians. If `val1` and `val2` are 0 then `errno` is set to `EDOM` (domain error).

`double ceil(` `  double val);`	math.h	Function: Rounds `val` up to the nearest whole number. Return: The nearest integer value converted to a `double`.
`double cos(` `  double val);`	math.h	Function: To determine the cosine of `val`. Return: Cosine value.
`double cosh(` `  double val);`	math.h	Function: To determine the hyperbolic cosine of `val`. Return: The hyperbolic cosine. If an overflow occurs the return value is `HUGE_VAL` and `errno` is set to `ERANGE` (out of range).
`double exp(` `  double val);`	math.h	Function: To determine the exponential e to the power of `val`. Return: The exponential power. If an overflow occurs the return value is `HUGE_VAL` and `errno` is set to `ERANGE` (out of range).
`double fabs(` `  double val);`	math.h	Function: To determine the absolute value of `val`. Return: Absolute value returned as a `double`.

`double floor(` `double val);`	math.h	Function: Rounds `val` down to the nearest whole number. Return: The nearest integer value converted to a `double`.

`double fmod(`
  `double val1,`
  `double val2);`
	math.h	Function: Determines the remainder of a divsion of `val1` by `val2` and rounds to the nearest whole number.
		Return: The nearest integer value converted to a `double`.

`double log(` `double val);`	math.h	Function: Determines the natural logarithm of `val`. Return: The natural logarithm. If the value passed into the function is less than or equal to 0 then `errno` is set to `EDOM` and the value passed back is `HUGE_VAL`.
`double log10(` `double val);`	math.h	Function: Determines the base-10 logarithm of `val`. Return: The base-10 logarithm. If the value passed into the function is less than or equal to 0 then `errno` is set with `EDOM` and the value passed back is `HUGE_VAL`.

`double pow(`
  `double val1,`
  `double val2);`
	math.h	Function: Determines `val1` to the power of `val2`.
		Return: The raised power. If an overflow occurs or the power is incalculable then the return value is `HUGE_VAL` and `errno` is set to `ERANGE` (out of range) or `EDOM` (domain error). If both arguments passed are 0 then the return is 1.

`int rand(void);`	math.h	Function: Generates a pseudo-random number from 0 to `val-1`. Return: The generated random number.
`double sin(` `double val);`	math.h	Function: To determine sine of `val`. Return: Sine value.
`double sinh(` `double val);`	math.h	Function: To determine hyperbolic sine of `val`. Return: The hyperbolic sine. If an overflow occurs the return value is `HUGE_VAL` and `errno` is set to `ERANGE` (out of range).
`double sqrt(` `double val);`	math.h	Function: Determines the square root of `val`. Return: The square root. If the value passed into the function is less than 0 then `errno` is set with `EDOM` and the value returned is 0.

`void srand(`
`   unsigned int`
`      seed);`
	stdlib.h	Function: Initializes the random-generator with seed.
		Return: None.

`double tan(` `   double val);`	math.h	Function: To determine the tangent of `val`. Return: The hyperbolic tangent. If an overflow occurs the return value is `HUGE_VAL` and `errno` is set to `ERANGE` (out of range).
`double tanh(` `   double val);`	math.h	Function: To determine the hyperbolic tangent of `val`. Return: The hyperbolic tangent.

Turbo Pascal Reference

The following is a list of Turbo Pascal procedures.

Append	GetIntVec	Release
Arc	GetLineSettings	RenameSetDate
Assign	GetMem	ResetSetFAttr
AssignCrt	GetPalette	RestoreCrtMode
Bar	GetTextSettings	Rewrite
Bar3D	GetTime	RmDir
BlockRead	GetVerify	RunError
BlockWrite	GetViewSettings	Sector
ChDir	GotoXY	Seek
Circle	Halt	SetActivePage
ClearDevice	HighVideo	SetAllPalette
ClearViewPort	Inc	SetAspectRatio
Close	InitGraph	SetBkColor
CloseGraph	Insert	SetCBreak
ClrEol	InsLine	SetColor
ClrScr	Intr	SetFillPattern
Dec	Keep	SetFillStyle
Delay	Line	SetFTime
Delete	LineRel	SetGraphMode
DelLine	LineTo	SetIntVec
DetectGraph	LowVideo	SetLineStyle
Dispose	Mark	SetPalette
Ellipse	MkDir	SetRGBPalette
Erase	Move	SetTextBuf
Exec	MoveTo	SetTextJustify
Exit	New	SetTextStyle
Fail	NormVideo	SetTime
FillChar	NoSound	SetUserCharSize
FillEllipse	OutText	SetVerify
FillPoly	OutTextXY	SetViewPort
FindFirst	OvrClrBuf	SetVisualPage
FindNext	OvrInit	SetWriteMode
FloodFill	OvrInitEMS	Str
Flush	OvrSetBuf	SwapVectors

FreeMem	OvrSetRetry	TextBackground
FSplit	PackTime	TextColor
GetArcCoords	PieSlice	TextMode
GetAspectRatio	PutImage	Truncate
GetCBreak	PutPixel	UnPackTime
GetDate	Randomize	Val
GetDir	Read (text)	Window
GetFAttr	Read (typed)	Write (text)
GetFillSettings	ReadLn	Write (typed)
GetFTime	Rectangle	WriteLn
GetImage		

The following is a list of Turbo Pascal functions.

Abs	GetGraphMode	ParamStr
Addr	GetMaxMode	Pi
ArcTan	GetMaxX	Pos
Chr	GetMaxY	Pred
Concat	GetModeName	Ptr
Copy	GetPaletteSize	Random
Cos	GetPixel	ReadKey
Cseg	GetX	Round
DiskFree	GetY	SeekEof
DiskSize	GraphErrorMsg	SeekEoln
DosexitCode	GraphResult	Seg
DosVersion	Hi	SetAspectRatio
Dseg	ImageSize	Sin
EnvCount	InstallUserDriver	SizeOf
EnvStr	InstallUserFont	Sound
Eof (text)	Int	SPtr
Eof (typed)	IOResult	Sqr
Eoln	KeyPressed	Sqrt
Exp	Length	SSeg
FExpand	Lo	Succ
FilePos	MaxAvail	Swap
FileSize	MemAvail	TextHeight
Frac	MsDos	TextWidth
FSearch	Odd	Trunc
GetBkColor	Ofs	TypeOf
GetColor	Ord	UpCase
GetDefaultPalette	OvrGetBuf	WhereX
GetDriverName	OvrGetRetry	WhereY
GetEnv	ParamCount	

Turbo Pascal accesses some standard procedures and functions through units, which are libraries of precompiled modules. For example, the Crt unit contains routines which access the text display. To use a unit the uses keyword must be included in a statement near the top of the program. For example, to use the clrscr function:

```
program test;

uses crt;

begin
   clrscr;        (* clears the screen *)
   textcolor(RED);
   textbackground(YELLOW);
   writeln('Hello');
end.
```

The listing of modules in the crt unit is:

AssignCrt	InsLine	TextBackground
ClrEol	KeyPressed	TextColor
ClrScr	LowVideo	TextMode
Delay	NormVideo	WhereX
DelLine	NoSound	WhereY
GotoXY	ReadKey	Window
HighVideo	Sound	

The listing of modules in the system unit is:

Abs	GetDir	ReadLn
Addr	GetMem	Release
Append	Halt	Rename
ArcTan	Hi	Reset
Assign	Inc	Rewrite
BlockRead	Insert	RmDir
BlockWrite	Int	Round
ChDir	IOResult	RunError
Chr	Length	Seek
Close	Ln	SeekEof
Concat	Lo	SeekEoln
Copy	Mark	Seg
Cos	MaxAvail	SetTextBuf
CSeg	MemAvail	Sin
Dec	MkDir	SizeOf
Delete	Move	Sptr
Dispose	New	Sqr
DSeg	Odd	Sqrt
Eof (text)	Ofs	Sseg
Eof (typed)	Ord	Str

Eoln	ParamCount	Succ
Erase	ParamStr	Swap
Exit	Pi	Trunc
Exp	Pos	Truncate
FilePos	Pred	UpCase
FileSize	Ptr	Val
FillChar	Random	Write (text)
Flush	Randomize	Write (typed)
Frac	Read (text)	WriteLn
FreeMem	Read (typed)	

The listing of modules in the dos unit is:

DiskFree	Fsplit	MsDos
DiskSize	GetCBreak	PackTime
DosExitCode	GetDate	SetCBreak
DosVersion	GetEnv	SetDate
EnvCount	GetFAttr	SetFAttr
EnvStr	GetFTime	SetFTime
Exec	GetIntVec	SetIntVec
Fexpand	GetTime	SetTime
FindFirst	GetVerify	SetVerify
FindNext	Intr	SwapVectors
Fsearch	Keep	UnpackTime

The listing of modules in the graph unit is:

Arc	GetMaxY	PutPixel
Bar	GetModeName	Rectangle
Bar3D	GetModeRange	RegisterBGIdriver
Circle	GetPalette	RegisterBGIfont
ClearDevice	GetPaletteSize	RestoreCrtMode
ClearViewPort	GetPixel	Sector
CloseGraph	GetTextSettings	SetActivePage
DetectGraph	GetViewSettings	SetAllPalette
Drawpoly	GetX	SetAspectRatio
Ellipse	GetY	SetBkColor
FillEllipse	GraphDefaults	SetColor
FillPoly	GraphErrorMsg	SetFillPattern
FloodFill	GraphResult	SetFillStyle
GetArcCoords	ImageSize	SetGraphBufSize
GetAspectRatio	InitGraph	SetGraphMode
GetBkColor	InstallUserDriver	SetLineStyle
GetColor	InstallUserFont	SetPalette
GetDefaultPalette	Line	SetRGBPalette
GetDriverName	LineRel	SetTextJustify

GetFillPattern	LineTo	SetTextStyle
GetFillSettings	MoveRel	SetUserCharSize
GetGraphMode	MoveTo	SetViewPort
GetImage	OutText	SetVisualPage
GetLineSettings	OutTextXY	SetWriteMode
GetMaxColor	PieSlice	TextHeight
GetMaxMode	PutImage	TextWidth
GetMaxX		

The following is a quick reference to commonly used functions.

Abs	`function Abs(x) :` `(Same type as parameter)`	Returns absolute value.
ArcTan	`function ArcTan(x : real) :` `real;`	Returns the arctangent of the argument. Turbo Pascal does not have a Tan function, but tangents can be calculated using the expression Sin(x) / Cos(x).
Chr	`function Chr(x : Byte) :` `Char;`	Returns a character with a specified ordinal number.

Concat	`function Concat(s1`
	`[, s2, ..., sn] : string):`
	`string;`
		Concatenates a sequence of strings.

Copy	`function Copy(s : string;`
	`index : Integer;count :`
	`Integer) : string;`
		Returns a substring of a string.

Cos	`function Cos(x : real) :` `real;`	Returns the cosine of the argument.
DiskFree	`function DiskFree(` `Drive: Byte) : Longint;`	Returns the number of free bytes of a specified disk drive. Unit name Dos.
DiskSize	`function DiskSize(` `Drive: Byte) : Longint;`	Returns the total size in bytes of a specified disk drive.
DosVersion	`function DosVersion : Word;`	Returns the DOS version number. The low byte of the result is the major version number, and the high byte is the minor version number. Unit name Dos.
Eof	`function Eof(var f)` `: Boolean;`	Returns the end-of-file status of a typed or untyped file.

Eoln	`function Eoln [` `  (var f : text) ] : Boolean;`	Returns the end-of-line status of a file.
Exp	`function Exp(x : real) :` `  real;`	Returns the exponential of the argument.
File	`function FilePos(var f) :` `  Longint;`	Returns the current file position of a file.
FileSize	`function FileSize(var f) :` `  Longint;`	Returns the current size of a file.
Frac	`function Frac(x : real) :` `  real;`	Returns the fractional part of the argument.

Fsearch	`function FSearch(Path:`
	`  PathStr;DirList:string) :`
	`  PathStr;`
		Searches for a file in a list of directories by DirList. The directories in DirList must be separated by semicolons. The PathStr type is defined in the Dos unit as string [79]. Unit Name Dos.

Int	`function Int(x : real) :` `  real;`	Returns the Integer part of the argument.
KeyPressed	`function KeyPressed :` `  Boolean;`	Returns True if a key has been pressed on the keyboard, and False otherwise. Unit name Crt.
Length	`function Length(s : string) :` `  Integer;`	Returns the dynamic length of a string.
MaxAvail	`function MaxAvail : Longint;`	Returns the size of the largest contiguous free block in the heap, corresponding to the size of the largest dynamic variable that can be allocated at that time.
MemAvail	`function MemAvail : Longint;`	Returns sum of all free blocks in the heap.
MsDos	`procedure MsDos(` `  var Regs : Registers);`	Executes a DOS function call. Unit name Dos.
Odd	`function Odd(x : Longint) :` `  Boolean;`	Tests if the argument is an odd number.
Ord	`function Ord(x) : Longint;`	Returns the ordinal number of an ordinal-type value.

SeekEoln	`function SeekEoln [` `  (var f : text) ]: Boolean;`	Returns the end-of-line status of a file.
Sin	`function Sin(x : real) :` `  real;`	Returns the sine of the argument.
UpCase	`function UpCase(ch : Char) :` `  Char;`	Converts a character to uppercase.
WhereX	`function WhereX : Byte;`	Returns the X-co-ordinate of the current cursor position, relative to the current window. Unit name Crt.
WhereY	`function WhereY : Byte;`	Returns the Y-co-ordinate of the current cursor position, relative to the current window. Unit name Crt.

The following is a quick reference to commonly used procedures.

Append	`procedure Append(var f :` `  text);`	Opens an existing file for appending.
Assign	`procedure Assign(var f;` `  name : string);`	Assigns the name of an external file to a file variable.

BlockRead	`procedure BlockRead(var f:`
	`  file; var buf;count : Word`
	`  [; var result: Word])`
		Reads one or more records into a variable, where f is an untyped file variable, buf is any variable, count is an expression of type Word, and result is a variable of type Word.

BlockWrite	`procedure BlockWrite(var f:`
	`  file;var buf; count: Word`
	`  [; var result: Word])`
		Writes one or more records from a variable, where f is an untyped file variable, buf is any variable, count is an expression of type Word, and result is a variable of type Word.

ChDir	`procedure ChDir(s : string);`	Changes the current directory.
Close	`procedure Close(var f);`	Closes an open file.
ClrEol	`procedure ClrEol;`	Clears all characters from the cursor position to the end of the line without moving the cursor. Unit name Crt.

ClrScr	`procedure ClrScr;`	Clears the active window and places the cursor in the upper left-hand corner. Unit name Crt.
Dec	`procedure Dec(var x` `[ ; n : Longint]);`	Decrements a variable.
Delay	`procedure Delay(MS : Word);`	Delays a specified number of milliseconds. Unit name Crt.

Delete	`procedure Delete(var s :`
	`  string; index : Integer;`
	`  count : Integer);`
		Deletes a substring from a string.

DelLine	`procedure DelLine;`	Deletes the line containing the cursor.
Erase	`procedure Erase(var f);`	Erases an external file.
Exec	`procedure Exec(Path,` `  CmdLine : string);`	Executes a specified program with a specified command line. Unit name Dos.
Exit	`procedure Exit;`	Exits immediately from the current block. If the current block is the main program, it causes the program to terminate.
FillChar	`procedure FillChar(var x;` `  count : Word; ch : Char);`	Fills a specified number of contiguous bytes with a specified value.

FindFirst	`procedure FindFirst(Path :`
	`  string; Attr : Word;`
	`  var S : SearchRec);`
		Searches the specified (or current) directory for the first entry matching the specified file name and set of attributes. Unit name Dos.

FindNext	`procedure FindNext(var S :` `  SearchRec);`	Returns the next entry that matches the name and attributes specified in an earlier call to FindFirst. Unit name Dos.
Flush	`procedure Flush(var f :` `  text);`	Flushes the buffer of a text file open for output.
FreeMem	`procedure FreeMem(var p :` `  pointer; size : Word);`	Disposes a dynamic variable of a given size. Should not be used with Mark or Release.

GetCBreak	`procedure GetCBreak(var Break: Boolean);`	Returns the state of Ctrl-Break checking in DOS. Unit Name Dos.
GetDate	`procedure GetDate(var Year, Month, Day, DayOfWeek : Word);`	Returns the current date set in the operating system. Unit name Dos.
GetDir	`procedure GetDir(d : Byte; var s : string);`	Returns the current directory of a specified drive.
GetFAttr	`procedure GetFAttr(var F; var Attr : Word);`	Returns the attributes of a file. Unit name Dos.
GetFTime	`procedure GetFTime(var F; var Time : Longint);`	Returns the date and time a file was last written. Unit name Dos.
GetIntVec	`procedure GetIntVec(IntNo : Byte; var Vector : pointer);`	Returns the address stored in a specified interrupt vector. Unit name Dos.
GetMem	`procedure GetMem(var p : pointer; size : Word);`	Creates a new dynamic variable of the specified size and puts the address of the block in a pointer variable.
GetTime	`procedure GetTime(var Hour, Minute, Second, Sec100 : Word);`	Returns the current time set in the operating system. Unit name Dos.
GotoXY	`procedure GoToXY(X, Y : Byte);`	Positions the cursor. Unit name Crt.
Halt	`procedure Halt [ (exitcode : Word ) ];`	Stops program execution and returns to the operating system.
HighVideo	`procedure HighVideo;`	Selects high intensity characters. Unit name Crt.
Inc	`procedure Inc(var x [ ; n : Longint ] );`	Increments a variable.
Insert	`procedure Insert(source : string; var s : string; index : Integer);`	Inserts a substring into a string.
InsLine	`procedure InsLine;`	Inserts an empty line at the cursor position. Unit name Crt.
Intr	`procedure Intr(IntNo : Byte; var Regs : Registers);`	Executes a specified software interrupt. Unit name Dos.

Keep	`procedure Keep(ExitCode : Word);`	Keep (or Terminate Stay Resident) terminates the program and makes it stay in memory. Unit name Dos.
MkDir	`procedure MkDir(s : string);`	Creates a subdirectory.
Move	`procedure Move(var source, dest; count : Word);`	Copies a specified number of contiguous bytes from a source range to a destination range.
NoSound	`procedure NoSound;`	Turns off the internal speaker.
Randomize	`procedure Randomize;`	Initializes the built-in random generator with a random value.
Read	`procedure Read(f , v1 [, v2,...,vn ] );`	Reads a file component into a variable.
ReadLn	`procedure ReadLn( [ var f : text; ] v1 [, v2,...,vn] );`	Executes the Read procedure, then skips to the next line of the file.
Rename	`procedure Rename(var f; newname : string);`	Renames an external file.
Reset	`procedure Reset(var f [ : file; recsize : Word ] );`	Opens an existing file.
Rewrite	`procedure Rewrite(var f : file [;recsize : Word ] );`	Creates and opens a new file.
RmDir	`procedure RmDir(s : string);`	Removes an empty subdirectory.
Seek	`procedure Seek(var f; n : Longint);`	Moves the current position of a file to a specified component.
SetCBreak	`procedure SetCBreak(Break: Boolean);`	Sets the state of Ctrl-Break checking in DOS. Unit Name Dos.
SetDate	`procedure SetDate(Year, Month, Day : Word);`	Sets the current date in the operating system. Unit name Dos.
SetFAttr	`procedure SetFAttr(var F; Attr : Word);`	Sets the attributes of a file. Unit name Dos.
SetFTime	`procedure SetFTime(var F; Time : Longint);`	Sets the date and time a file was last written. Unit name Dos.
SetIntVec	`procedure SetIntVec(IntNo : Byte; Vector : pointer);`	Sets a specified interrupt vector to a specified address. Unit name Dos.

SetTime	`procedure SetTime(Hour, Minute, Second, Sec100 : Word);`	Sets the current time in the operating system. Unit name Dos.
Str	`procedure Str(x [ : width [ : decimals ]]; var s : string);`	Converts a numeric value to the same string representation that would be output by Write.
TextColor	`procedure TextColor(Color : Byte);`	Selects the foreground character colour. Unit name Crt.
TextMode	`procedure TextMode(Mode : Integer);`	Selects a specific text mode. Unit name Crt.
Truncate	`procedure Truncate(var f);`	Truncates the file size at the current file position.
Val	`procedure Val(s : string; var v; var code : Integer);`	Converts the string value to its numeric representation, as if it were read from a text file with Read. s is a string-type variable; it must be a sequence of characters that form a signed whole number. v is an Integer-type or real-type variable.
Window	`procedure Window(X1, Y1, X2, Y2 : Byte);`	Defines a text window on the screen. Unit name Crt.
Write	`procedure Write(f, v1 [, v2,...,vn ] );`	Writes a variable into a file component.
WriteLn	`procedure WriteLn( [ var f : text; ] v1 [, v2,...,vn ] );`	Executes the Write procedure, then outputs an end-of-line marker to the file.

Assembly Language Reference

D.1 Assembly language mnemonics

Table D.1 outlines the Assembly Language mnemonics (in column 1) and the equivalent encoded bit values (in column 3). It also shows the number of cycles for a 8086 processor and a 80386 processor (columns 4 and 5). The explanation of the encoded bit values is given after the table.

Table D.1 Assembly Language reference

Mnemonic	Description	Encoding	8086	386
AAA	Adjust after addition	00110111	8	4
AAD	Adjust before division	11010101 00001010	60	19
ADC *accum,immed*	Add immediate with carry to accumulator	0001010w *mod,reg,r/m*	4	2
ADC *r/m,immed*	Add immediate with carry to operand	100000sw *mod,reg,r/m*	4	2
ADC *r/m,reg*	Add register with carry to operand	000100dw *mod,reg,r/m*	3	2
ADC *reg,r/m*	Add operand with carry to register	000100dw *mod,reg,r/m*	3	2
ADD *accum,immed*	Add immediate to accumulator	0000010w	4	2
ADD *r/m,immed*	Add immediate to operand	100000sw *mod,000,r/m*	4	3
ADD *r/m,reg*	Add register to operand	0000010w *mod,reg,r/m*	4	2
ADD *reg,r/m*	Add operand to register	0000010w *mod,reg,r/m*	9	6
AND *accum,immed*	Bitwise AND immediate with operand	0010010w	4	2
AND *r/m,immed*	Bitwise AND register with operand	100000sw *mod,100,r/m*	4	2
AND *r/m,reg*	Bitwise AND operand with register	001000dw *mod,reg,r/m*	3	2
CALL *label*	Call instruction at label	11101000	19	7
CALL *r/m*	Call instruction indirect	11111111	16	7
CBW	Convert byte to word	10011000	2	3
CLC	Clear carry flag	11111000	2	2
CLD	Clear direction flag	11111100	2	2
CLI	Clear interrupt flag	11111010	2	3
CMC	Complement carry flag	11110101	2	2
CMP *accum, immed*	Compare immediate with accumulator	0011110w	4	2
CMP *r/m,immed*	Compare immediate with operand	100000sw	4	2
CMP *reg,r/m*	Compare register with operand	001110dw	3	2

CMPS arc,dest	Compare strings	1010011w	22	10
CMPSW	Compare strings word by word	1010011w	22	10
CMPSB	Compare string byte by byte	1010011w	22	10
CWD	Convert word to double word	10011001	5	2
DAA	Decimal adjust for addition	00100111	4	4
DAS	Decimal adjust for subtraction	00101111	4	4
DEC *r/m*	Decrement operand	1111111w	3	2
DEC *reg*	Decrement 16-bit register	01001 *reg*	3	2
DIV *r/m*	Divide accumulator by operand	1111011w	80	14
ESC *immed,r/m*	Escape with 6-bit immediate and operand	11011TTT		
HLT	Halt	11110100	2	5
IDIV	Integer divide accumulator by operand	1111011w		
IMUL	Integer multiply accumulator by operand	1111011w		
IN *accum,immed*	Input from port	1110010w	10	12
IN *accum,*DX	Input form port given by DX	1110110w	8	13
INC *r/m*	Increment operand	1111111w *mod,*000*,r/m*	3	2
INC *reg*	Increment 16-bit register	01000 *reg*	3	2
INT *immed*	Software interrupt	11001101	51	37
INTO	Interrupt on overflow	11001110	53	35
IRET	Return from interrupt	11001111	32	22
JA *label*	Jump on above	01110111	4	3
JAE *label*	Jump on above or equal	01110011	4	3
JBE *label*	Jump on below	01110110	4	3
JC *label*	Jump on carry	01110010	4	3
JCXZ *label*	Jump on CX zero	11100011	4	3
JE *label*	Jump on equal	01110100	4	3
JG *label*	Jump on greater	01111111	4	3
JGE *label*	Jump on greater or equal	01111101	4	3
JL *label*	Jump on less than	01111100	4	3
JLE *label*	Jump on less than or equal	01111110	4	3
JMP *label*	Jump to label	11101011	15	7
JMP *r/m*	Jump to instruction directly	111111 *mod,*110*,r/m*	11	7
JNA *label*	Jump on not above	01110110	4	3
JNAE *label*	Jump on note above or equal	01110010	4	3
JNB *label*	Jump on not below	01110011	4	3
JNBE *label*	Jump on not below or equal	01110111	4	3
JNC *label*	Jump on no carry	01110011	4	3
JNE *label*	Jump on not equal	01110101	4	3
JNG *label*	Jump on not greater	01111110	4	3
JNGE *label*	Jump on not greater or equal	01111100	4	3
JNO *label*	Jump on not overflow	01110111	4	3
JNP *label*	Jump on not parity	01111011	4	3
JNS *label*	Jump on not sign	01111001	4	3
JNZ *label*	Jump on not zero	01110101	4	3

Instruction	Description	Opcode		
JO *label*	Jump on overflow	01110000	4	3
JP *label*	Jump on parity	01111010	4	3
JPE *label*	Jump on parity even	01111010	4	3
JPO *label*	Jump on parity odd	01111011	4	3
JS *label*	Jump on sign	01111000	4	3
JZ *label*	Jump on zero	01110100	4	3
LAHF	Load AH with flags	10011111	4	2
LDS *r/m*	Load operand into DS			
LEA *r/m*	Load effective address of operand			
LES *r/m*	Load operand into ES			
LOCK	Lock bus			
LODS arc	Load string			
LODSB	Load byte from string into AL			
LODSW	Load word from string into AL			
LOOP *label*	Loop	11100010	17	11
LOOPE *label*	Loop while equal			
LOOPNE *label*	Loop while not equal			
LOOPNZ *label*	Loop while not zero			
LOOPZ *label*	Loop while zero			
MOV *accum*,mem	Move memory to accumulator	101000dw	10	4
MOV mem, *accum*	Move accumulator to memory	101000dw	10	3
MOV *r/m,immed*	Mover immediate to operand	1100011w *mod*,000,*r/m*	10	2
MOV *r/m,reg*	Move register to operand	100010dw *mod,reg,r/m*	2	2
MOV *r/m,segreg*	Mover segment *reg*ister to operand	100011d0 *mod,sreg,r/m*	2	2
MOV *reg,immed*	Move immediate to register	1011w *reg*	4	2
MOV seg*reg,r/m*	Move operand to segment register	100011d0 *mod,sreg,r/m*	2	2
MOVS dest,src	Move string	1010010w	18	7
MOVSB	Move string byte by byte	1010010w	18	7
MOVSW	Move string word by word	1010010w	18	7
MUL *r/m*	Multiply accumulator by operand	1111011w *mod*,100,*r/m*	70	9
NEG *r/m*	Negate operand	1111011w *mod*,011,*r/m*	3	2
NOP	No operation	10010000	3	3
NOT *r/m*	Invert bits	1111011w *mod*,010,*r/m*	3	2
OR *accum,accum*	Bitwise OR immediate with accumulator	000010dw *mod,reg,r/m*	3	2
OR *r/m,immed*	Bitwise OR immediate with operand	100000sw	4	2
OR *r/m,reg*	Bitwise OR register with operand	000010dw *mod,reg,r/m*	3	2
OR *reg,r/m*	Bitwise OR operand with register	000010dw *mod,reg,r/m*	3	2
OUT DX,*accum*	Output to port given by DX	1110111w	8	11
OUT *immed,accum*	Output to port	1110011w	10	10
POP *r/m*	Pop 16-bit operand	10001111 *mod*, 000,*r/m*	17	5
POP *reg*	Pop 16-bit register from stack	01011 *reg*	8	4
POPF	Pop flags	10011101	8	5
PUSH *r/m*	Push 16-bit operand	11111111 mem,110,*r/m*	16	5
PUSH *reg*	Push 16-bit register onto stack	010101 *reg*	11	2
PUSHF	Push flags	10011100	10	4
RCL *r/m*,1	Rotate left through carry by 1 bit	1101000w *mod*,010,*r/m*	2	3

RCL *r/m*,CL	Rotate left through carry by CL bits	1101001w *mod*,010,*r/m*	8+4n	3
RCR *r/m*,1	Rotate right through carry by 1 bit	1101000w *mod*,011,*r/m*	2	3
RCR *r/m*,CL	Rotate right through carry by CL bits	1101001w *mod*,011,*r/m*	8+4n	3
REP	Repeat	11110010	9	8
REPE	Repeat if equal	11110011		
REPNE	Repeat if not equal	11110011		
REPNZ	Repeat if not zero	11110011		
RET [*immed*]	Return after popping bytes from stack	11000010		
ROL *r/m*,1	Rotate left by 1 bit	1101000w *mod*,000,*r/m*	2	3
ROL *r/m*,CL	Rotate left by CL bits	1101001w *mod*,000,*r/m*	8+4n	3
ROR *r/m*,1	Rotate right by 1 bit	1101000w *mod*,001,*r/m*	2	3
ROR *r/m*,CL	Rotate right by CL bits	1101001w *mod*,001,*r/m*	8+4n	3
SAHF	Store AH into flags	10011110	4	3
SAL *r/m*,1	Shift arithmetic left by 1 bit	1101000w *mod*,100,*r/m*	2	3
SAL *r/m*,CL	Shift arithmetic left by CL bits	1100000w *mod*,100,*r/m*	8+4n	3
SAR *r/m*,1	Shift arithmetic right by 1 bit	1101000w *mod*,101,*r/m*	2	3
SAR *r/m*,CL	Shift arithmetic right by CL bits	1100000w *mod*,101,*r/m*	8+4n	3
SBB *accum,immed*	Subtract immediate and carry flag	0001110w	4	2
SBB *r/m,immed*	Subtract immediate and carry flag	100000sw *mod*,011,*r/m*	4	2
SBB *r/m,reg*	Subtract register and carry flag	000110dw *mod*,reg,*r/m*	3	2
SBB *reg,r/m*	Subtract operand and carry flag	000110dw *mod*,reg,*r/m*	3	2
SCAS dest	Scan string	1010111w	15	7
SCASB	Scan string for byte in AL	1010111w	15	7
SCASW	Scan string for word in AX	1010111w	15	7
SHL *r/m*,1	Shift left by 1 bit	1101000w *mod*,100,*r/m*	2	3
SHL *r/m*,CL	Shift left by CL bits	1100000w *mod*,100,*r/m*	8+4n	3
SHR *r/m*,1	Shift right by 1 bit	1101000w *mod*,101,*r/m*	2	3
SHR *r/m*,CL	Shift right by CL bits	1100000w *mod*,101,*r/m*	8+4n	3
STC	Set carry flag	11111001	2	2
STD	Set direction flag	11111101	2	2
STI	Set interrupt flag	11111011	2	3
STOS dest	Store string	1010101w	11	4
STOSB	Store byte in AL at string	1010101w	11	4
STOSW	Store word in AX at string	1010101w	11	4
SUB *accum,immed*	Subtract immediate from accumulator	0010110w	4	2
SUB *r/m,immed*	Subtract immediate from operand	100000sw *mod*,101,*r/m*	4	2
SUB *r/m,reg*	Subtract register from operand	001010dw *mod*,reg,*r/m*	3	2
SUB *reg,r/m*	Subtract operand from register	001010dw *mod*,reg,*r/m*	3	2
TEST *accum,immed*	Compare immediate bits with accumulator	1010100w	4	2
TEST *r/m,immed*	Compare immediate bits with operand	1111011w *mod*,000,*r/m*	5	2
TEST *reg,r/m*	Compare register bits with operand	1000011w *mod*,reg,*r/m*	3	2
WAIT	Wait	10011011	4	6
XCHG *accum,reg*	Exchange accumulator with register	100011w *mod*,reg,*r/m*	4	3
XCHG *r/m,reg*	Exchange operand with register	100011w *mod*,reg,*r/m*	17	5
XCHG *reg,accum*	Exchange register with accumulator	100011w *mod*,reg,*r/m*	4	3
XCHG *reg,r/m*	Exchange register with operand	100011w *mod*,reg,*r/m*	17	5

XOR *accum,immed*	Bitwise XOR immediate with accumulator	001110dw *mod,reg,r/m*	4	2
XOR *r/m,immed*	Bitwise XOR immediate with operand	001100dw *mod,reg,r/m*	4	2
XOR *r/m,reg*	Bitwise XOR register with operand	001100dw *mod,reg,r/m*	3	2
XOR *reg,r/m*	Bitwise XOR operand with register	001100dw *mod,reg,r/m*	3	2

Syntax:

reg   A general-purpose register of any size.

seg*reg*   A segment registers, such as DS, ES, SS or CS.

accum   An accumulator of any size: AL or AX (or EAX on 386/486).

mem   A direct or indirect memory operand of any size.

label   A labeled memory location in the code segment.

src,dest   A source of destination memory operand used in a string operand.

immed   A constant operand.

and the bits are specified by:

d   **direction bit**. If set (1) then the transfer is from memory to register or register to register, and the destination is a reg field. If not set then the source is a register field and the transfer is from register to memory.

w   **word/byte bit**. If set the 16-bit operands are used, else 8-bit operands are used.

s   **sign bit**. If set then the operand has a sign-bit.

mod   **mode**. Identifies the register/memory mode. These are:

            00   If r/m is 110 then direct memory is used, else the displacement is 0 and an indirect memory operand is used.

            01   An indirect memory operand is used with an 8-bit displacement.

            10   An indirect memory operand is used with a 16-bit displacement.

            11   A two-register instruction is used; the reg field specifies the destination and the r/m field specifies the source.

reg   **register**. Specifies one of the general-purpose registers. These are:

reg	16-bit, if w=1	8-bit, if w=0
000	AX	AL
001	CX	CL
010	DX	DL
011	BX	BL
100	SP	AH
101	BP	CH
110	SI	DH
111	DI	BH

r/m   **register/memory**. Specifies a memory of register operand. If the mod file is 11 then the register is specified with the reg field (as given above), else it has the following settings:

reg	Operand address
000	**DS:[BX+SI+*disp*]**
001	**DS:[BX+DI+*disp*]**
010	**SS:[BP+SI+*disp*]**
011	**SS:[BP+DI+*disp*]**
100	**DS:[SI+*disp*]**
101	**DS:[DI+*disp*]**
110	**DS:[BP+*disp*]**
111	**DS:[BX+*disp*]**

The instruction encoding has the form:

OPCODE	*mod,reg,r/m*	*disp*	*immed*
(1–2 bytes)	(0–1 byte)	(0–2 bytes)	(0–2 bytes)

where:

disp   **displacement**. Specifies the offset for memory operands.

immed   **register/memory**. Specifies the actual values for constant values.

D.2 Assembler directives

Table D.2 outlines some Assembly Language directives.

Table D.2   Assembly Language reference

Directive	Description
`.386`	Enables assembly of 386 code.
`.8086`	Enables assembly of 8086 code.
`ASSUME` *segreg:name*	Selects *segreg* to be the default segment register for all symbols in the named segment.
`.CODE`	Defines the code segment.
`COMMENT`	Defines a comment.
`.CONST`	Defines a constant.
name `DB` *init*	Allocations and optionally initializes a byte of storage for each init.
name `DW` *init*	Allocations and optionally initializes a word of storage for each init.
name `DD` *init*	Allocations and optionally initializes a double word of storage for each init.
name `DF` *init*	Allocations and optionally initializes a far word (6 bytes) of storage for each init.
name `DQ` *init*	Allocations and optionally initializes a quad word (8 bytes) of storage for each init.
`.DATA`	Define the data segment.
`ELSE`	Defines an alternative block of the `IF` directive.
`END`	Defines the end of a module.
`ENDIF`	Defines the end of the `IF` directive.
`EXTERN` *names*	Defines one of more external variables, labels or systems which are called *names*.
`IF` *expression* *ifstatements* `ENDIF`	Defines the `IF` directive.
`INCLUDE` *filename*	Includes source code.
`.MODEL` *mem_model*	Defines memory model.

NAME *module_name*	Defines the module name.
ORG *express*	Organizes the program in memory.
.STACK *size*	Defines the stack size.
TITLE *text*	Defines title.

D.3 C and Pascal interrupts

In Turbo/Borland C there are four main functions to interrupt the processor: int86x(), intdos(), intr() and int86(). These functions are prototyped in the header file dos.h. This header file also contains a structure definition that allows a C program to gain access to the processor's registers. Parameters are passed into and out of the interrupt sevice routines via these registers. The format of the structure is:

```
struct   WORDREGS
{
   unsigned int ax;
   unsigned int bx;
   unsigned int cx;
   unsigned int dx;
   unsigned int si;
   unsigned int di;
   unsigned int cflag;
}
struct   BYTEREGS
{
   unsigned char al,ah;
   unsigned char bl,bh;
   unsigned char cl,ch;
   unsigned char dl,dh;
}
union REGS
{
   struct   WORDREGS x;
   struct   BYTEREGS h;
}
```

Registers are accessed either as 8-bit registers (such as AL, AH) or 16-bit registers (such as AX, BX). If a structure name regs is declared, then:

regs.h.al   accesses the 8-bit AL register
regs.x.ax   accesses the 16-bit AX register.

The syntax of the function int86() takes the form of

```
int86(intno,&inregs,&outregs);
```

where the first argument of the parameter list is the interrupt number, the input registers are passed as the second argument and the output registers the third. Parameters are passed to the interrupt routine by setting certain input registers and parameters are passed back from the interrupt in the output registers.

In a similiar way Turbo Pascal provides access throught the routine named `Intr()`. To gain access to this procedure the *uses dos;* statement is placed near the top of the program. A data type named `Registers` has also been predefined, as shown below. Note that it is possible to use either the 16-bit registers (such as AX, BX) or 8-bit (such as AL, AH):

```
type
   Registers = record
                  case Integer of
                     0: (AX,BX,CX,DX,BP,SI,DI,DS,ES,Flags: Word);
                     1: (AL,AH,BL,BH,CL,CH,DL,DH: Byte);
   end;
```

In Program D.1, the DOS interrupt 21h and function 02h (write character to the output) is used to display the character 'A'. In this case, the function number 02h is loaded into AH and the character to be displayed is loaded into DL.

📄 Program D.1
```
#include <dos.h>

int    main(void)
{
union REGS inregs,outregs;

   inregs.h.ah=0x02;
   inregs.h.dl='A';

   int86(0x21,&inregs,&outregs);
   return(0);
}
```

📄 Program D.1
```
program DOS1;
{       Program to display 'A' using DOS interrupt          }

uses dos;

var REGS:registers;
begin

     regs.ah:=$02;
     regs.dl:=65;

     intr($21,regs);
end.
```

Program D.2 shows how a program can gain access to the system date. The function used in this example is 2Ah. Test run D.1 gives a sample run.

📄 **Program D.2**

```c
#include <stdio.h>
#include <dos.h>

int   main(void)
{
int   day,month,year,day_of_week;

union REGS inregs,outregs;

    inreqs.h.ah=0x2A;
    int86(0x21,&inregs,&outregs);

    day=outregs.h.dl;
    month=outregs.h.dh;
    year=outregs.x.cx;
    day_of_week=outregs.h.al;

    printf("\nDate is %d/%d/%d day of week %d\n",
                            day,month,year,day_of_week);
    return(0);
}
```

📄 **Program D.2**

```pascal
program DOS2;
uses dos;

var REGS:registers;
    day,month,year,day_of_week:integer;

begin
    regs.ah:=$2a;
    intr($21,regs);

    day:=regs.dl;
    month:=regs.dh;
    year:=regs.cx;
    day_of_week:=regs.al;
    writeln('Date is ',day,'/',month,'/',year);

    writeln('Day of week is ',day_of_week);
end.
```

🖥 Test run D.1

```
Date is 8/9/1993
Day of week is 3
```

ASCII Character Set

ANSI defined a standard alphabet known as ASCII. This has since been adopted by the CCITT as a standard, known as IA5 (International Alphabet No. 5). The following tables define this alphabet in binary, as a decimal, as a hexadecimal value and as a character.

Binary	Decimal	Hex	Character	Binary	Decimal	Hex	Character
00000000	0	00	NUL	00010000	16	10	DLE
00000001	1	01	SOH	00010001	17	11	DC1
00000010	2	02	STX	00010010	18	12	DC2
00000011	3	03	ETX	00010011	19	13	DC3
00000100	4	04	EOT	00010100	20	14	DC4
00000101	5	05	ENQ	00010101	21	15	NAK
00000110	6	06	ACK	00010110	22	16	SYN
00000111	7	07	BEL	00010111	23	17	ETB
00001000	8	08	BS	00011000	24	18	CAN
00001001	9	09	HT	00011001	25	19	EM
00001010	10	0A	LF	00011010	26	1A	SUB
00001011	11	0B	VT	00011011	27	1B	ESC
00001100	12	0C	FF	00011100	28	1C	FS
00001101	13	0D	CR	00011101	29	1D	GS
00001110	14	0E	SO	00011110	30	1E	RS
00001111	15	0F	SI	00011111	31	1F	US

Binary	Decimal	Hex	Character	Binary	Decimal	Hex	Character
00100000	32	20	SPACE	00110000	48	30	0
00100001	33	21	!	00110001	49	31	1
00100010	34	22	"	00110010	50	32	2
00100011	35	23	#	00110011	51	33	3
00100100	36	24	$	00110100	52	34	4
00100101	37	25	%	00110101	53	35	5
00100110	38	26	&	00110110	54	36	6
00100111	39	27	/	00110111	55	37	7
00101000	40	28	(	00111000	56	38	8
00101001	41	29	)	00111001	57	39	9
00101010	42	2A	*	00111010	58	3A	:
00101011	43	2B	+	00111011	59	3B	;
00101100	44	2C	,	00111100	60	3C	<
00101101	45	2D	–	00111101	61	3D	=
00101110	46	2E	.	00111110	62	3E	>
00101111	47	2F	/	00111111	63	3F	?

Binary	Decimal	Hex	Character	Binary	Decimal	Hex	Character
01000000	64	40	@	01010000	80	50	P
01000001	65	41	A	01010001	81	51	Q
01000010	66	42	B	01010010	82	52	R
01000011	67	43	C	01010011	83	53	S
01000100	68	44	D	01010100	84	54	T
01000101	69	45	E	01010101	85	55	U
01000110	70	46	F	01010110	86	56	V
01000111	71	47	G	01010111	87	57	W
01001000	72	48	H	01011000	88	58	X
01001001	73	49	I	01011001	89	59	Y
01001010	74	4A	J	01011010	90	5A	Z
01001011	75	4B	K	01011011	91	5B	[
01001100	76	4C	L	01011100	92	5C	\
01001101	77	4D	M	01011101	93	5D	]
01001110	78	4E	N	01011110	94	5E	`
01001111	79	4F	O	01011111	95	5F	_

Binary	Decimal	Hex	Character	Binary	Decimal	Hex	Character
01100000	96	60		01110000	112	70	p
01100001	97	61	a	01110001	113	71	q
01100010	98	62	b	01110010	114	72	r
01100011	99	63	c	01110011	115	73	s
01100100	100	64	d	01110100	116	74	t
01100101	101	65	e	01110101	117	75	u
01100110	102	66	f	01110110	118	76	v
01100111	103	67	g	01110111	119	77	w
01101000	104	68	h	01111000	120	78	x
01101001	105	69	i	01111001	121	79	y
01101010	106	6A	j	01111010	122	7A	z
01101011	107	6B	k	01111011	123	7B	{
01101100	108	6C	l	01111100	124	7C	:
01101101	109	6D	m	01111101	125	7D	}
01101110	110	6E	n	01111110	126	7E	~
01101111	111	6F	o	01111111	127	7F	DEL

The standard ASCII character set is a 7-bit character and ranges from 0 to 127. This code is rather limited as it does not contain symbols such as Greek letters, lines, and so on. For this purpose the extended ASCII code has been defined. This fits into character number 128 to 255. The following 4 tables define a typical extended ASCII character set.

Binary	Decimal	Hex	Character	Binary	Decimal	Hex	Character
10000000	128	80	Ç	10010000	144	90	É
10000001	129	81	ü	10010001	145	91	æ
10000010	130	82	é	10010010	146	92	Æ
10000011	131	83	â	10010011	147	93	ô
10000100	132	84	ä	10010100	148	94	ö
10000101	133	85	à	10010101	149	95	ò
10000110	134	86	å	10010110	150	96	û
10000111	135	87	ç	10010111	151	97	ù
10001000	136	88	ê	10011000	152	98	ÿ
10001001	137	89	ë	10011001	153	99	Ö
10001010	138	8A	è	10011010	154	9A	Ü
10001011	139	8B	ï	10011011	155	9B	¢
10001100	140	8C	î	10011100	156	9C	£
10001101	141	8D	ì	10011101	157	9D	¥
10001110	142	8E	Ä	10011110	158	9E	₧
10001111	143	8F	Å	10011111	159	9F	ƒ

Binary	Decimal	Hex	Character	Binary	Decimal	Hex	Character
10100000	160	A0	á	10110000	176	B0	░
10100001	161	A1	í	10110001	177	B1	▒
10100010	162	A2	ó	10110010	178	B2	▓
10100011	163	A3	ú	10110011	179	B3	│
10100100	164	A4	ñ	10110100	180	B4	┤
10100101	165	A5	Ñ	10110101	181	B5	╡
10100110	166	A6	ª	10110110	182	B6	╢
10100111	167	A7	º	10110111	183	B7	╖
10101000	168	A8	¿	10111000	184	B8	╕
10101001	169	A9	⌐	10111001	185	B9	╣
10101010	170	AA	¬	10111010	186	BA	║
10101011	171	AB	½	10111011	187	BB	╗
10101100	172	AC	¼	10111100	188	BC	╝
10101101	173	AD	¡	10111101	189	BD	╜
10101110	174	AE	«	10111110	190	BE	╛
10101111	175	AF	»	10111111	191	BF	┐

Binary	Decimal	Hex	Character	Binary	Decimal	Hex	Character
11000000	192	C0	└	11010000	208	D0	╨
11000001	193	C1	┴	11010001	209	D1	╤
11000010	194	C2	┬	11010010	210	D2	╥
11000011	195	C3	├	11010011	211	D3	╙
11000100	196	C4	─	11010100	212	D4	╘
11000101	197	C5	┼	11010101	213	D5	╒
11000110	198	C6	╞	11010110	214	D6	╓
11000111	199	C7	╟	11010111	215	D7	╫
11001000	200	C8	╚	11011000	216	D8	╪
11001001	201	C9	╔	11011001	217	D9	┘
11001010	202	CA	╩	11011010	218	DA	┌
11001011	203	CB	╦	11011011	219	DB	█
11001100	204	CC	╠	11011100	220	DC	▄
11001101	205	CD	═	11011101	221	DD	▌
11001110	206	CE	╬	11011110	222	DE	▐
11001111	207	CF	╧	11011111	223	DF	▀

Binary	Decimal	Hex	Character	Binary	Decimal	Hex	Character
11100000	224	E0	α	11110000	240	F0	Ξ
11100001	225	E1	ß	11110001	241	F1	±
11100010	226	E2	Γ	11110010	242	F2	≥
11100011	227	E3	π	11110011	243	F3	≤
11100100	228	E4	Σ	11110100	244	F4	⌠
11100101	229	E5	σ	11110101	245	F5	⌡
11100110	230	E6	μ	11110110	246	F6	÷
11100111	231	E7	τ	11110111	247	F7	≈
11101000	232	E8	Φ	11111000	248	F8	°
11101001	233	E9	Θ	11111001	249	F9	•
11101010	234	EA	Ω	11111010	250	FA	·
11101011	235	EB	δ	11111011	251	FB	√
11101100	236	EC	φ	11111100	252	FC	^{n}
11101101	237	ED	φ	11111101	253	FD	^{2}
11101110	238	EE	E	11111110	254	FE	■
11101111	239	EF	Λ	11111111	255	FF	□

 **Index**

Lightning Source UK Ltd.
Milton Keynes UK
UKOW06f0103170913

217328UK00017B/582/A